Understanding the Global Energy Crisis

Purdue Studies in Public Policy

Understanding the Global Energy Crisis

EDITED BY EUGENE D. COYLE AND
RICHARD A. SIMMONS

Published on behalf of the
Global Policy Research Institute
by Purdue University Press
West Lafayette, Indiana

This book is licensed under a Creative Commons CC-BY-NC License. To view a copy of this license, visit http://creativecommons.org/licenses/by-nc/2.0/.

Understanding the global energy crisis / edited by Eugene D. Coyle and Richard A. Simmons.
    pages cm. -- (Purdue studies in public policy)
 Includes bibliographical references and index.
 ISBN 978-1-55753-661-7 (pbk. : alk. paper) -- ISBN 978-1-61249-309-1 (epdf) -- ISBN 978-1-61249-310-7 (epub)
 1. Energy consumption. 2. Energy policy. 3. Energy development. 4. Renewable energy sources. I. Coyle, Eugene D.
 HD9502.A2U495 2014
 333.79--dc23

An electronic version of this book is freely available, thanks to the support of libraries working with Knowledge Unlatched. KU is a collaborative initiative designed to make high quality books Open Access for the public good. The Open Access ISBN for this book is 978-1-55753-701-0. More information about the initiative and links to the Open Access version can be found at www.knowledgeunlatched.org.

*To our wives,
Lucy Dendy Coyle and Michelle Simmons,
for their abiding encouragement and inspiration*

Table of Contents

Foreword	xi
Preface	xiii
Introduction	1

PART 1: THE GLOBAL ENERGY CRISIS IN CONTEXT

Chapter 1: Reflections on Energy, Greenhouse Gases, and Carbonaceous Fuels — 11
   1.1. Introduction: Man's Quest for Energy — 12
   1.2. Earth's Atmosphere and Greenhouse Gases — 13

Chapter 2: Global Energy Policy Perspectives — 27
   2.1. Introduction: Energy Demand and Expected Growth — 28
   2.2. United States Energy and Climate Policy — 34
   2.3. Energy and Climate Policy in the European Union — 42
   2.4. China Energy and Climate Change Policy — 50
   2.5. Russia Energy and Climate Change Policy — 52
   2.6. Brazil Energy and Climate Change Policy — 54
   2.7. India Energy and Climate Change Policy — 58

Chapter 3: Social Engagement by the Engineer — 73
   3.1. Introduction — 74
   3.2. Social Systems — 75
   3.3. Common Authentic Values and Principles — 78

PART 2: ENERGY CONVERSION TECHNOLOGY

Chapter 4: Harnessing Nature: Wind, Hydro, Wave, Tidal, and Geothermal Energy — 91
   4.1. Introduction — 91
   4.2. Wind Energy — 92
   4.3. Hydroelectric Energy — 99
   4.4. Wave and Tidal Energy — 102
   4.5. Geothermal Energy — 114
   4.6. Impact of Renewable Technologies on Electricity Grid Developments — 117

Chapter 5: Solar Power and the Enabling Role of Nanotechnology — 125
- 5.1. Introduction — 125
- 5.2. Solar Power Overview — 126
- 5.3. Nanotechnology — 128
- 5.4. Solar Photovoltaics — 132
- 5.5. Thermoelectrics — 135
- 5.6. Nanotechnology in Other Energy Applications — 140

Chapter 6: Biofuel Prospects in an Uncertain World — 151
- 6.1. Biofuels History — 152
- 6.2. First Generation Biofuels — 153
- 6.3. Second Generation Biofuels — 155
- 6.4. Conclusions: Major Challenges and Opportunities — 162

Chapter 7: A Future Role for Nuclear Energy? — 167
- 7.1. Introduction: Essentials of Nuclear Energy — 168
- 7.2. History of Nuclear Engineering — 171
- 7.3. Current Status of Nuclear Energy — 171
- 7.4. Nuclear Energy Safety — 175
- 7.5. Nuclear Accidents and Impacts — 177
- 7.6. Challenges in Nuclear Waste Management — 180
- 7.7. Future Role for Nuclear Power — 182
- 7.8. Social Engagement — 185
- 7.9. Future for Nuclear Fusion — 188

PART 3: ENERGY DISTRIBUTION AND USE

Chapter 8: Taking Emerging Renewable Technologies to Market — 193
- 8.1 Introduction — 194
- 8.2 Economic Factors — 194
- 8.3 Political Factors — 196
- 8.4 Social Factors — 199
- 8.5 Maintainability Factors — 202
- 8.6 Economics of Energy — 202
- 8.7 Some Challenges for Emerging Wave Energy Technologies — 204
- 8.8 Conclusion — 209

Chapter 9: Transportation and Energy — 215
- 9.1 Transportation Energy Overview — 216
- 9.2 Electric and Hybrid Vehicles — 225
- 9.3 Aviation Fuels and Regulation — 239

Chapter 10: Policy Challenges for the Built Environment: The Dilemma of the Existing Building Stock — 255
- 10.1 Introduction — 255
- 10.2 Energy Conserving Building Retrofit Technologies — 257

10.3 Complexity of Energy Efficiency Retrofit Strategies   *261*
10.4 Policy Challenges to Energy Efficiency Retrofit Success   *266*
10.5 Building Energy Reduction Programs Recently Employed in the US and EU   *272*
10.6 Recommendations and Opportunities for Future Solutions   *276*

Epilogue: Reflections on Our Path Forward   283

Index   291

Foreword

I believe that this book will be regarded as a classic in describing the emerging global energy crises and alternate approaches for addressing them. It will not only serve several purposes in the technology and policy worlds but will also appeal to a broader audience. As the authors intended, it is an excellent technical source book for engineers and technologists. It provides a comprehensive review of the history of energy conversion and use; current and emerging technologies to achieve energy sustainability in a highly stressed planet; and contemporary international efforts to find solutions to the complex issues involved. Accordingly, it is also a must read for all young individuals of social consciousness, who see themselves as inheritors of *grand challenge* world issues and have a keen desire to contribute to their solution.

The book is organized such that each chapter begins with an abstract of the subject matter and ends with a summary of key points. The language is aimed at a *Popular Science* level of technical exposition and is relatively jargon-free considering the wide spectrum of technologies presented. Each chapter includes an extensive list of references to assist the reader in finding sources and additional details of the referenced content.

However, the most important aspect of the book, which is relatively unique, is the way the subject matter is organized. Energy sustainability is presented as a complex issue (or a *wicked problem*) that is interrelated with other complex issues, such as environmental sustainability, economic sustainability, and in the case of biomass-derived fuels, water and soil sustainability. For such problems there exists not a single solution but a multiplicity of solutions, which must then be judged by equally complex interrelationships among technical, social, and economic factors. These in turn vary regionally throughout the world based on different histories, cultures, social norms, etc. The book carefully addresses these complexities for each energy policy topic presented.

The editors have carefully selected expert authors to explain the technical, social, and economic factors for each topic and present alternative approaches to a solution. The book intentionally avoids advocacy and attempts to be an honest broker to the readers so that they can draw their own conclusions based on the relative advantages and disadvantages presented.

I am exceedingly proud that this book originated in Professor Coyle's Fulbright Fellowship at Purdue University's Global Policy Research Institute. I also applaud the members of the faculty at Purdue and at the Dublin Institute of Technology who contributed to it, exemplifying a successful trans-Atlantic partnership. This product is a glowing example of the vision of the GPRI to engage more faculty and students at Purdue and elsewhere to conduct research and careful analysis of grand challenge global issues to inform the nation's policy and decision makers.

Arden L. Bement, Jr.

Biographical Sketch

Arden Bement, Jr. retired from his position as the founding Director of the Global Policy Research Institute at Purdue University in 2013. Prior to that position, he was the Director of the National Science Foundation from 2004 to 2010. He served as a member of the U.S. National Commission for UNESCO and as the vice-chair of the Commission's Natural Sciences and Engineering Committee. He is a member of the U.S. National Academy of Engineering, a fellow of the American Academy of Arts and Sciences, and a fellow of the American Association for the Advancement of Science.

Preface

This book brings together experts in energy policy, social science, power systems, solar energy, agronomy, renewable energy technologies, nuclear engineering, transportation, and the built environment from both sides of the Atlantic to explore the future of energy production and consumption from technological, political, and sociological perspectives. The volume is not intended to serve as complete in-depth coverage of all energy sector technologies, nor to cover energy policy comprehensively for all world regions. It is, however, hoped that the topics selected and questions addressed will encourage further engagement and debate among not only students, but anyone with interest in energy sustainability, climate change, and related challenges.

These issues are multi-dimensional and complex in nature; "wicked problems" with no easy answers. The book explores issues such as financial outlay and tariff support, the readiness of emerging technologies such as wave and tidal energy converters, the degree of wind energy that may be accommodated on national networks, the extent to which solar energy may be deployed, challenges and uncertainties in the production of advanced biofuels, concerns about natural gas extraction via hydraulic fracture (hydrofracking), and whether nuclear energy should become more widely used or taken out of the generation mix.

In many quarters there is a sense of a race against time in trying to undo the current and introduce the new technologies that will help reduce carbon emissions back to within acceptable levels and, in so doing, offset further increases in global average temperature. It is also important to remain focused and seek agreement on practical steps that may be taken in both the short term, through research and innovation for renewable technologies and efficiency in energy use, and longer term through replacement of coal, oil, and gas by commercially viable renewable technologies in much greater proportions than are achievable today.

We the editors are strong proponents of a growing dialogue between the technology and policy communities, and attest to the value of a broader exchange among stakeholders. Through our respective participation in programs such as the Fulbright Scholarship and the AAAS Science and Technology Policy Fellowship, we have witnessed ways in which this dialogue can both inspire and be transformed into action.

We wish to extend our thanks to the Dublin Institute of Technology, Purdue University, and both the Irish and US Fulbright Commissions for facilitating the faculty exchanges between DIT and Purdue that were the origin of this book. We would also like to thank Dr. Arden Bement, Emeritus Director of the Global Policy Research Institute and his team at Purdue University; Yvonne Desmond and Amy Van Epps, librarians at DIT and Purdue respectively; and the staff at Purdue University Press, especially editor Jennifer Lynch and director Charles Watkinson. Last but not least, we wish to thank Dr. Marek Rebow for his energy and dedication to research at DIT and to Dr. Melissa Dark for her role in research collaboration between Purdue University and DIT.

Eugene D. Coyle, Military Technological College, Sultanate of Oman

Richard A. Simmons, Purdue University, West Lafayette, Indiana

> Then I say the Earth belongs to each generation during its course, fully and in its own right . . . [but] no generation can contract debts greater than may be paid during the course of its own existence.
>
> Thomas Jefferson (1743–1826)

Introduction

Energy is everywhere and drives everything. Our modern lives, both individual and societal, have come to depend on its abundance, convenience, and potential. It is the motive force within our bodies, propelling our vehicles, lighting our world. Consider a power outage, or a dead cell phone battery; living without energy, for even ten minutes, demonstrates how indelible its imprint is on daily activities. At the same time, we inhabit an amazing ecosystem, as resilient as it is fragile. Our energy comes from and returns to a global environment. The world is in a predicament, yet this is no book of gloom and doom, but rather of technology and policy. These tools help us not only understand the energy and climate context of our world, but allow us to begin solving its challenges. History has shown that when technology and policy are not aligned or well-proportioned, they fail on their collective promises. But taken together, in an intentional, practical, and coordinated manner, they can be the stimulus behind a new and far superior energy future. And the world has never needed that more than it does today.

Planet Earth is facing an energy crisis owing to an escalation in global energy demand, continued dependence on fossil-based fuels for energy generation and transportation, and an increase in world population, exceeding seven billion people and rising steadily. Excessive burning of fossil fuels is not only depleting natural resources, but is resulting in a steady increase of carbon dioxide emissions, which experts believe is responsible for increasing average global temperatures. While natural cyclical variations do occur in regional and global climates, there is now widespread agreement among scientific communities and governments that recent climate change is accelerating as a result of human intervention and that rapid and profound measures will be required to reduce harmful impacts. Concentration levels of greenhouse gases are rising steadily and are now greater than at any time in the past eight hundred thousand years. If concentration levels are not reversed, major changes to the world climate may result, bringing significant effects on people, industry, and the world economy. The International Energy Agency (IEA) has outlined critical steps that, if implemented quickly, can help reduce the upward trend in atmospheric emissions. To reduce traditional fuel use and CO_2 emissions, major countermeasures include increased energy efficiency and conservation, efforts to advance alternative energy technologies, and efforts to control future energy demand.

Prior to modern industrialization, concentration levels of carbon dioxide in the atmosphere remained relatively stable at 280 ppm. Over recent decades there has been a steady rise in emissions, with levels now approaching 400 ppm (a forty percent increase) and rising an average of 2.3 percent per annum. The Intergovernmental Panel on Climate Change (IPCC) has coined the now widely accepted "hockey stick graph" characteristic to describe atmospheric pollutant increase. The graph has been used for numerous reconstructions of Northern Hemisphere temperatures for the last 600 to 1,000 years. Reconstructions have consistently shown that late 20th century and early 21st century temperatures are rising sharply in tandem with concentrations of greenhouse gases, in particular carbon dioxide (Figure 1).

It is now believed that a doubling of atmospheric carbon dioxide to 560 ppm, projected by IPCC to occur by mid-century, will yield a global average temperature increase of at least 4°F (2.2°C). To gain an appreciation of world average temperature statistics it is noted that the twentieth century average global temperature for the month of June was 60°F (15.5°C).

Even an increase of 2°C over pre-industrial levels may result in significant world climatic change with detrimental social, human and economic impact. Therefore, to the extent such temperature increases can be avoided, it behooves governments and concerned members of civil society to implement appropriate, yet practical policies and actions in response.

To that end, IEA in 2009 proposed a plan entitled the *450 Scenario* with an aggressive timetable of actions that would be required to limit the long-term concentration of greenhouse gases in Earth's atmosphere to 450 parts per million of carbon-dioxide

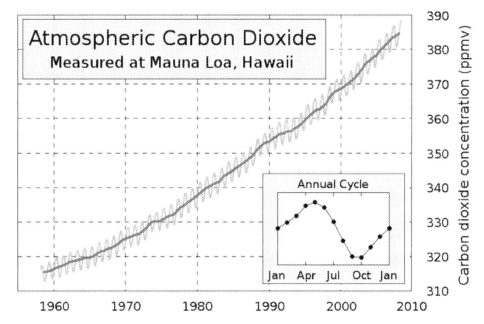

Figure 1. Keeling Curve of Atmospheric CO_2 Concentrations, Measured at Mauna Loa Observatory[1]

equivalent, setting a limit on global temperature rise to around 2°C above pre-industrial levels. The plan outlines a timeline to 2030 with actions that include the introduction of energy efficient technologies, low-carbon energy technologies, enhanced generation integration through renewable energy resources, increase in nuclear energy as a base load provider, and incorporation of energy plants fitted with carbon capture and storage capabilities. In road transportation, the plan advocates a shift from the current balance of greater than 99 percent combustion-powered vehicles to at least 60 percent hybrid and electric vehicles.

In addition to the 450 Scenario, IEA proposed a range of policy scenarios in the 2012 *World Energy Outlook*, which if followed could result in very different outcomes in global climate.[2] In researching projections and likely outcomes it is clear that the grand challenges presenting in energy and climate are global in nature and require concerted action and coordination across state, country and continental borders. Commendable inroads have been achieved through work of the United Nations Framework Convention on Climate Change (UNFCCC), the Intergovernmental Panel on Climate Change, the Environmental Protection Agency and related organizations. By the end of 2011, 191 countries had become signatories to the Kyoto Protocol, and in so doing committed to reaching designated national targets for reduction in greenhouse gas emissions.

A study on global climate change, commissioned by the National Research Council in the United States and conducted by a wide ranging team of experts, resulted in a comprehensive report etitled *America's Climate Choices* in 2011. The world currently emits upwards to thirty billion tons of CO_2 per year from the combustion of fossil fuels. Twenty percent of these emissions are created by the United States. In *America's Climate Choices* it is acknowledged that limiting climate change will necessitate global participation and contribution, noting that greenhouse gases do not observe national boundaries:

> A molecule of CO_2 emitted in India or China has the same effect on the climate system as a molecule emitted in the United States. There is wide agreement that limiting the magnitude of climate change will require substantial action on behalf of all major GHG-emitting nations, including both the industrialized nations and the rapidly developing countries whose relative share of global emissions is rapidly increasing.[3]

The report proffers that development of strong credible policies by the US for reducing emissions will ultimately help advance similar response by individual nations and help facilitate greater international cooperative engagement.

Western industrialized countries carry a much greater responsibility for past emissions and continue to emit large quantities of carbon dioxide; however many developing countries through rapid expansion and growth are now significant greenhouse gas polluters. In recent years China has surpassed the US as the world's largest emitter of greenhouse gases. China, US, the European Union, Brazil, Indonesia, Russia, and India account collectively for approximately 60 percent of emissions. This group of countries also accounts for approximately 55 percent of world population. More than 75 percent of carbon dioxide emissions derive from burning of fossil fuels, principally coal, oil and natural gas.

Aside from concerns about fossil fuel emissions, there is increasing concern regarding global supply to meet market demand for crude-oil. In "Oil's tipping point has passed," Murray and King argue that since 2005 conventional crude oil production has not risen to match increasing demand. Prior to 2005, production increased in line with growing demand, however supply has been relatively constant over the ensuing eight years to the present day:

> In 2005, global production of regular crude oil reached about 72 million barrels per day. From then on, production capacity seems to have hit a ceiling at 75 million barrels per day. Analysis of prices against production from 1998 to today . . . shows this dramatic transition, from a time when supply could respond elastically to rising prices caused by increased demand, to when it could not. As a result, prices swing wildly in response to small changes in demand.[4]

In a special report entitled "Golden Rules for a Golden Age of Gas," IEA explores the case for exploitation of unconventional gas (in particular shale gas, tight gas, and coalbed methane) and questions whether natural gas is poised to enter a golden age. Some view wider deployment of natural gas as a way to provide increased energy security, while others remain concerned about potential environmental damage which may result from hydraulic fracturing (fracking). Additional complexities surrounding availability and supply of crude oil, coal and natural gas will be further addressed in ensuing chapters.

The authors of this book are primarily engineers, social scientists, and policy specialists and do not claim in-depth expertise in the related sciences of climatology and greenhouse gas emissions. The book will therefore not attempt to explain the complex relationship between energy, emissions and climate nor further argue in favor or against the case for accepting a particular projection. As concerned citizens and educators of student engineers, scientists, and technologists, this book rather seeks to question how greenhouse gas emissions can be reduced, address challenges in bringing advanced technologies to market, and identify steps to be taken that will facilitate more diversified and sustainable global energy systems.

Engineers will need to be engaged in solving these issues to the same extent as they have, however unwittingly, contributed to their creation. Major steps forward in the adaptation, development, and use of technology will be required. Greater investment in energy efficient technologies, low carbon technologies, renewable technologies, nuclear energy, and carbon capture and storage technologies is now needed. In transportation, increased vehicle efficiency along with a gradual shift from conventional petroleum-fueled technology to hybrids and other advanced vehicles promises to extend fuel and diversify to new sources of energy. Applying biofuels to both air and ground based transportation, which have notably different fuel specifications, is also of crucial importance. Greater momentum is now evident in applied research with a focus on product-to-market renewable energy technologies including solar photovoltaic, wind, marine, and geothermal. The future of nuclear energy and the development of smart grids and super grids are seminal research questions facing today's engineers and policy makers. Adaptations to existing residential and commercial buildings to more passive, energy efficient, less fossil-fuel dependent dwellings is an area of growing concern in many western countries today. Research on energy distribution and future interconnected grid networks is of growing interest to national energy utilities, with emerging opportunities for greater cooperation between nations in both energy trading and in ensuring energy availability and security of supply.

In a modern book discussing the interactions of energy and climate, technology and policy, an attempt to integrate all aspects of these issues would be daunting at best. Therefore we have organized the narrative by selecting target technologies, representative policy concepts, and instructive case studies. Specific technologies, policies, and countries have necessarily been chosen in order to expand the reader's breadth and depth of understanding and stimulate additional discussion and investigation. No attempt has been made to introduce every conceivable energy solution, or even to suggest a priority for those most promising today. Rather, it is believed that technology and policy remain flexible, and that a truly robust energy and climate strategy should be adaptable to achieve multiple objectives in the face of changing variables, evolving economies and electorates, and new scientific data and discoveries.

In order to achieve solutions of the required scale and magnitude within a limited timeline, it is essential that engineers, scientists, and technologists be not only technologically adept, but also aware of the wider social and political issues that governmental policy-makers face. Likewise, it is imperative that policy makers work closely with the academic community to interpret data and chart the way to achieve bold, timely, and lasting change. This book is designed to bridge the gap between these two communities. Central issues in global energy will be discussed through interdisciplinary dialogue and contribution by a host of experts in their respective fields.

Book Layout

The book is organized in three parts.

Part 1: Global Energy Crisis in Context. Chapter 1 considers man's dependence on carbonaceous fuels for survival through time. The technological and economic developments of the industrial revolution are recalled, with a focus on the detrimental effects resulting from excessive burning of coal to meet energy requirements. This coincides with emerging scientific awareness in the eighteenth century of the nature of Earth's atmosphere and the delicate balance of its constituent elements. The history of society's growing dependence on coal, oil, and natural gas, the emergence of new methods of extraction such as hydraulic fracking, and the introduction of clean technologies and the proposed capture and storage of carbon are reviewed in context.

Chapter 2 explores current global energy demand and expected demand growth in the coming decades. Demand has more than doubled in the last four decades; with reliance still heavily weighted on the traditional fossil fuels: coal, oil, and natural gas. A review and comparison of energy policy in the US and the EU is made, including the important 1997 Kyoto Protocol and subsequent UN energy and climate conventions. This is followed by an exploration of energy policy directives and trends in China, Russia, Brazil, and India. These nations all have large populations as well as significant energy demands and resources, and will be critical players in global efforts to align energy and climate trade and policies.

As educators, our primary objective is to equip graduates with the necessary skill sets to understand social context and to help them contribute to the solutions of energy's challenges. Chapter 3 explores social engagement by the engineer, through understanding the social environment and awareness of common authentic values and principles. Themed case studies are included to address how social environment influences engineering practice.

Part 2: Energy Conversion Technology. In harnessing the forces of nature Chapter 4 reviews a range of renewable energy technologies including wind, hydro, marine wave and tidal, and geothermal energy. A discussion of recent developments and growth in both onshore and offshore wind energy is followed by an appraisal of historical developments in hydropower. The case for wave and tidal energy is made with a review of emerging technologies and the challenges engineers continue to face. The chapter concludes with an investigation of geothermal energy and its place in the energy mix.

Solar energy is emerging as an important source of renewable energy with potential for increased grid penetration. Developments in nanotechnology have enabled the study of materials at an atomic level, opening up an exciting frontier in materials science. Applications of nanotechnology to solar energy devices are resulting in improvements in solar energy conversion efficiency. Emerging technologies are also enabling improved robustness to thermal variation and environmental degradation of solar devices. Chapter 5 addresses the current status of research in nanotechnology in association with solar photovoltaic, solar concentration, and thermoelectric devices. The chapter also explores future opportunities for nanotechnology in energy conversion and storage.

Bioenergy is a forefront research frontier. Chapter 6 provides a history of first and second-generation biofuel production, and explores policy which has enabled developments in biofuels in the US, Brazil, and the EU. Feedstocks, conversion processes, and end products in advanced biofuel technologies are explored. An examination is made of five uncertainties associated with the industrial development of biofuels and other challenges and opportunities facing the industry are explored. The chapter includes with a technology update of advanced biofuel conversion projects.

In chapter 7 we shift focus from renewable energy technologies, to consider the role of nuclear engineering. We examine the social, environmental, technological, and power capacity capability of nuclear fission reactors. An exploration of the historical development of nuclear engineering, the nuclear fuel cycle and nuclear energy as a provider of baseload generation is followed by a review of nuclear energy safety. Nuclear accidents and their effect on public perception are explored through scenario discussions of Three Mile Island, Chernobyl, and Fukushima. Challenges in handling waste with current policy, including disposal or storage of nuclear fuel stockpile, are explored and quantified. The chapters closes with a brief discussion on nuclear fusion and where it might lead.

Part 3: Energy Distribution and Use. Chapter 8 explores policy perspectives and challenges presenting in taking emerging renewable energies to market. The chapter first explores a range of influential factors including economic, political, social, environmental, and maintainability. A brief appraisal is then made of the economics of energy together with a study of levelized costs of new generation energy resources, both dispatchable (including coal, gas, nuclear, and biomass) and non-dispatchable (such as wind, solar, and hydro). An exemplar study of challenges for emerging wave energy technologies concludes the chapter.

In chapter 9, attention turns to consideration of energy used for transportation, noting the sector's disproportionate reliance on oil. The resulting geopolitical, economic, and environmental consequences present difficult near and long-term challenges. The chapter is divided into three parts; part 1 is an introduction and overview of current transportation energy issues. Part 2 explores the specific challenges facing the automotive transportation sector. A brief history of automotive technology is followed by a classification of modern vehicle configurations, including internal combustion engine and hybrid electric driveline configuration developments. In Part 3, we turn our attention to the aviation transportation sector by exploring aviation fuels and regulations, followed by a discussion of challenges to the development and production of alternative aviation fuels and fuel emissions.

Noting that energy use in the built environment accounts for approximately 40% of the energy consumed in developed countries, chapter 10 is devoted to this highly important sector. In 2004, the emissions resulting from direct energy use in the built environment were estimated at close on 9 Gt of CO_2 per year. There is general agreement that through the use of mature technologies, building energy usage can be reduced substantially. The chapter begins with a thorough introduction to the magnitude of energy consumed by existing buildings in the developed world. Practical and currently available retrofit technologies with long-term potential for building energy reduction are described along with variables that influence the choice and effectiveness of these technologies. A discussion ensues of the challenges and barriers to implementing these technologies. This is followed by an exploration of policy challenges that confront energy efficient retrofits. A review of building energy reduction programs employed in the US and EU is followed with recommendations and opportunities for future solutions.

Epilogue

The epilogue provides a pivotal synthesis of questions posed, lessons learned, and insights gained, and of the continued challenges in both meeting future energy demands and helping reduce manmade carbon emissions.

Notes

1. The Keeling curve, available from the Scripps Institution of Oceanography at http://keelingcurve.ucsd.edu/, measures the concentration of carbon dioxide in the atmosphere. Measurements are recorded on top of Hawaii's Mauna Loa. This work commenced in 1958 under the tutelage of Charles David Keeling, and is the longest running such measurement in the world.
2. Energy Information Administration, *Annual Energy Outlook 2011 with Projections to 2035* (Washington, D.C.: U.S. Department of Energy, 2011).
3. Committee on America's Climate Choices, Board on Atmospheric Sciences and Climate, and National Research Council of the National Academies, *America's Climate Choices* (Washington, D.C.: National Research Council, 2011), 32.
4. Murray, James, and David King, "Climate Policy: Oil's Tipping Point Has Passed," *Nature International Weekly Journal of Science* 481 (2012): 434. http://dx.doi.org/10.1038/481433a.

Bibliography

Committee on America's Climate Choices, Board on Atmospheric Sciences and Climate, and National Research Council of the National Academies. *America's Climate Choices*. Washington, D.C.: The National Academies Press, 2011.

Dow, Kirstin, and Thomas E. Downing. *The Atlas of Climate Change: Mapping the World's Greatest Challenge*. 3rd ed. Abingdon: Earthscan, 2011.

Energy Information Administration. *Annual Energy Outlook 2011 with Projections to 2035*. Washingon, D.C.: U.S. Department of Energy, 2011. http://www.eia.gov/forecasts/archive/aeo11/.

Murray, James, and David King. "Climate Policy: Oil's Tipping Point Has Passed." *Nature International Weekly Journal of Science* 481 (2012): 433–435. http://dx.doi.org/10.1038/481433a.

PART 1

THE GLOBAL ENERGY CRISIS IN CONTEXT

Chapter 1

Reflections on Energy, Greenhouse Gases, and Carbonaceous Fuels

EUGENE D. COYLE, WILLIAM GRIMSON,
BISWAJIT BASU, AND MIKE MURPHY

Abstract

In this chapter, we review the history of man's dependence on carbonaceous fuels for survival, beginning with pre-industrial civilizations, during which charcoal was processed for thousands of years to smelt iron and copper. In the eighteenth and nineteenth centuries, however, coke and coal became prime energy resources which powered the engine rooms of the industrial revolution. Accompanying the economic and societal benefits of this period was the recognition of the damage resulting from smog owing to excessive burning of coal, which affected both human health and the natural environment. These pivotal centuries laid the foundation for the advancement of scientific knowledge and discovery which underpinned both engineering developments and the sciences of the natural world, including earth science, atmospheric science, and meteorology. These developments in turn led to our modern understanding of climate change and the effect of greenhouse gases.

Today coal, petroleum, and natural gas still play a vital role in our global energy mix. While scientists and engineers have developed clean coal technologies such as carbon capture and storage, it is important to question whether such technologies can offset the growing carbon footprint caused by the use of carbonaceous fuels. This challenge is complicated by the growth in scale of total global world energy demand, the scale of economic investment required to implement such technologies, and the race against time to minimize the damage resulting from continued use of fossil fuel energy.

1.1. Introduction: Man's Quest for Energy

Humankind has always needed energy, and while the source and usage of energy have changed over time some patterns have remained constant. In earlier times food was the key source of energy for people and their livestock. This form of energy not only allowed our race to survive but dictated in part how civilization developed. Societies worldwide focused on developing new and sustainable food sources. The storage of food and its distribution was a factor in how groups learned to organize themselves communally, best survive periods of shortage, and also benefit from occasional abundances. The discovery of methods of processing and preserving food meant that new sources of food could be used with increased efficiency and increasingly less waste. People migrated across continents, seas, and oceans in response to sometimes complex social pressures, but certainly the search for food and reliable sources of food was a common factor in their movements. There may be a greater urgency today than heretofore to identifying sustainable sources of energy, increasing the efficiency of energy usage, and finding new sources of energy due to expanding world population, depletion of energy resources, and growing environmental concerns; but there is no question that similar patterns have been in evidence for thousands of years. And there is something timeless and circular about modern society growing crops that once would have been considered food, but now are solely intended to produce energy as biofuels.

The history of how energy is and was used illustrates how competing usages dictate the exploitation of resources, often to the detriment of the original but less powerful first adopters. Charcoal as fuel for cooking has a long history and is still in demand today for use in barbecues. Yet more than five thousand years ago, people found that it was useful in smelting of iron and in the Bronze Age applied it to the production of copper and more valuably, bronze. These and subsequent developments caused the clearing of woodlands and competed with land once intended only for agricultural purposes. The use of banks to divide land facilitated the retention of some trees which were then coppiced to provide a source of charcoal. By the thirteenth century Europeans had learned of the Chinese explosive gunpowder, which created a new demand for charcoal yet again. The military use of gunpowder necessitated the casting of cannons, requiring a considerable amount of charcoal. These factors put pressure on supplies of wood suitable for charcoal production, leading to the introduction of restrictions in certain countries. By the eighteenth century the demand for charcoal to support the iron industry was so high that an alternative was desirable, and this was found in the form of coke. Not only could coke replace charcoal for many industrial purposes, but a byproduct of coke production was a combustible gas that could be used in households. Not surprisingly coal and coke producers encouraged the use of their products, further reducing the demand for charcoal. The historical relationship between coke and charcoal demonstrate how a single energy source can have many interacting uses and drivers for its exploitation, and that the resultant interrelationships between users and suppliers are complex.

During World Wars I and II and their aftermath, the world witnessed both the horror of the destructive power of nuclear energy and the potential promise of an efficient, reliable and clean source of electrical energy. The debate on the future mix of nuclear power in global energy provision, which had to address such issues as nuclear waste disposal, nuclear power plant accidents and their environmental and social consequence, and the continued development and dependence on nuclear energy from an armaments perspective, continues today (these issues are explored further in chapter 7). Furthermore, the general argument that environmental factors are not the only ones that influence decisions on energy production also applies to what might be called green or clean technologies. Lobby groups pushing their own agendas have not always supported their stances with high quality economic and environmental data. As a result, the informed public has rightly become more robust in questioning the latest projects to harness power through renewable and sustainable sources, whether those involve estuary barrages, wave power, offshore wind, solar power, or bioenergy. Apart from searching for new solutions and developing new methods of production, energy engineers have a clear responsibility to help inform policy makers and the general public of the pros and cons of each means of energy production.

The world has truly become a global village. The challenges to achieving global economic security and sustainable living—in a world of increasing population and multivariable levels of wealth and social inequality—are complex and vast. The relationship between man and machine, productivity and industrial development, marches on. Whether in cities of the so-called developed nations or in the rapidly expanding urban population centers of the developing world, concern for the atmosphere that sustains Earth's ecosystem is of growing importance. Air pollution affects the overall balance and ultimate health of the ecosystem. It is instructive to briefly review the nature and composition of Earth's atmosphere and to explore the important role played by carbonaceous fuels throughout human history.

1.2. Earth's Atmosphere and Greenhouse Gases

1.2.1. Climate Variability

Climate variability is one of the great discussion points and climate change one of the great concerns of humankind today. Research in climate science and meteorology is long established and it is therefore fitting to briefly review the writings of a selected band of pioneering thought leaders of the nineteenth century in their contemplations of Earth's atmosphere and its makeup.

In the 1820s, Jean Baptiste Joseph Fourier calculated that, based on its size and distance from the sun, planet Earth should be considerably cooler than it actually is, assuming it is warmed only by the effects of incoming solar radiation. He examined various possible sources of the additional observed heat, and ultimately concluded that the Earth's atmosphere acts in some way as an insulator, thus retaining quantities of incoming solar heat. This observation may be considered the earliest scientific contribution to what today is commonly known as the greenhouse effect.[1]

Forty years later John Tyndall identified the radiative properties of water vapor and CO_2 in controlling surface temperatures. In 1861, after two years of painstaking experiments, Tyndall published a lengthy paper packed with results. Among the findings, he reported that moist air absorbs thirteen times more heat than dry, purified air.[2] Tyndall observed that:

> The waves of heat speed from our earth through our atmosphere towards space. These waves dash in their passage against the atoms of oxygen and nitrogen, and against the molecules of aqueous vapor. Thinly scattered as these latter are, we might naturally think meanly of them as barriers to the waves of heat.[3]

In the early twentieth century, Swedish scientist Svante Arrhenius asked whether the mean temperature of the ground was in any way influenced by the presence of the heat-absorbing gases in the atmosphere. This question was debated throughout the early part of the twentieth century and is still a main concern of earth scientists today. Arrhenius went on to become the first person to investigate the effect that doubling atmospheric carbon dioxide would have on global climate and was awarded the 1903 Nobel Prize for Chemistry.[4]

It is well understood that Earth's atmosphere comprises a layer of gases surrounding the planet and retained by gravity.[5] Extending from Earth's surface, the atmosphere protects life on Earth by absorbing ultraviolet solar radiation, warming the surface through heat retention (the *greenhouse effect*), and reducing temperature extremes between day and night through a process called *diurnal variation*. The air we breathe contains approximately 78.1% nitrogen, 20.9% oxygen, 0.9% argon, 0.04% carbon and small amounts of other gases. These other gases, often referred to as trace gases, also comprise the *greenhouse gases*.

An atmospheric greenhouse gas (GHG) can absorb and emit radiation within the thermal infrared (IR) range of the electromagnetic spectrum of light.[6] The primary greenhouse gases of Earth's atmosphere are water vapor, carbon dioxide, methane, nitrous oxide, and tropospheric ozone.[7] Solar radiation passing through the atmosphere heats the surface of the Earth. Some of the energy returns to the atmosphere as longwave heat energy radiation, some energy is captured by the layer of gases that surrounds the Earth, and the remainder passes into space. The concentration and proportional mix of these gases in the atmosphere influence climate stability and changes in composition can result in climate change. Since the commencement of the industrial revolution, human activity such as the burning of fossil fuels, the release of industrial chemicals, the removal of forests that would otherwise absorb carbon dioxide, and their replacement with intensive livestock ranching, has changed the types and quantities of gases in the atmosphere. This in turn has substantially increased the capacity of the atmosphere to absorb heat energy and emit it back to Earth. Some greenhouse gases stay in the atmosphere for only a few hours or days, while others remain for decades, centuries, or even millennia. Greenhouse gases emitted today will drive climate change long into the future, and the process cannot be quickly reversed.[8]

1.2.2. Carbonaceous Fuels

Carbon dioxide emissions come from combustion of carbonaceous fuels such as coal, oil and natural gas. Carbon dioxide has an atmospheric lifetime of about one hundred years; methane, twelve years; and nitrous oxide, one hundred fourteen years. Methane is up to twenty-five times more effective than carbon dioxide in the capture of heat in the atmosphere and its radiative effect is approximately seventy times larger, however it exists in much smaller concentrations and therefore its overall environmental impact is significantly less. In addition to its production through farming livestock, rice cultivation, and coal mining, there are large quantities of methane in arctic permafrost ice[9] and below ocean sediments. Release of such gas could result in major environmental damage; large-scale release has not occurred in recent history, but remains a point of genuine concern.

Isn't it ironic that the natural elements of coal, gas, and oil, having sustained human life over thousands of years, are now viewed to a certain degree as offenders, responsible for the pollution that has upset the balance of nature? It is, of course, mankind that has created the current instability through insatiable exploitation of Earth's resources. It is therefore mankind's responsibility to ensure every effort be made to redress the damage done and to work toward a more sustainable eco-environment.

1.2.3. Fossil Fuels Through History

Fossil fuels are formed by natural processes such as the anaerobic decomposition of buried dead organisms, through exposure to heat and pressure in the Earth's crust over time periods of typically millions of years. Containing high percentages of carbon, fossil fuels include coal, petroleum, and gas. They range from volatile materials with low carbon to hydrogen ratios, such as methane (CH_4), to liquid petroleum, to nonvolatile materials composed of almost pure carbon, such as anthracite coal.[10] George Agricola is credited as the first scientist to have articulated the biogenic of fossil fuel creation. His most famous work, the *De re metallica libri xii*, a treatise on mining and extractive metallurgy, was published in 1556. Agricola described and illustrated how ore veins occur in and on the ground, making the work an early contribution to the developing science of geology.[11]

In 2011, fossil fuel consumption in the United States totaled eighty quadrillion British thermal units (Btu). The US Energy Information Administration (EIA) estimated that 80% of that energy was derived from fossil fuels, specifically 35.3% from petroleum, 19.6% from coal, and 26.8% from natural gas. Nuclear energy and renewable energy accounted for 8.3% and 9.1%, respectively.[12]

Fossil fuels are non-renewable resources because they take millions of years to form, and reserves are being depleted much faster than new ones are being made. The burning of fossil fuels produces over twenty-two billion tonnes of carbon dioxide (CO_2) per year, but it is estimated that natural processes can only absorb about half of that amount. This causes a net increase of eleven billion tonnes of atmospheric carbon dioxide per year.

1.2.3.1. Coal

One of Earth's most valued natural resources, coal has been a provider of warmth and energy to humankind for hundreds, if not thousands of years. Resulting from decaying woodland vegetation, compressed by rain water and repeatedly added to through further additional mineral vegetation deposit over hundreds of thousands of years, peat was formed which over time hardened to lignite (brown coal) and then to coal, a dark colored sedimentary rock made of both inorganic and organic matter. With many different classifications of grade and composition, also referred to as coal rank, coal is primarily composed of carbon, while also containing elements of hydrogen, oxygen, nitrogen, aluminum, silicon, iron, sulfur and calcium. Coal can in fact contain as many as one hundred twenty inorganic compound trace elements with over seventy of the naturally occurring elements of the periodic table. Designated coal types range from lignite to flame coal, sub-bituminous, bituminous through to nonbaking coal and anthracite, classified in accordance with percentage element composition. The particles of organic matter in coal are referred to as macerals, indicative of plants or parts of plants including bark, roots, spores and seeds, which originally contributed to a particular coal formation. Coal rank is determined by the percentage of fixed carbon, moisture, volatile matter, and calorific value in British thermal units after the sulfur and mineral-matter content have been subtracted.[13]

Coal is the world's most abundant and widely distributed fossil fuel, accounting for more than one quarter of global primary energy demand. With global proven reserves totaling nearly one trillion tonnes it remains one of the most important sources of energy for the world, particularly for power generation.[16] Coal fuels high percentages of electricity to the United States (49%), India (69%), China (79%), Poland (92%), and South Africa (97%), and supplies in excess of forty percent of the global electricity generation requirements, including Germany and much of central Europe. More than twenty-three percent of total world energy and thirty-six percent of world electricity is produced by coal, with a projected growth of 2.4% annually in the consumption of electricity between 2005 and 2030. Over the last decade demand for coal has outpaced that for gas, oil, nuclear power, and renewable energy sources. North America, the former Soviet Union, and Pacific Asia combined account for more than eighty percent of proven coal reserves. Global coal production in 2009 topped 6.9 billion tonnes, with China producing approximately 46 percent, the United States 16 percent, and Australia and India equal producers at roughly 6 percent. Bituminous coal dominates world production, followed by lignite and coking coal. Sixty percent of coal is produced through underground mining. Australia and Indonesia are the two main coal exporting countries. Most coal-producing nations produce for their home markets exclusively, and import the balance required to meet national demand. In spite of environmental concerns, coal is expected to continue to be the second greatest global source of energy through 2030.[17]

Coal-fired power plants, however, are facing new challenges owing to increased competition from natural gas and new air pollutant regulations advanced by the EPA in 2011, requiring in particular reduced emissions of mercury, acid gases, and soot.[18] Average CO_2 emissions from coal-fired power plants are roughly double those from natural gas plants, approximately 2,250 and 1,135 pounds per megawatt hour, respectively. Coal fired plant retirements are projected to rise to nine thousand megawatts by 2014, with a reduction in generating capacity from coal of well in excess of ten percent.[19]

Carbon and the Industrial Revolution

Socio-techno-economic factors all played their part in how industrial revolutions originated, developed, transformed and then eventually evolved to a post-revolution industrial society. One of the key factors undoubtedly was the availability of energy and invariably that source of energy was coal. In Great Britain Matt Ridley noted that it was not just the availability of coal but that for other and existing sources "there was never going to be enough wind, water or wood in England to power the factories, let alone in the right place."[14] Of course this comment has to be qualified in that we now know that there was and is sufficient energy available from wind and water, but the technology and know-how did not exist to harness the levels of energy required by industries such as iron and transport. Ridley refers exclusively to windmills, water mills, and charcoal. Another point Ridley makes is that the widespread use of horses required a huge amount of food (itself an energy source) that required up to one-third of the available arable land—land required to feed a growing population. It was therefore necessary to abandon renewable sources of energy if the Industrial Revolution was to take off. But the picture was complex, as technological innovation was required in order to exploit coal at an economic advantage. Effective water pumps were required for mining, and new transport solutions were needed to deliver coal to where it was to be used. Steam engines for both pumps and early trains, as well as the rapidly expanding rail network, required machines to manufacture and shape the necessary parts, and it was coal that ultimately provided the power.

Coal as it was used during the Industrial Revolution came at an additional cost in terms of a set of disadvantages. First, it was dirty, resulting in huge amounts of ash. Second, it produced a range of toxic flue gases as well as carbon dioxide. Third, the production of coal left its mark on the landscape and more importantly on the men and their families who carried out the mining. And because coal was abundant, there were few incentives to investigate alternatives or even to be much concerned with efficiency. A plentiful supply of coal replaced a number of largely clean energy sources, but this was not considered an issue as long as profits were increasing. Of course today clean—or more correctly cleaner—coal technology (such as flue gas scrubbing) has been developed, but the long-term damage cannot be undone.

Some of the problems that accompany coal mining include:
- *Acid mine drainage* results when coal beds and surrounding strata containing medium to high amounts of sulfur (sulphide compounds) are disrupted by mining, thereby exposing sulphides to air and water.
- Atmospheric sulfur oxides (SO_x) and subsequent acid decomposition, such as acid rain, result from the burning of medium to high-sulfur coal.
- The quality of surface and ground water may be adversely affected by the disposal of the ash and sludge that result from the burning of coal and flue gases.
- Environmental greenhouse gas emissions result from the release of carbon dioxide (CO_2) and nitrogen oxides (NO_x) through the burning of coal.

> - Additional trace elements are released through burning of coal.
>
> Developments both during and subsequent to the Industrial Revolution also resulted in great benefits to society. The Enlightenment provided the stimulus for creative development and innovation, an increased interchange of knowledge coupled with a new entrepreneurial vigour.[15]

1.2.3.2. Petroleum[20]

Petroleum, also termed crude oil, contains hydrocarbons and other organic compounds. It is found in natural formations beneath the Earth's surface. It is derived from ancient organic materials such as zooplankton and algae. Petroleum is recovered mostly through oil drilling. Colonel Edwin Laurentine Drake is credited as the first person to have successfully drilled for oil in the United States. Employed by the Seneca Oil Company in 1858 to investigate oil deposits in Pennsylvania, Drake devised a 10-foot long cast-iron drive pipe which struck bedrock at 32 feet. The following morning crude oil was seen to be rising up and oil was brought to the surface using a hand-pitcher pump.[21]

The discovery of oil triggered an oil rush in America, fueled by an American law which conferred ownership of underground resources to the landowner. During this period, crude oil was refined into kerosene and was used to light homes and businesses. In 1863 John D. Rockefeller entered the fray and concentrated his business on the refining, transportation, and distribution of petroleum. After founding the Standard Oil Company in 1870, Rockefeller became the dominant figure in the late nineteenth century petroleum industry. Exploration in other parts of the United States, in particular Texas, and in countries including Russia, Dutch East Indies, Indonesia, Venezuela, Trinidad, and Mexico opened up the market with competition from companies including the Royal Dutch Company and Shell. It wasn't until around 1910 that oil began to overtake coal as the primary global energy driver. This was largely due to the rapid growth of the automobile industry and the necessity for widespread availability and supply of gasoline.[22]

Today, petroleum is refined and separated into a range of consumer products, including gasoline (petrol), kerosene (paraffin), asphalt (bitumen), and chemical reagents used to make plastics and pharmaceuticals. Petroleum is used in manufacturing a wide variety of materials, and it is estimated that the world consumes up to eighty-eight million barrels per day. The term petroleum strictly refers to crude oil, however in common usage it includes all liquid, gaseous, and solid hydrocarbons. Under certain surface pressure and temperature conditions, lighter hydrocarbons including methane, ethane, propane, and butane exist as gases, while pentane (C_5H_{12}) and heavier organic compounds exist in the form of liquids or solids.[23]

An oil well produces predominantly crude oil, with some natural gas dissolved in it. Because the pressure is lower at the surface than underground, some of the gas will come out of solution and be recovered (or burned) as *associated gas* or *solution gas*. A gas well produces predominantly natural gas, however because the underground temperature and pressure are higher than at the surface, the gas may contain heavier hydrocarbons such as pentane, hexane, and heptane in the gaseous state. At surface conditions these will condense out of the gas to form natural gas condensate. The proportion of light hydrocarbons in the petroleum mixture varies greatly among different oil fields, ranging from as much as 97 percent by weight in the lighter oils to as little as 50 percent in the heavier oils and bitumens.

Within the Organization of Petroleum Exporting Countries (OPEC), the five Middle Eastern countries Saudi Arabia, Iran, Iraq, Kuwait, and United Arab Emirates accounted for seventy percent of oil reserves in 2009, with Saudi Arabia alone accounting for twenty-six percent of total OPEC reserves.[24] Non-OPEC production accounts for about sixty percent of current global oil supply. China has emerged as the largest oil consuming country in the world with annual growth rate of about seven percent.

Global crude oil consumption grew by 0.6 million barrels per day (b/d) in 2012, reaching 88 million b/d. OECD consumption actually declined by 1.2% in line with trends over recent years. Non-OECD consumption grew by 2.8%. China had the largest consumption growth at 5.5% (505,000 b/d).

1.2.3.3. Natural Gas

Natural gas is found in deep underground natural rock formations or associated with other hydrocarbon reservoirs in coal beds and as methane clathrates. As discussed above, petroleum is also found in proximity to and with natural gas. Most natural gas was created over time by either a biogenic or thermogenic mechanism. *Biogenic gas* is created by *methanogenic* organisms (microorganisms that produce methane as a metabolic byproduct in anoxic conditions) in marshes, bogs, landfills, and shallow sediments. Deeper in the earth, at greater temperature and pressure, *thermogenic gas* is created from buried organic material.[25]

Natural gas is often informally referred to simply as *gas*, especially when compared to other energy sources such as oil or coal. In the nineteenth century, natural gas was obtained as a byproduct of producing oil, since the small, light gas carbon chains came out of solution as the extracted fluids underwent pressure reduction from the reservoir to the surface. If unwanted, natural gas was burned off at source in the oil field. Today, unwanted gas may be returned to the reservoir through injection wells. Where economical, gas may be transported using a network of pipelines. By converting gas into a form of liquid gasoline or diesel, it may also be exported as a liquid, commonly referred as *gas to liquid* (GTL), or to a jet fuel by applying the *Fischer-Tropsch process* (a collection of chemical reactions that converts a mixture of carbon monoxide and hydrogen into liquid hydrocarbons).

Before natural gas can be used as a fuel, it must undergo processing to remove impurities, including water, to meet the specifications of marketable natural gas. The byproducts of processing include ethane, propane, butanes, pentanes, and higher molecular weight hydrocarbons, hydrogen sulfide (which may be converted into pure sulfur), carbon dioxide, water vapor, and sometimes helium and nitrogen.

Natural gas extracted from oil wells is called *casinghead* gas or *associated* gas. The natural gas industry is extracting an increasing quantity of gas from challenging resource types, including *sour gas, tight gas, shale gas,* and *coalbed methane*.

The world's largest proven gas reserves are located in Russia, at approximately 48 terra (10^{12}) cubic meters. Russia is frequently the world's largest natural gas extractor. Other major proven resources (in billion cubic meters) exist in Iran (26,370 in 2006), Qatar (25,790 in 2007), Saudi Arabia (6,568 in 2006) and the United Arab Emirates (5,823 in 2006).

The world's largest gas field is Qatar's offshore North Field, estimated to have twenty-five trillion cubic meters of gas in place.[26] The second largest natural gas field is the South Pars Gas Field in Iranian waters in the Persian Gulf. Located next to Qatar's North Field, it has an estimated reserve of eight to fourteen trillion cubic meters of gas.

1.2.3.4. Shale Gas

Shale is a fine-grained, clastic sedimentary rock composed of mud that is a mix of flakes of clay minerals and tiny fragments of other minerals, notably quartz, calcite, feldspar, and dolomite. The ratio of clay to other minerals is variable. Shale is characterized by breaks along thin laminae or parallel layering or bedding less than one centimeter in thickness, called *fissility*.[27] Shale is easy to break or split, slate-like, into smaller planar sheets. Oil shale occurs where organic material is present in the process of producing the sedimentary rock. Extraction of gas and oil is carried out by heating the shale to temperatures in the region of 475°C. It is estimated that there are about nine hundred trillion cubic meters of unconventional gas such as shale gas, of which one hundred eighty trillion may be recoverable.[28]

Shale gas was first extracted as a resource in Fredonia, New York in 1821, in shallow, low-pressure fractures.[29] In the mid-1800s James Young, a Scottish chemist, devised a method of extracting from shale an oil product which he then distilled to yield kerosene, naphtha, heavier lubricating oils, and paraffin wax. Fuel (kerosene) to provide lighting was an important use of the shale oil at the end of the industrial revolution. The shale residue gave rise to shale bings (small hills or tips) which are still a feature of the landscape today in West Lothian, Scotland.

Horizontal drilling began in the 1930s, and in 1947 a well was first fracked[30] in the United States. Work on industrial-scale shale gas production did not begin until the 1970s, when declining production potential from conventional gas deposits in the United States spurred the federal government to invest in research, development, and demonstration projects that ultimately led to directional and horizontal drilling, microseismic imaging, and massive hydraulic fracturing. Up until the public and private demonstration projects of the 1970s and 1980s, drilling in shale was not considered to be commercially viable.

Approximately thirty countries have oil shale in quantities that are economically extractable and they include Brazil, China, Germany, Russia, Sweden, and the United States. Worldwide peak production occurred in the 1980s (forty-six million metric tons).[31] Allix et al. note that "current estimates of the volumes recoverable from shale oil deposits are in the trillions of barrels, but recovery methods are complicated and expensive . . . but may soon become economically viable."[32]

As with many sources of energy, the use of shale gives rise to environmental issues. Two major disadvantages of oil recovery from shale is the huge amounts of waste rock (spent shale) and the requirement for large quantities of water in the post-extraction treatment. To mitigate the waste problem, one company, Shell, has investigated *in situ* extraction of oil based on a method first developed in Sweden using electric heaters. Trials have also investigated the use of high temperature injected steam. As the United States has the largest oil shale deposits in the world and bearing in mind its policy of being as independent as possible of other oil suppliers, it is likely that the exploitation of these deposits and the development of extraction methods will intensify.

There are growing fears among concerned citizens about the social and environmental impacts of hydrofracking, not least in the United States. The New York Marcellus Shale formation extends from West Virginia to southern New York. Lobby groups such as the Nature Conservancy[33] point out that high volume horizontal fracturing (hydrofracking) would necessitate use of millions of gallons of water per fracking treatment. The water used also contains oil, grease, and small amounts of other chemicals and it is estimated that up to forty percent of this water will return to the surface, resulting in various degrees of environmental pollution.

1.2.4 Clean Fossil Fuels: Future Challenges and Prospects

1.2.4.1. Overview

While there has been a substantial growth in generation of power from renewable and green sources, coal will remain a significant source of fuel for power generation due to the requirements of availability, security and diversity of supply. It is therefore important to review the state of the art in the field of Clean Coal Technologies (CCTs) and how such technologies may help reduce carbon emissions going forward.

The primary drawback associated with the use of traditional coal is that modern coal-fired power plants operate at low efficiencies and emit large amounts of pollutants. This drawback can be circumvented by instead using clean coal. CCT is a product of several generations of technological advances. Since the process of combustion is the key for energy generation, CCT has led to more efficient combustion of coal with reduced emissions of greenhouse gases (GHGs), including carbon dioxide, sulfur dioxide, and nitrogen oxide. The market for CCT is steadily growing owing to the imminent need to reduce GHGs and improve upon power plant efficiency.

One of the biggest challenges in the implementation of CCT is that the quality of coal is extremely variable and that coal combustion structurally produces more pollutants than other fossil fuels. Coal is also a major ingredient in the production of steel. A further concern is China; the world's largest and fastest growing economy ranks number one in coal production, accounting for more than forty percent of global production. The extensive use of coal worldwide, coupled with a large number of old, inefficient power plants lacking proper emission control equipment, adds to the pollution generated through burning coal.

Given the likelihood that coal will continue to feature prominently in the energy mix for decades to come, adaptation and deployment of CCT in both new and existing

plants, where possible, is essential. Power plants being built today are more efficient and emit ninety percent less pollutants (SO_2, NOx, particulates and mercury) than plants built in the 1970s. There are three stages to achieving clean coal:
1. Controlling and reducing pollutants (excluding CO_2),
2. Deploying advanced technologies, and
3. Installing CO_2 capture and storage.

1.2.4.2. Advanced Technologies

To control and reduce pollutants it is necessary to remove the source of pollution before burning, avoid production of pollutions during combustion, and remove pollutants prior to stack emission.

Plant efficiency and pollutant emission reduction can be improved upon by deploying advanced technologies and improving the thermodynamic cycle of power generation. For example, there is a modern shift from steam-cycle plants to gas-cycle plants. Advanced ultra-supercritical (USC) parameters for steam are used in some plants. Further to the use of these parameters for steam conditions, other advanced technologies incorporated include several clean air technologies; innovative design of burners, new schemes for combustion in the boiler furnace, new design of steam superheaters and systems for gas cleaning.

In fluidized bed combustion technology, limestone and dolomite are added during the coal combustion process to mitigate sulfur dioxide formation.

An integrated gasifier combined cycle uses heat and pressure in the thermodynamic cycle to convert coal into a gas/liquid phase. The coal in this transformed phase can be further refined, resulting in reduced environmental impact. The heat energy from the gas turbine is also used to power a steam turbine. This technology has the potential to improve the thermodynamic system efficiency of a coal plant to fifty percent.

Flue gas desulfurization, or scrubber technology, removes large quantities of sulfur, particulate matter, and other impurities from the emissions. Low Nitrogen Oxide (NOx) burners help reduce the generation of NO_x, a set of gases which contribute to ground-level ozone. This is achieved by restricting oxygen and manipulating the combustion process. Selective Catalytic Reduction (SCR) achieves NO_x reductions of between eighty and ninety percent. Electrostatic Precipitators remove particulates from emissions by electrically charging particles and then capturing them on collection plates.

1.2.4.3. Carbon Capture and Storage

There is a long term view toward achieving effective capture and storage of carbon dioxide. CO_2 emissions from burning of coal are calculated based on the *emission factor*, EF_C, where

$$EF_C = (CR \times CC \times CE \times 44)/(HV \times 12)$$

HV is the heating value of the fuel (12–32 MJ/kg), CC is the carbon content of the coal (60–90%), CE is the combustion efficiency (0.9–0.95), and CR is an opportune conversion rate (0.2778 in the case of kWh).[34]

Carbon capture and storage requires capturing CO_2 emissions and then storing them either in geologic formations or deep in the ocean bed where the gas is then dissolved under pressure. CCS technologies under development include:

- Post-combustion capture: This involves capture from the flue gases and necessitates use of an amine as solvent and chilled ammonia.
- Pre-combustion capture: Here, integrated gasifier combined cycling is used to isolate and capture CO_2 before it is released to the atmosphere.
- Oxy-Coal combustion process: This is an improved combustion process using pure oxygen in the boiler, resulting in significant reduction in the dilution of CO_2 in the exhaust gas stream.

Improved efficiency of power plants will reduce the levels of CO_2 emissions, however, carbon capture and storage would be a more effective solution. Unfortunately, CCS technologies require energy to implement and operate, reducing overall plant performance. Other pollutant emissions are also created in the CCS process, including limestone and ammonia. This technology is still at an early stage of development and can be considered as a future technology.

The development of CCT is growing worldwide with active research and development in both the US and Europe. If successful, CCT will play a vital role in allowing the continued worldwide use of abundant coal resources, in an affordable and sustainable manner. Such advanced technologies can contribute significantly to the areas of mercury control and carbon capture and storage, while also assisting in the reduction of SO_2 and NO_x emissions. Zero emissions through carbon sequestration, is a long-term objective.

While developments in clean coal technologies mark a welcome phase in the history of coal as a power fuel, there are considerable challenges relating to economic cost, plant refurbishment, effective utilization, and wide scale global deployment of such technologies. There is a sense of a race against time in the proposition that clean coal technologies may significantly offset the damaging effects of carbon emissions or that deployment will justify the further and continued exploitation of coal as a principal source of energy going forward. Even if CCT doesn't allow current levels of coal utilization to be sustainable, effective development of clean technologies could result in justification for the use of coal in reduced quantities in the energy mix until such time that carbon free renewable technologies reach mature status and can be shown to be effective for deployment. The future for nuclear energy, debated in chapter 7, will also influence the viability for significant investment in clean coal and related technologies.

Notes

1. Spencer R. Weart, "The Discovery of Global Warming," last modified February 2013, http://www.aip.org/history/climate/index.htm.
2. John Tyndall, "On Radiation Through the Earth's Atmosphere," *Notices of the Proceedings of the Meetings of the Members of the Royal Institution of Great Britain with Abstracts of the Discourses of the Evening Meetings*, vol. 4: 1862–1866 (London: W. Nicol, 1866), 4–8.
3. Ibid., 5.

4. "Svante Arrhenius (1859–1927)," NASA Earth Observatory, last modified 2013, http://earthobservatory.nasa.gov/Features/Arrhenius/.
5. Atmosphere: Greek *atmos* (vapor) and *sphaira* (sphere).
6. With longer wavelengths than those of visible light, IR extends from the nominal red edge of the visible spectrum at 0.74 micrometres (μm) to 300 μm. This range of wavelengths corresponds to a frequency range of approximately 1 to 400 THz. The existence of infrared radiation was first discovered in 1800 by astronomer William Herschel.
7. Ozone (O_3), a triatomic allotrope of oxygen, is much less stable than the diatomic allotrope, O_2. When occurring in the lower atmosphere, O_3 is an air pollutant. The ozone layer in the upper atmosphere, however, is a protector against excess ingress of electromagnetic radiation.
8. Kristin Dow and Thomas E. Downing, *The Atlas of Climate Change: Mapping the World's Greatest Challenge*, 3rd ed. (Abingdon: Earthscan, 2011).
9. Permafrost, or cryotic soil, is soil at or below the freezing point of water for two or more years.
10. Paul Mann, Lisa Gahagan, and Mark B. Gordon, "Tectonic Setting of the World's Giant Oil and Gas Fields," in *Giant Oil and Gas Fields of the Decade: 1990–1999*, ed. Michel T. Halbouty (Tulsa, OK: American Association of Petroleum Geologists, 2003).
11. "Höfflichkeit und Bergkgeschrey, Georgius Agricola 1494–1555," ETH-Bibliothek Zürich, last modified 2005, http://www.library.ethz.ch/exhibit/agricola/index.html.
12. Energy Information Administration, "Total Energy," United States Government, 2011, http://www.eia.gov/totalenergy/.
13. Stanley P. Schweinfurt, "Coal: A Complex Natural Resource," USGS Science for a Changing World, last modified January 11, 2013, http://pubs.usgs.gov/circ/c1143/html/text.html.
14. Matt Ridley, *The Rational Optimist: How Prosperity Evolves* (New York: Harper, 2010).
15. Chris Freeman and Francisco Louçã, *As Time Goes By: From the Industrial Revolutions to the Information Revolution* (Oxford: Oxford University Press, 2001).
16. International Energy Agency, *World Energy Outlook 2010* (Paris: International Energy Agency, 2010).
17. Subhes C. Bhattacharyya, *Energy Economics* (London: Springer-Verlag, 2011), 383–87. http://dx.doi.org/10.1007/978-0-85729-268-1.
18. Environmental Protection Agency, "Mercury and Air Toxics Standards," United States Government, last modified March 27, 2012. http://epa.gov/mats.
19. Eric B. Svenson, Jr., "EPA's Utility Air Regulations: New Jersey Clean Air Council," PSEG, last modified October 12, 2011. http://www.pseg.com/info/media/newsreleases/2011/attachments/njclean_air_council10-12-2011.pdf.
20. Latin *petroleum*, from Greek: *petra* (rock) and Latin: *oleum* (oil).
21. M. S. Vassiliou, *Historical Dictionary of the Petroleum Industry* (Lanham, MD: Scarecrow Press, 2009).
22. Bhattacharyya, *Energy Economics*.
23. Milton Beychok, "Petroleum Refining Processes," in *The Encyclopedia of Earth*, last modified February 12, 2012, http://www.eoearth.org/view/article/169791/.
24. Bhattacharyya, *Energy Economics*.
25. Energy Information Administration, "Natural Gas Explained," United States Government, last modified April 4, 2013. http://www.eia.gov/energyexplained/index.cfm?page=natural_gas_home.
26. Robert L. Braun and Alan K. Burnham, *Chemical Reaction Model for Oil and Gas Generation from Type I and Type II Kerogen* (UCRL-ID-114143) (Livermore, CA: Lawrence Livermore National Laboratory, 1993).
27. Harvey Blatt and Robert J. Tracy, *Petrology: Igneous, Sedimentary and Metamorphic*, 2nd ed. (New York: W. H. Freeman, 1996).
28. Helen Knight, "Wonderfuel: Welcome to the Age of Unconventional Gas," *New Scientist*, no. 2764 (2010): 44–47.
29. Paul Stevens, *The 'Shale Gas Revolution': Developments and Changes* (EERG BP 2012/04) (London: Chatham House, 2012).

30. Hydraulic fracturing, or fracking, is the fracturing of various rock layers by a pressurized liquid.
31. Pierre Allix et al., "Coaxing Oil from Shale," *Oilfield Review* 22, no. 4 (Winter 2010/2011): 6.
32. Ibid., 5.
33. Nature Conservancy, http://www.nature.org/.
34. Alessandro Franco and Ana R. Diaz, "The Future Challenges for 'Clean Coal Technologies': Joining Efficiency Increase and Pollutant Emission Control," *Energy* 34, no. 3 (2009): 348–354. http://dx.doi.org/10.1016/j.energy.2008.09.012.

Bibliography

Allix, Pierre, Alan Burnham, Tom Fowler, Michael Herron, Robert Kleinberg, and Bill Symington "Coaxing Oil from Shale." *Oilfield Review* 22, no. 4 (2011): 4–15.

Beychok, Milton. "Petroleum Refining Processes." *The Encyclopedia of Earth*. Last modified February 12, 2012. http://www.eoearth.org/view/article/169791.

Bhattacharyya, Subhes C. *Energy Economics*. London: Springer-Verlag, 2011. http://dx.doi.org/10.1007/978-0-85729-268-1.

Blatt, Harvey, and Robert J. Tracy. *Petrology: Igneous, Sedimentary and Metamorphic*. 2nd ed. New York: W. H. Freeman, 1996.

Braun, Robert L., and Alan K. Burnham. *Chemical Reaction Model for Oil and Gas Generation from Type I and Type II Kerogen* (UCRL-ID-114143). Livermore, CA: Lawrence Livermore National Laboratory, 1993. http://ds.heavyoil.utah.edu/dspace/bitstream/123456789/4949/1/222043.pdf.

Dow, Kristin, and Thomas E. Downing. *The Atlas of Climate Change: Mapping the World's Greatest Challenge*. 3rd ed. Abingdon: Earthscan, 2011.

Energy Information Administration. "Natural Gas Explained." United States Government. Last modified April 4, 2013. http://www.eia.gov/energyexplained/index.cfm?page=natural_gas_home.

Energy Information Administration. "Total Energy." United States Government. Last modified 2013. http://www.eia.gov/totalenergy/.

Environmental Protection Agency. "Mercury and Air Toxics Standards." United States Government. Last modified March 27, 2012. http://epa.gov/mats.

ETH-Bibliothek. "Höflichkeit und Bergkgeschrey, Georgius Agricola 1494–1555." ETH-Bibliothek Zürich. Last modified 2005. http://www.library.ethz.ch/exhibit/agricola/index.html.

Franco, Alessandro, and Ana R. Diaz. "The Future Challenges for 'Clean Coal Technologies': Joining Efficiency Increase and Pollutant Emission Control." *Energy* 34, no. 3 (2009): 348–354. http://dx.doi.org/10.1016/j.energy.2008.09.012.

Freeman, Chris, and Francisco Louçã. *As Time Goes By: From the Industrial Revolutions to the Information Revolution*. Oxford: Oxford University Press, 2001.

International Energy Agency. *World Energy Outlook 2010*. Paris: International Energy Agency, 2010. http://www.worldenergyoutlook.org/media/weo2010.pdf.

International Energy Agency. *Key World Energy Statistics 2010*. Paris: International Energy Agency, 2010. http://dx.doi.org/10.1787/9789264095243-en.

Irby-Massie, Georgia L., and Paul T. Keyser. *Greek Science of the Hellenistic Era: A Sourcebook*. London: Routledge, 2002.

Knight, Helen. "Wonderfuel: Welcome to the Age of Unconventional Gas." *New Scientist*, no. 2764 (2010): 44–47.

Mann, Paul, Lisa Gahagan, and Mark B. Gordon. "Tectonic Setting of the World's Giant Oil and Gas Fields." In *Giant Oil and Gas Fields of the Decade, 1990–1999*, edited by Michel T. Halbouty, 15–106. Tulsa, OK: American Association of Petroleum Geologists, 2003.

Mattusch, Carol. "Metalworking and Tools." In *The Oxford Handbook of Engineering and Technology in the Classical World*, edited by John Peter Oleson, 418–38. Oxford: Oxford University Press, 2008.

NASA Earth Observatory. "Svante Arrhenius (1859–1927)." National Aeronautics and Space Administration. Accessed 2013. http://earthobservatory.nasa.gov/Features/Arrhenius/.

Ridley, Matt. *The Rational Optimist: How Prosperity Evolves*. New York: Harper, 2010.

Schweinfurt, Stanley P. "Coal: A Complex Natural Resource." USGS Science for a Changing World. Last modified January 11, 2013. http://pubs.usgs.gov/circ/c1143/html/text.html.

Smith, A. H. V. "Provenance of Coals from Roman Sites in England and Wales." *Britannia* 28 (1997): 297–324. http://dx.doi.org/10.2307/526770.

Stevens, Paul. *The 'Shale Gas Revolution': Developments and Changes* (EERG BP 2012/04). London: Chatham House, 2012. http://www.chathamhouse.org/sites/default/files/public/Research/Energy,%20Environment%20and%20Development/bp0812_stevens.pdf.

Svenson, Eric B., Jr. "EPA's Utility Air Regulations: New Jersey Clean Air Council." PSEG. Last modified October 12, 2011. http://www.pseg.com/info/media/newsreleases/2011/attachments/njclean_air_council10-12-2011.pdf.

Tyndall, John. "On Radiation Through the Earth's Atmosphere." *Notices of the Proceedings of the Meetings of the Members of the Royal Institution of Great Britain with Abstracts of the Discourses of the Evening Meetings,* vol. 4: 1862–1866. London: W. Nicol, 1866. http://archive.org/details/noticesproceedi00britgoog.

Vassiliou, M. S. *Historical Dictionary of the Petroleum Industry*. Lanham, MD: Scarecrow Press, 2009.

Weart, Spencer R. The Discovery of Global Warming. Last modified February, 2013. http://www.aip.org/history/climate/index.htm.

Wrigley, E. A. *Energy and the English Industrial Revolution*. Cambridge: Cambridge University Press, 2010.

Chapter 2
Global Energy Policy Perspectives

RICHARD A. SIMMONS, EUGENE D. COYLE,
AND BERT CHAPMAN

Abstract

The global demand for energy and the enhanced quality of life it affords is strong and growing. In this chapter a review is made of developments and interactions of energy, environment and climate policy in the United States, Europe and globally. Contemporary demand for energy as well as initiatives promoting diversity of energy supplies, efficiency, and policies aimed at curbing emissions are reviewed. Legislation for energy and the environment is explored along with a discussion of challenges facing the world's most developed nations and regions. Viable technology and policy solutions adapted for introduction throughout the world are investigated. Through this dialogue the authors sense that cross-discipline partnerships of a global nature, social conscience, and compelling market factors will be critical in driving tomorrow's energy and climate trends.

Through exploration of in-depth perspectives for national energy and environment policy mechanisms a review and comparison of energy policy in the US and the EU is made, with inclusion of major historical milestones, the seminal 1997 Kyoto Protocol, and subsequent UN energy and climate conventions. The chapter concludes with analysis of energy and climate policy directives and trends in several of the world's largest economies including China, Russia, Brazil and India. All have large populations, significant energy demands and resources, and will be critical players in global efforts to align energy and climate trade and policies.

2.1. Introduction: Energy Demand and Expected Growth

Richard A. Simmons

Energy has been an enabling driver of unprecedented levels of economic growth, prosperity, and globalization, particularly during the past century. Throughout this period, a variety of primary energy sources have enjoyed eras of popularity, including traditional biomass, coal, oil, and natural gas. Due to a complex combination of factors, including the prospects of resource constraint, security of supply, and heightened environmental concern, a host of alternatives to traditional fossil fuels including renewable and unconventional sources of energy have been introduced to the global energy matrix in recent decades. However, the demand for energy and the enhanced quality of life it affords is strong and growing. Appropriately managing this global reality is the primary motivation of numerous energy and climate policy measures that are being analyzed, developed, and implemented across the globe.

In this chapter we review the interaction of energy, environment, and climate policy in the United States, Europe, and other major energy markets. We consider the current demand for energy and review initiatives developed to promote diversity of energy supplies, efficiency, and policies and regulations aimed at curbing emissions. We explore reasons why comprehensive energy and environment legislation has presented major challenges in the world's most developed regions, review global perspectives on energy and environment policies, and discuss mechanisms being used to promote broader dialogue on energy policy issues.

A brief glance at history provides insight into the link between energy and economic growth. Critical eras such as the Industrial Revolution, the post-WWII boom, and the oil crisis of the mid 1970s come quickly to mind. More recently, the economic crisis that began in 2007 has has resulted in intense volatility and price fluctuations for oil and natural gas, renewed concerns over the use of nuclear energy for power generation, as well as polarization over the near-term promise of many renewable energy technologies. Since the dawn of the industrial revolution some 250 years ago, historical evidence indicates that major energy transitions take longer, are more complicated, and often cost more than initially expected.[1] Power, heat, and electricity produced from traditional biomass gave way to coal, which has given way to oil, natural gas, and even nuclear fuels, albeit over intervals closer to fifty years, not ten or twenty. Like a massive mechanical flywheel, once major energy infrastructure has been adopted and integrated, it has great inertia owing to its cost and complexity. This makes it difficult to adapt quickly to new fuels and technologies. Now we are equipped with more advanced data, tools, and resources than at any previous point in history, and we aspire to understand how new transitions will be implemented over the coming decades. The future will likely bring periods of uncertainty, including growth and recession, but continued economic progress will hinge upon sustainable supplies of energy. It is vital to both understand these challenges and develop a plan to address them.

The wealth and economic status that has been amassed by much of the developed world is now available in varying degrees to developing countries and emerging economies. Some of the opportunities arising from globalization can be a double-edged sword. Individuals have benefited tremendously from increased quality of life, medical services, and opportunities for social mobility as developing economies boom; yet many countries are confounded in their attempts to keep pace with growing resource demands by rapid population expansion. Frequently, the infrastructure for energy and other critical resource services associated with clean air and water, waste management, and the transportation of people and goods is taxed beyond intended design limits; initial assumptions and conventional approaches can be ill-equipped to address projected population growth trends. Undesirable side effects include congestion, harmful air quality, price gouging, and electrical blackouts. Methodical approaches will be required to optimize resources, improve efficiency, encourage conservation, and reduce waste. Given appropriate implementation, such actions may enable the delivery of critical services, alleviate the risks of scarcity, and sustain trends toward a greater quality of life.

Energy is at the nexus of people, environment, and economic development, and energy supply and management requires careful implementation in order to navigate many of these challenges. This is not an issue to be delegated to or solved exclusively by policy makers or by any single group of stakeholders. Globally, more than eighty percent of the world's energy requirement is derived from fossil fuels, with oil (thirty-three percent), coal (twenty-eight percent), and natural gas (twenty-one percent) the principal constituents.[2] Combustion of these fossil fuels releases greenhouse gases directly into Earth's atmosphere. Scientific and economic experts are in increasing agreement that our current energy paradigm is no longer tenable, not least due to reserve and supply uncertainties, price volatility, and fiscal and environmental strains on the world's major markets and ecosystems. Numerous studies on this topic highlight aspects of the present challenges and discuss a range of viable technology and policy solutions, often concurring there is no one-size-fits-all model. Furthermore, it is unlikely a single technology or a single country will swing the needle entirely by itself. More likely, combined efforts such as global public and private partnerships, increased social conscience, and compelling market factors will continue to drive tomorrow's energy and climate trends. Prior to delving into potential solutions to this problem, it is important to assess the current reality, including energy demand and scenarios of expected growth, in addition to the social, economic, and environmental impacts such growth may have.

In addition to monitoring major energy supply disruptions and advising member countries regarding appropriate and timely responses, a primary mission of the International Energy Agency (IEA) is to compile and analyze historical energy data. This data is used to estimate future supply and demand scenarios and to develop policy advice based in part upon these projected trends. The IEA provides a series of outlook scenarios based on assumptions including the availability and reliability of the energy supply, energy consumption, growth, and the uptake of alternative energy sources. While an exhaustive review of this data is not the intent of this text, an overview of key global demands and trends is certainly revealing and instructive. Comparing energy data between countries is not a trivial task, given the obvious differences in energy infrastructure, modernization, and regulatory policies, let alone variances in the quality and reliability of the data itself. Despite this, the IEA's publically available assessment of energy supply and consumption is a robust database that conveys a sense of gravity and context for the energy challenge. Consider the following snap-shot in Table 2.1 of major energy indices, indicating gross domestic product, total primary energy supply, and estimated carbon dioxide emissions for countries and regions inclusive of US, EU, China, Russia, Brazil, India. Summed totals for world figures are also provided, indicating the aforementioned countries are responsible for approximately sixty-six percent of global energy-related CO_2 emissions.

In many ways, this type of data speaks for itself and is useful to help frame the global energy situation at a particular instant in time. That said, such summaries do not adequately capture either the strategic agendas of individual countries or the trends in these key indicators. Important questions are therefore raised: where is energy demand growing, at what rates, with what resources and technologies, and why? This text will explore the technical and geopolitical aspects of some of these urgent questions.

Data for both energy and climate has been collected since before utilities began electrifying the world. In terms of tracking energy metrics, analysis has progressed with greater rigor since the 1970s and the Arab oil embargo. From a climate perspective, attempts to understand and quantify the links and potential impacts between emissions and the combustion of fossil fuels are more recent, with research commencing in the 1980s and evolving quickly over the past two decades. It would appear that trends have motivated growing research interest: annual greenhouse gas emissions from the combustion of fuel between 1971 and 2012 have more than doubled from about fourteen to thirty-one Gt CO_2e (billion metric tonnes of carbon dioxide equivalent).[4] Recent transitions of fuel source from coal to natural gas and more stringent regulations on aging coal power plants have helped mitigate emissions to some degree; however, these have been largely outpaced by increased overall energy growth rates. In 2010, total global primary energy supply was estimated to be in excess of 12,700 Mtoe (million tonnes of oil equivalent), with an expected increase of at least one hundred percent by the year 2035. These figures are largely based on anticipated population growth in the developing world. Even in an optimistic

Country or Region	Population	Gross Domestic Product (GDP)	Total Primary Energy Supply (TPES)	TPES per capita	CO_2 Emissions	CO_2 per capita
	(million)	(2005 USD)	(Mtoe)	(toe/capita)	(Mt)	(t/capita)
USA	312	13,226	2,191	7.02	5,287	16.94
EU*	503	12,626	1,654	3.29	3,543	7.04
China	1,344	4,195	2,728	2.03	7,955	5.92
Russia	142	947	731	5.15	1,653	11.65
Brazil	197	1,127	270	1.37	408	2.07
India	1,241	1,317	749	0.60	1,745	1.41
World	6,958	52,486	13,113	1.88	31,342	4.50

Table 2.1. Key World Energy Statistics for 2011.[3]

scenario in which leading world economies implement aggressive new policies aimed at limiting carbon dioxide emissions to 450 parts per million (PPM) (believed to correlate to a temperature rise of two degrees Celsius above pre-industrial levels) by 2035, energy consumption is still projected to increase by at least twenty percent with respect to 2010 levels. A less aggressive scenario, allowing an approximate temperature rise of 3.5°C, projects energy consumption to increase thirty-six percent over this same period.[5] The global share of renewable-based energy consumption is projected to increase from about eighteen percent of total energy use in 2010 to between twenty and thirty-five percent by 2035. While this transition from fossil to non-fossil fuel energy resources will constitute a step in the right direction for Earth's climate, fossil fuel energy sources may still constitute a sizable majority of world energy supply by mid-century unless significant shifts in policy, increases in alternative technology uptake, and large scale capital investments are implemented.

Public awareness of the negative impacts of greenhouse gas emissions has grown significantly in recent years and this concern is beginning to translate into effective policy action. While the potential positive impact of large scale countermeasures and low carbon energy deployment may still be decades away, there is cause for hope. Through climate science research and dissemination, the links between energy consumption and associated environmental impacts are becoming more widely accepted and understood. In the twenty-five years from the inception of the Intergovernmental Panel on Climate Change (IPCC) in 1988 to the present, researchers have acquired new evidence complemented by powerful new modeling capabilities. This obviously provides decision makers a greater database of reliable information. Unfortunately, during this same twenty-five year period, the global community has consumed more exajoules (EJ) of fossil fuels than in the previous forty years (1948–1988).[6] These accelerating trends combined with the sobering message depicted in recent energy and climate data has given stakeholders ample reason for pause. It has also served to sound an alarm. Public opinion, far from unanimous on either energy or environmental policy, is beginning to reflect a growing sensitivity to select issues. Whether this has been driven home by higher oil or gasoline prices, volatile heating bills, or a more nuanced reading of energy and climate trends, it is occurring. This may usher in an era no less complicated but characterized by critical focus on meaningful long term strategic action and enabled by the clear interpretation of science-based climate and energy data. Such action may include a range of steps, including individual consumer behavioral shifts, industrial responses to economic and market factors, commercialization of innovative technologies, and broad policy measures taken by state and federal governments.

Understanding these issues for developing countries becomes exponentially more complex, as are efforts to expand real time learning of energy and climate in a perpetually evolving, increasingly globalized world. The age of two-way trade between major superpowers has given way to a global matrix of producers and consumers, the models for which, be they economic, environmental, societal or geopolitical, are in constant flux. In a January 2013 speech, then-US Secretary of State Hillary Clinton explained the energy and climate change balance succinctly: "Managing the world's energy supplies in a way that minimizes conflict and supports economic growth while protecting the future of our planet is one of the greatest challenges of our time."[7] Consider the implications of energy and climate policy for a few of the world's most rapidly growing economies: China, Russia, Brazil, and India. The significance and complexity of energy and climate issues in these regions are markedly different than they are in the more developed and established domains of the United States and the European Union.

Two differences, in particular, stand out. First, increased global trade and urbanization throughout the world (fueled primarily by fossil fuels) are now hitting full stride in several of the world's most populous countries. By contrast, energy consumption in the developed world has more or less plateaued, been augmented by larger shares of cleaner energy (including nuclear, natural gas, and hydropower), and stabilized on a per capita basis.

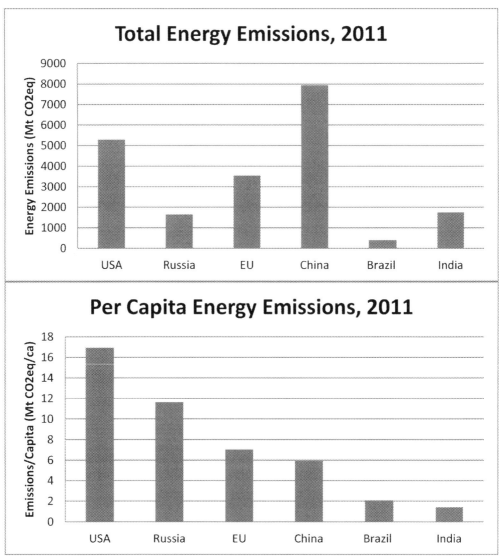

Figure 2.1. (a) Total Energy Emissions for Selected Countries, 2011; (b) Energy Emissions Per Capita for Selected Countries, 2011.[8]

Second, from the standpoint of industry and emissions, the light regulation and inexpensive supply typical of the past have set precedents that weigh heavily in economic and business modeling. Not surprisingly, developing countries commonly rely upon these precedents in strategic planning and public policy. It is argued that the rates of growth for energy and emissions should not be subject to sudden change or regulation, as the developed world fueled much of its own growth relatively unchecked by environmental constraints or international opinions. For example, China has argued that developing states should be afforded some leniency in emissions as they are currently in critical stages of economic development.[8]

Consider, for example, the two energy emissions charts shown as Figure 2.1. Is it more appropriate to measure CO_2 emissions by country or per capita? The answer obviously depends on your perspective. China and India can leverage their large populations in this debate to argue that their CO_2 intensity per capita is much lower than the developed world, yet China as a nation leads in overall emissions. India has formally announced during climate negotiations that their per capita CO_2 emissions will not exceed that of developed countries, falling far short of negotiators' aspirations but sending a salient and sobering message to the West.

In terms of recent trends, total energy-related CO_2 emissions have actually continued to fall slightly for the US and EU, for example between 2010 and 2011, but continued to increase between 4% and 8% for Brazil, Russia, Indian and China over the same one year period.[9] These emission trajectories are qualitatively consistent with a recent six-year period between 2005 and 2011 as illustrated in Figure 2.2.

Leading economies of the world recognize their future will be in part defined by how they create a sustainable balance between the supply of energy, its environmental impact, and the prevailing pursuit of economic prosperity and growth. Like other monumental challenges of our times, this is much more easily stated than solved. To achieve the greatest global impact in a world of increasing globalization and population, national

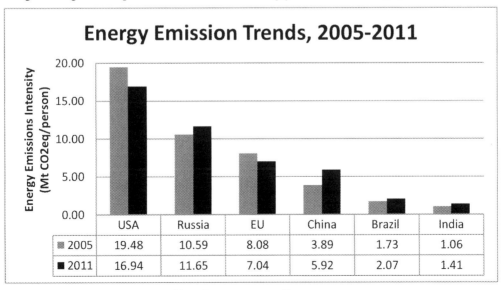

Figure 2.2. Energy Emissions Trends, 2005–2011.[10]

efforts should not occur in a vacuum but rather, to the extent possible, in a coordinated, informed manner. Major consumer nations are reminded frequently and acutely that the world has finite resources, and there will be increasing competition for them.

It is upon this energy and climate backdrop that members of productive society will strive to confront epic challenges and sustain recent positive trends in health, economic development, and quality of life. Lasting solutions will require not just revolutionary technology, but also an understanding of some very disparate perspectives, productive global discourses about the nature of the problem, and effective and pragmatic policy implementation on an unprecedented scale.

2.2. United States Energy and Climate Policy

Bert Chapman

2.2.1. United States Energy Policy

Although the United States' federal government has been engaged in energy and climatic policymaking for many decades, the emergence of modern US energy policy commenced with the 1973 Arab oil embargo. The embargo was a reaction by many petroleum producing countries against US military support for Israel during the October 1973 Yom Kippur war. It was a vivid illustration of the dangers of increasing US dependence on oil imports. This embargo included reductions in oil output by five to ten percent per month, and a total ban on oil exports to the United States which resulted in significant energy supply shortages. It also increased retail gasoline prices in the United States by approximately fifty percent from $0.37 per gallon in 1973 to nearly $0.57 per gallon in 1975. Crude oil prices rose from $3.18 per barrel in 1970 to $7.67 in 1975.[11]

The embargo caught the United States and other governments unprepared, leading to a variety of policy responses as western countries became more fully aware of their vulnerability to abrupt energy supply disruptions.[12]

The eventual creation of a centralized Department of Energy (DOE) in 1977 from existing energy policymaking entities, including the Federal Power Commission, was a direct result of these developments. The DOE would have multiple policymaking arms with responsibility for overseeing the domestic US energy agenda, including fossil and renewable resources; conducting energy policy research through its national laboratories; managing the US nuclear weapons arsenal; and analyzing energy sector trends and developments both domestically and internationally.[13]

Another federal response was the establishment of the Strategic Petroleum Reserve (SPR) in 1975 to create a store of up to one billion barrels of petroleum to meet domestic economic needs in the event of another supply disruption. The SPR is located in salt caverns on the Louisiana and Texas coasts and had an inventory of about 695 million barrels at the end of 2012. This equates to approximately sixty days of import protection, and more than one hundred days' supply if private stocks are included. Presidential administrations of both parties have generally resisted calls to tap into the SPR as a response to rising domestic energy prices, though releases occurred in

1991, 2005, and 2011, usually to mitigate short term supply disruptions from global conflicts or natural disasters.[14]

Through a cooperative effort between the United States and other member countries of the Organization of Economic Cooperation and Development (OECD), the International Energy Agency (IEA) was also formed in response to the 1973 oil crisis. The IEA's initial role was to help countries coordinate a collective response to major disruptions in oil supply through the release of emergency oil stocks to the markets. The IEA has since broadened its charter to encompass energy analyses, technology surveys, projections, and policy recommendations, but remains a primary international mechanism for monitoring national petroleum reserves against global oil supply and demand and coordinating multi-lateral policy responses to international supply disruptions.[15]

Inflation in the late 1970s combined with geopolitical tension arising from the Iranian revolution and the Iranian seizure of US diplomats as hostages resulted in subsequent energy price increases. US retail gasoline prices nearly doubled to $1.19 per gallon between 1978 and 1980, while crude oil prices more than tripled from $9.00 to $31.77 per barrel between 1978 and 1981.[16]

Another seminal event affecting US energy policy was the partial meltdown of a reactor at the Three Mile Island nuclear power plant near Harrisburg, Pennsylvania in March 1979. This incident was contained and various regulatory reforms were instigated by the Nuclear Regulatory Commission (NRC) including enhanced safety and training protocols and reactor design modifications at US nuclear power plants. However, the accident damaged the then-burgeoning nuclear industry's reputation and resulted in an effective moratorium on US nuclear power plant construction that persisted for more than three decades. Although the number of operating nuclear units has increased from 70 in 1979 to 103 in 2013, the NRC issued its first new construction permits in thirty-three years in 2012 for new generation reactors slated to be constructed in Georgia and South Carolina.[17] These two separate projects, administered by regional utilities, are underway with the construction of Westinghouse AP1000 pressurized water reactors. Two such reactors at each site, of 1100 MW nameplate capacity, are expected to come online by 2018.[18] When the construction permit was issued, the NRC Commissioner stated, "If they are built as proposed and in accordance with NRC requirements, [the reactors] will represent a new era of enhanced nuclear safety."[19] Prior to the recent construction permit issuances, less than forty percent of nuclear power plant operating licenses had been issued since 1979, with the last of these licenses issued in 1996.[20]

Key similarities in US energy policy response can be observed by comparing the 1970s with a multi-year period beginning in about 2007. Not surprisingly, both eras were characterized by dramatic spikes in oil price, global economic stress, and geopolitical instability. During both periods, efforts were made to improve energy security, increase US domestic energy production, and institutionalize efficiency and conservation measures. For example, the Corporate Average Fuel Economy (CAFE) program, inaugurated in 1975, set standards for increasing car and light truck fuel efficiency. Actual regulatory standards experienced no change between 1992 and 2011, but CAFE has returned as a significant piece of recent energy policy, and new rules strive to double fuel economy in new vehicles (between 2011 and 2025) while exploiting efficiency benefits such as reduced oil consumption and emissions (Please see chapter 9 for more on CAFE). Similarly, measures have been enacted to increase the use of alternative fuels, for example via the Energy Independence and Security Act (EISA 2007) and the Renewable Fuels Standard (RFS 2005 and annual updates), which mandate the use of increasing quantities of alternative fuels such as cellulosic ethanol and other advanced biofuels through 2022.

Natural gas prices were decontrolled by the Natural Gas Policy Act of 1978 and this process was completed in 1985. Oil prices were decontrolled upon issuance of an Executive Order by President Reagan (Order 12287) on January 28, 1981. These events would produce significant increases in natural gas and oil supplies and associated reductions in energy commodity prices. Retail gasoline would fall from $1.35 per gallon in 1981 to $0.96 in 1988 and industrial natural gas would fall from $6.98 per thousand cubic feet in 1982 to $3.32 by 1995.[21]

More recently, natural gas has re-emerged as a central issue in US energy policy. The so-called shale gas revolution has been made possible by hydraulic fracturing techniques, which allow access to and retrieval of gas trapped in source rock via horizontal drilling. Just as conventional on- and offshore gas wells began a slow decline in the United States, these new technologies reached commercial scale around 2005. The potential is astounding. The US Department of Energy's Energy Information Administration (DOE-EIA) predicts that the share of shale gas as a percentage of all natural gas will grow from five percent to about forty-five percent between 2005 and 2020. This turns an otherwise declining natural resource into one projected to grow approximately fifty percent to an annualized twenty-seven trillion cubic feet of gas by 2020.[22]

The flood of shale gas is on; and the abundant supplies caused commodity prices to drop two to threefold in US markets between 2008 and 2012, creating a host of opportunities, challenges and unexpected consequences. On the positive side, natural gas, with about half the carbon content of coal, can truly be a bridge fuel as high carbon alternatives are phased out and carbon neutral options get deployed, reducing the environmental impact of energy generation. In addition to adding jobs in the energy sector, cheap natural gas has rekindled the competitiveness of US manufacturing, enabling key industries to regain global market share. Unfortunately, many fear a reliance on another, albeit cleaner, fossil fuel, may significantly impair the commercial viability and deployment of renewable energy technologies. Others fear loose state-controlled regulation and weak oversight may result in unintended environmental risks such as water contamination or un-combusted emissions of methane during extraction and transport. The debate will continue, as the economic benefits appear compelling and the environmental benefits and risks, thus far at least, relatively manageable. A new decision impacting foreign policy may be on the near horizon as well, as the United States grapples with whether to leverage newfound energy abundance strictly for domestic gain, or to begin exporting lower carbon energy resources to a willing global marketplace.

By way of advising policymakers about technical implications relevant to future energy trends in the United States, EIA summarizes key analysis in its Annual Energy Outlook. In its 2013 projection to 2040, EIA predicts:

- Growth in energy production will outstrip consumption growth;
- Crude oil production will rise sharply over the next decade;
- Motor gasoline consumption will reflect more stringent fuel economy standards;
- Renewable fuel use will grow at a much faster rate than fossil fuel use;
- The United States will become a net exporter of natural gas by the early 2020s; and
- US energy-related carbon dioxide emissions will remain below their 2005 level through 2040.[23]

2.2.2. United States Policy and Climate Change

An important emerging strand affecting US energy policy is that of global warming and climate change. The idea that human-created emissions are responsible for warming temperatures remains controversial in the United States. The phrase *global warming* first appeared in a January 1986 Environmental Protection Agency (EPA) stratospheric ozone protection plan, when it was proposed that chlorofluorocarbons (CFCs) are infrared-absorbing gases acting like carbon dioxide which can result in raised global surface temperatures.[24] The remainder of the 1980s and early 1990s would see analysis, controversy, policy proposals, and scrutiny of this polarizing issue play out in literature produced by many federal agencies and the Congress.[25]

Climate change has become a critical aspect of US foreign policy as other nations have sought to coordinate plans and perspectives as well as to propose multi-lateral action to address it. These efforts culminated in a 1992 United Nations sponsored Conference on Environment and Development (UNCED), informally titled Earth Summit, that was held in Rio de Janeiro and drew one hundred twenty heads of state, accompanied by numerous government officials. Topics addressed included personal energy consumption, energy resource availability, forest production and deforestation, population impact on energy, and various commitments to addressing climate change. The Summit's message—"that nothing less than a transformation of our attitudes and behaviour would bring about the necessary changes"—was transmitted by almost ten thousand on-site journalists and heard by millions around the world. The message reflected the complexity of the problems facing us: that poverty as well as excessive consumption by affluent populations place damaging stress on the environment. Governments recognized the need to redirect international and national efforts to ensure that economic decisions account for environmental impacts.[26] President George H. W. Bush, however, refused to sign the Convention on Biodiversity due to compensation requirements for countries providing animal and plant sources for biotechnology inventions. Summit efforts also faltered due to wide differences of opinion between developed and developing countries over emission reduction targets and financial liability for enforcing climate change countermeasures.[27]

During the Clinton administration, the United States participated in the 1997 Kyoto Protocol Conference in Japan as a member of the United Nations Framework Convention on Climate Change. In the Kyoto agreement, signatory countries agreed to emission reductions of greenhouse gases such as carbon dioxide, methane, nitrous oxide, and sulfur hexafluoride. Although most countries have signed Kyoto, the US Senate did not ratify this agreement due to concerns that it would negatively impact US economic growth and weaken national sovereignty. The United States has also expressed concern that the Protocol's loose emission restrictions on China and India would not achieve significant global emissions reductions, and that it would increase US dependence on foreign oil and adversely impact US fuel and energy prices.

On March 28, 2001, President George W. Bush announced that the US would not implement Kyoto, and on December 12, 2011 Canada, an initial Kyoto signatory, withdrew from the Kyoto Protocol to avoid monetary penalties for failure to comply with its emission targets and its promotion of domestic energy industries such as Alberta's oil sands.[28]

The 2005 Energy Policy Act was a significant piece of energy legislation that mirrored widespread sentiment in the United States, European Union, and globally, in response to climate and energy related issues. It coincided with the introduction in the European Union by the European Council of a Mandatory Energy Policy which is expounded upon in the next section. General statute provisions of the 2005 US Energy Policy Act encompassed a broad remit including energy efficiency, renewable energy such as hydro and geothermal power, alternative fuels such as ethanol and hydrogen, oil and gas, coal, nuclear power, energy tax incentives, and climate change technology. Specific provisions include sustainable design principles in new federal buildings, tax credits for improved energy efficiency in homes and to help meet Energy Star requirements, provisions to increase domestic energy production and reduce dependence on foreign energy, expansion of unconventional fuel resources such as oil sands; implementation of new regulatory tools, and the development initiatives to promote alternative fuels and new vehicle technologies.[29]

The American Reinvestment and Recovery Act (ARRA) was signed into law in February 2009 in response to the global financial crisis, and included significant federal investments in energy efficiency and renewable energy. Under ARRA, the Obama administration allocated more than ninety billion dollars (of nearly eight hundred billion dollars total) in grants and tax incentives for a host of clean energy programs. It is of particular note that nearly half of the award grants went to energy efficiency initiatives, such as programs to assist in the weatherization of homes and buildings. Efficiency improvements under ARRA would utilize available technologies, be deployed rapidly, and could therefore result in the greatest economic and environmental benefits in the near term. The balance of energy investments and provisions were spread across a wide suite of programs including renewable energy pilot projects, alternative fuels, smart grid, environmental clean-up, and carbon capture and storage. Some ventures failed on either technical or financial grounds, for example Beacon Power, Solar Trust, and Solyndra, which resulted in hundreds of millions of federally funded losses. Critics cite such failures as evidence of political cronyism and add that poor business performance by these companies can cast doubts on the financial viability of certain clean tech ventures and worthiness of government support.[30] Amidst failures, other successful projects continue to achieve stated objectives. Most energy projects by nature are longer-term, and by dispatching funds quickly, the administration may have created an unreasonable expectation that all funded programs would succeed and/or generate immediate and measurable impacts. This situation led to controversy over whether the federal government should pick economic winners and losers, with complaints that investment had not yielded sufficient returns.[31] Though the intent to stimulate the energy economy was genuine, even the most successful projects may progress at a deliberate pace and achieving "positive" outcomes may be challenging in the short term.

Political controversy has also ensued over the Environmental Protection Agency's (EPA) intent to regulate carbon dioxide emissions as a public health threat contributing to global warning. In 2009, the outgoing Bush administration resisted this on economic grounds, but the April 2007 US Supreme Court decision *Massachusetts v. EPA* ruled by a 5–4 margin that the EPA could promote the development of auto tailpipe greenhouse gas emissions standards under Section 202 of the Clean Air Act (CAA). This decision was reinforced by the Supreme Court's 2011 ruling in *American Electric Power v. Connecticut* and the EPA has since worked to promote various standards for reducing CO_2 emissions such as capture and sequestration.

In March 27, 2012 the EPA issued proposed carbon pollution standards for new power plants which would limit the amount of carbon pollution these plants can emit and ensure that these facilities adopt new cleaning technologies. The fate of the proposed regulations is uncertain.[32] Among the objections are that possible geographic costs and benefits from climate change do not align with Congressional intent under the CAA, that the Court was forcing the EPA to regulate these emissions under a law never intended to cover climate change, and that unilateral US emission limits will be ineffective and may become a disincentive for China and other countries to reduce their emissions. Another critique of EPA's policy in this area maintains that the proposed rule has not monetized costs or benefits to the electric power sector nor by extension to electric power consumers.[33]

The Obama administration sought to use the December 2009 Copenhagen Summit to recognize the acute global environmental challenge resulting from climate change by limiting temperature increases to two degrees Celsius. The administration also committed the United States to reducing its emissions seventeen percent by 2020. The 2009 draft document was not legally binding and contained no enforcement mechanisms for reducing CO_2 emissions. However, UN climate negotiation efforts are ongoing, with a 2015 goal to formalize a climate agreement that would take effect in 2020 (for more, please see the section on EU climate policy). Despite this, consensus has been elusive within the US government and among the international community over exactly how to proceed in reducing emissions and combating climate change. As one of the more complicated issues of our times, the future of US participation in international climate change initiatives remains uncertain.[34]

Energy and climate change factors heavily into US national security policymaking. The outgoing Bush administration issued a policymaking document stressing the Arctic's importance on a range of grounds including US national and homeland security, as well as natural resource management, environmental protection and economic development with other Arctic nations.[35] The Obama administration's May 2010 National Security Strategy stressed the importance of developing new energy sources to reduce dependence on foreign oil as critical to national security as are efforts to transition to low carbon energy sources and combat climate change.[36] Recent trends would indicate selected policies are taking hold, as the share of US oil imports in 2013 fell below fifty percent for the first time since 1995. Furthermore, Canada has become the largest oil importer to the United States, and along with Mexico and Venezuela, the western hemisphere presently accounts for a majority of US imports.[37] In 2009, the Central Intelligence Agency established a Center on Climate Change and National Security whose missions include examining the national security impact of desertification, natural resources competition, population shifts, and rising sea levels to policymakers and the scientific community through imagery and other means.[38] Resources describing national security implications resulting from climate change are produced by multiple military and civilian agencies.[39]

In 2011, the Obama administration issued the *Blueprint for a Secure Energy Future* with proposed solutions to major US energy problems. It outlined means of developing and securing America's energy supplies; providing consumers with choices to reduce costs and save energy; and innovating the nation's way to a clean energy future. In 2013 the *President's Climate Action Plan* was released. These objectives included specific policies supporting the responsible and safe expansion of domestic oil and gas development and production; more fuel and energy efficient cars, trucks, homes, and buildings; and the promotion of clean energy research and development.[40]

A 2012 supplement to the *Blueprint* contended that 2011 saw US domestic oil and natural gas production reach their highest levels since 2003. Meanwhile, the Interior Department announced a proposed expansion of the Outer Continental Shelf Oil and Gas Leasing Program, and light duty vehicle fuel economy standards increased to 54.5 miles per gallon by 2025 through advanced and clean energy vehicles.[41] Some energy policy critics assert, however, that the administration has been slow to issue drilling leases and permits and to promote drilling expansion in areas of potential promise. Following an extensive review of economic and environmental impacts, the administration deferred a permit for the Keystone XL Pipeline that would bring crude oil from Alberta, Canada, to Nebraska and, eventually, to refineries in Texas. In 2013, the State Department issued a supplemental environmental impact statement for a new permit with an alternate, lower impact route, but final authorization of national interest is still pending.

The need for low carbon solutions combined with increasing fuel prices and controversy over the Obama administration's green energy programs have renewed interest in nuclear power as an important element of a diverse US energy strategy. Nuclear energy accounts for about nineteen percent of US electricity generation, and the industry has been boosted by 2005 Energy Policy Act incentives such as construction loan guarantees, which are helping underwrite the aforementioned new generation projects. The EIA expects nuclear power output to increase at approximately half the rate of total electricity generation.[42] Despite its aging nuclear facility infrastructure, the United States remains the world's leading nuclear power producer in total capacity with 101 GW, though France leads in nuclear as a share of total electrical generation at seventy-five percent.⁴³ However, there is still significant public skepticism about nuclear energy, which has been exacerbated by the 2011 tsunami which caused the nuclear accident at Japan's Fukushima Dai-ichi plant. Polarization on the issue was reflected in a March 2012 Gallup opinion poll, in which fifty-seven percent of respondents said they strongly or somewhat favored using nuclear energy to provide electricity, with forty percent strongly or somewhat opposed to using nuclear energy for this purpose.⁴⁴

The Obama administration has also sought to promote expanded international energy cooperation through multilateral initiatives such as the UN's Sustainable Energy for All campaign, which was delineated at the "Rio+20" Summit in June 2012. This initiative calls on the international community to reach three aspirational goals by 2030: providing universal access to modern energy services, doubling the global rate of improvement in energy efficiency, and doubling the share of renewable energy in the global energy mix.⁴⁵ If implemented, the US would provide nearly two billion dollars worth of grants, loans, and loan guarantee resources to help developing countries create sustainable energy development, participate in clean energy technology partnerships, promote US energy technology exports, and finance and mobilize private capital to help developing world investors through debt financing, risk insurance, and new coverage for power purchases. Critics of the Rio+20 Summit contend that its policymaking objectives create expanded and unaccountable global international energy bureaucracies, as well as enhance international control over energy development and other economic activities.⁴⁶

US energy policymaking involves the complex and continually evolving interaction of federal, state, and local government authorities with civil society, the private sector, non-governmental organizations, and independent agencies. A diverse array of congressional and state legislative committees are also involved in this process, adding to its merit, but also its complexity, inefficiency, and cost. Regulatory authority may fall under federal or state jurisdiction, and certain differences between states have been known to complicate regional alignment. With such a diverse array of authorities and affected stakeholders from both the public and private sectors, the open and democratic US energy policy process is charged with the daunting task of equitably weighing and integrating all manner of input, optimizing effectiveness, while controlling costs. Policymaking efforts can be understandably more complex in the international domain, as US energy policy objectives intersect those of foreign governments and stakeholders.

2.2.3. Conclusion

Ideological differences on energy between the two largest political parties in the United States have been a source of uncertainty and tension, not least for the American voter. While some cite scientific evidence for human-caused global warming, skeptics cite leaked emails from East Anglia University's climate unit as justifying their charges of fraudulent scientific behavior by global warming proponents. These critics express concern that some proposed climate change solutions could injure the US economy and limit US national sovereignty. As mentioned, other critics take issue with direct government financial support of clean energy companies.⁴⁷

The questions and responses surrounding the extraction of oil, natural gas and unconventional fossil fuels have sharply divided US national interests. Environmental implications, energy demand growth, economic development, climate change and strategies to reduce the carbon footprint of the domestic energy matrix demand appropriate consideration and integration into the national energy dialogue. Common-ground outcomes must be agreed upon and pursued, despite traditional polarization of certain issues. While some contend that government energy policies are intrusive and counterproductive for both individuals and the private sector, others feel that government policies are excessively solicitous of individual and private sector interests. Nevertheless, the proper roles of fossil energy and renewable energy, protracted federal budget deficits, and public resistance to higher energy prices or reductions in energy consumption are hallmark issues of recent US energy policy and will remain part of the US energy policymaking landscape for the foreseeable future.[48]

It is clear that the process of optimizing energy resources and developing effective policies can be complicated, costly and time consuming. Throughout its history, the United States has pioneered significant technical, commercial, and even political energy innovations. Along the way, it has had to navigate major uncertainties, disasters, and challenges. As we look forward, accompanying unprecedented levels of globalization and economic growth looms the unknown risk of climate change. And while no system yet devised is perfect, the United States has both the responsibility and the privilege as the world's largest economy to contribute productively to global progress in the new energy era.

2.3. Energy and Climate Policy in the European Union

Eugene D. Coyle

2.3.1. Underpinnings of Modern European Energy Policy

The seeds of the European Union share a rich and interwoven history with energy, and the multi-national stage upon which energy related events have unfolded. Twentieth century energy policy legislation among European nations formally commenced during a post World War II period that coincided with the establishment of the Council of Europe in 1949[49] alongside efforts to reconstruct the economy and establish a lasting period of peace. In May 1950, French Foreign Minister Robert Schuman made his Schuman Declaration at the Quai d'Orsay, where he proposed that "Franco-German production of coal and steel as a whole be placed under a common High Authority, within the framework of an organization open to participation of other countries of Europe."[50] This led to the pooling of Franco-German coal and steel production and the formation of the European Coal and Steel Community (ECSC). The treaty was entered into force in July 1952, bringing France, Germany, Italy, Belgium, Luxembourg and the Netherlands together in a community with the aim of organizing free movement of coal and steel and free access to sources of production. This set a significant precedent: a common high authority supervised the market, with the aim of respecting and ensuring rules for competition and price transparency.

In similar fashion, the Messina Conference, held at Messina, Sicily in May 1955, was charged with preparation of a report on the creation of a common European market. The resulting Spaak Report was presented at the Intergovernmental Conference on the Common Market and EURATOM, at Val Duchesse, Brussels. It was agreed that two new communities would be established, the European Economic Community (EEC) and the European Atomic Energy Community (EURATOM). The two new high authorities (Commissions) would be separate to the ECSC's Council of High Authority. In March 1957, the Treaties of Rome were signed, with the new authorities coming into force on January 1, 1958. EURATOM was founded with the purpose of creating a specific market for nuclear power in Europe. Although legally distinct from the European Union (EU), EURATOM has nevertheless the same membership and is governed by the EU's institutions.[51]

Long before these developments, the strategic hundred-mile-long Suez Canal opened in 1869. Financed jointly by the French and Egyptian governments, it created a much needed shipping route and land bridge between Africa and Asia, the Mediterranean and the Indian Ocean, and enabled trade between Asia, the Middle East, Europe and the United States.[52] Owing to financial difficulties, Egypt was forced to sell its shares in ownership of the Suez Canal to the United Kingdom in 1875. However, an international convention in 1888 opened the Canal to all shipping from any nation. In 1956 tensions arose when the newly inaugurated Head of Government in Egypt, Gamal Abdel Nasser, declared his intent to place the Suez Canal under Egyptian control. The United States and United Kingdom withdrew previously committed financial support for the construction of the Aswan High Dam in the Nile, and in retaliation, Egypt nationalized the Canal. This became the Suez Crisis, also referred to as the Tripartite Aggression, Suez War, or Second Arab-Israeli War. The crisis became a diplomatic and military confrontation pitting Egypt against the combined interests of Britain, France and Israel. Intervention by the United States, the Soviet Union, and the United Nations brought the occupation to an end by late December 1956.[53] The fight over the canal also sowed the seeds for the eventual outbreak of the Six Day War in 1967 due to an inadequate peace settlement following the 1956 war.

The conflict is significant in that it lead to a severe oil shortage and financial crisis in the United Kingdom and Western Europe. Prior to the conflict approximately 1.5 million barrels of oil per day transited the canal, of which 1.2 million barrels were destined for Western Europe, equating to two thirds of total oil supplies. A third of the ships that passed through the Canal at the time were British, and three-quarters belonged to NATO countries.

In the aftermath of the turbulence surrounding the Suez Crisis, the Organization of the Petroleum Exporting Countries (OPEC), a permanent intergovernmental organization, was created at the Baghdad Conference on September 10–14, 1960, representing oil producing nations Iran, Iraq, Kuwait, Saudi Arabia, and Venezuela. The five founding members were later joined by nine other members: Qatar (1961–present); Indonesia (1962–2009); Libya (1962–present); United Arab Emirates (1967–present); Algeria (1969–present); Nigeria (1971–present); Ecuador (1973–1992, 2007–present); Angola (2007–present);

and Gabon (1975–1994). Originally headquartered in Geneva, Switzerland, OPEC has been based in Vienna, Austria, since 1965.[54] OPEC's stated principal objective is to "coordinate and unify petroleum policies among Member Countries, in order to secure fair and stable prices for petroleum producers; an efficient, economic and regular supply of petroleum to consuming nations; and a fair return on capital to those investing in the industry."[55]

The October 1973 Yom Kippur War and subsequent oil embargo had a profound impact on Europe as well as the United States. The Organization of Arab Petroleum Exporting Countries (OAPEC), comprising the Arab members of OPEC in addition to Egypt, Syria, and Tunisia, were the dominant supplier of crude oil to the European Union member states. When OAPEC proclaimed an oil embargo in October 1973, economic turmoil in the European Union ensued. The first effect of the crisis was a shortage of oil, which led to a number of measures to restrict consumption. As shortage fears diminished, increasing oil prices and their resulting financial consequences became the paramount concern. Prices for crude oil rose for many years, reaching twelve times their pre-crisis level (thirty-six dollars per barrel compared to three) after a second oil shock provoked by the Iran-Iraq war of 1980. This exorbitant increase in crude oil prices over the span of six years dealt a serious blow to the economies in several regions of the world, including Europe. The community member states, accustomed to trade surpluses, were now in a weakened position. Recession began to bite in nearly all the European countries.[56] In addition to the economic consequences, the 1973 crisis created a sense of insecurity among European countries, exposing the vulnerability of EU economies to their dependence on abundant supplies of cost-competitive oil. As noted earlier, such insecurity led to the formation of the Paris-based International Energy Agency (IEA) to help coordinate member country responses, track markets, and eventually advise on energy technologies and global policies.

The automotive industry was one of western Europe's most affected industries in the wake of the 1973 oil crisis. After the second World War most west European countries applied heavy taxes to imported automobiles and related accessories, and as a result most cars made in Europe were small and hence more economical to both purchase and operate. However, by the late 1960s, as individual wealth increased, vehicles and engines began to increase in size. The oil crisis reversed this trend in Europe, and convinced many people to revert to smaller and more efficient hatchback vehicles. This trend continued until the late 1980s, by which time hatchbacks dominated most European small and medium car markets and gained a substantial share of the larger family car market as well.

2.3.2. Energy Policy in Twenty-First-Century Europe

In meeting its current energy demands, the European Union is heavily dependent on imports of fossil fuels, with up to eighty percent imports of oil and sixty percent natural gas. Almost ninety-seven percent of uranium used in European nuclear reactors is imported from countries including Russia, Canada, Australia, Niger, and Kazakhstan, with only three percent mined in Europe.[57] The basic principles of European energy policy were laid down in 2006 with the release of the Commission's green paper *A European Strategy for Sustainable, Competitive and Secure Energy.*[58] In launching the strategy it was noted that Europe requires the importation of fifty percent of its energy for fuel and that global hydrocarbon reserves are being depleted. Investment of one trillion euros is required by 2020 in order to meet the expected energy demand and replace aging infrastructure. It was also accepted that global warming has already made the world 0.6°C hotter.

Intent on limiting global average temperature increase to less than two degrees Celsius above pre-industrial levels, key proposals of the European strategy include:

1. A cut of at least twenty percent in greenhouse gas emissions from all primary energy sources by 2020 (compared to 1990 levels), while pushing for an international agreement to succeed the Kyoto Protocol aimed at achieving a thirty percent cut by all developed nations by 2020.
2. A cut of up to ninety-five percent in carbon emissions from primary energy sources by 2050, compared to 1990 levels.
3. A minimum target of ten percent for the use of biofuels by 2020.
4. Unbundling of the energy supply and generation activities of energy companies from their distribution networks to further increase market competition.
5. Improving energy relations with the European Union's neighbors, including Russia.
6. The development of a European Strategic Energy Technology Plan to develop technologies in areas including renewable energy, energy conservation, low-energy buildings, fourth generation nuclear power, clean coal, and carbon capture.
7. Developing an Africa-Europe Energy partnership, to help Africa leap-frog to low-carbon technologies and to help develop the continent as a sustainable energy supplier.

While these goals were considered ambitious, subsequent developments have set the scene for change and have imparted responsibility to individual EU member states to advance, implement and achieve targets. Mechanisms for doing so include regular strategic energy reviews, introduction of a European Emissions Trading Scheme (EU-ETS, endeavoring to achieve cost-effective carbon dioxide emissions reductions), and the pursuit of targets to be achieved initially by year 2020. Among these are efforts to: 1) reduce greenhouse gas emissions by twenty percent, 2) increase energy efficiency to achieve a twenty percent savings in energy consumption, 3) achieve integration of renewable energy sources for twenty percent of total energy consumption, and 4) achieve ten percent integration of biofuels into the total consumption of vehicle fuels by 2020.

2.3.3. EU Policy and Climate Change

European policy strategies with respect to energy are arguably more closely linked to climate policy and international dialogue than perhaps the energy policy approaches that have traditionally been taken in the United States. As evidenced by the implementation of market-based emissions trading in the European Union, the motivation and nature of policy action can be quite different among developed countries. Here, we will explore

SET-Plan: The European Strategic Energy Technology Plan

The European Commission presented a strategic energy technology plan (SET Plan) on 22 November 2008, to accelerate the development and deployment of cost-effective low carbon technologies. The plan comprises measures relating to planning, implementation, resources and international cooperation in the field of energy technology. Set-Plan was introduced through communication from the Commission to the Council of the European Parliament on 22 November 2007, entitled: "A European strategic energy technology plan (SET-Plan)—Towards a low carbon future."[59]

Adopted in 2008, the aim through SET-Plan was to establish an energy technology policy for Europe. It is the principal decision-making support tool for European energy policy, with a goal of accelerating knowledge development, technology transfer and uptake, providing industrial leadership on low-carbon energy technologies, fostering science for transforming energy technologies to achieve the 2020 Energy and Climate Change goals, and contributing to the worldwide transition to a low carbon economy by 2050.

Implementation of the SET-Plan commenced with the establishment of the European Industrial Initiatives (EIIs) which bring together industry, the research community, the member states, and the Commission in risk-sharing, public-private partnerships aimed at the rapid development of key energy technologies at the European level. In parallel, the European Energy Research Alliance (EERA) has been working since 2008 to align the research and development activities of individual research organizations to the needs of the SET-Plan priorities, and to establish a joint programming framework in the European Union. A projected budget for the SET-Plan was estimated in excess of seventy billion euros.[60]

Six EIIs in total were established with a focus on data exchange on low-carbon energy technologies, including the European Industrial Bioenergy Initiative, European CO_2 Capture, Transport and Storage Initiative, European Electricity Grid Initiative, Sustainable Nuclear Initiative, Solar Europe Initiative, and European Wind Initiative.

European Technology Platforms (ETPs) were also established to liaise with SETIS and to examine how to reach Europe's energy targets through major technological advances. The ETPs, led by industry, help define research and development objectives and lay down concrete goals for achieving them. The ETPs in fields covered by the SET-Plan are aligned with the EIIs.

In some sectors of strategic importance to Europe, public-private funded European Joint Technology Initiatives (JTIs) were also established under the Seventh Framework Program (FP7) for large-scale initiatives. One such JTI, Fuel Cells and Hydrogen (FCH) was created to deliver hydrogen energy and fuel cell technologies developed to the point of commercial take-off.

more fully the role and involvement of the European Union following the adoption of the UN Framework Convention on Climate Change (UNFCCC) in 1994.[61] As mentioned, the Kyoto Protocol was adopted at the third conference of the parties to the UNFCCC in Kyoto, Japan in December 1997 and entered into force on 16 February 2005.[62] The detailed rules for the implementation of the Protocol were adopted at COP 7 (Conference of Parties) in Marrakesh in 2001, and are also called the *Marrakesh Accords*.[63] A major distinction between the convention and the protocol is that the convention encouraged developed countries to stabilize greenhouse gas emissions, while the protocol committed them to do so.

The major feature of the Protocol is that it sets binding targets for signatory countries and the European Community for reducing greenhouse gas (GHG) emissions. To be considered compliant, signatory countries had to reduce national greenhouse gas emissions an average of five percent compared to a 1990 baseline over the five year period from 2008 to 2012. Calculated estimates of GHG emissions in Mt CO_2e (million tonnes of carbon dioxide equivalent) would establish the baselines for participating countries. As of September 2011, 191 states had signed and ratified the Protocol. Individual countries in the European Union are responsible for developing annual national emission projections for greenhouse gases for all key sectors of their economy and for complying with EU reporting obligations and projections. Official submissions to the European Commission are required under Council Decision 280/2004.[64]

Under the Treaty, countries were encouraged to meet their targets primarily through national measures. This, incidentally, has interesting implications for the European Union, which is obviously comprised of member states that exhibit significant national diversity with respect to energy and emissions. The Protocol also offered additional means of meeting national targets by way of three market-based mechanisms. The Kyoto mechanisms are: 1) emissions trading, known as the *carbon market*, 2) clean development mechanism (CDM), and 3) joint implementation (JI). These mechanisms were proposed to help stimulate green investment and help Parties meet their emission targets in a cost-effective way.[65]

Participating countries were obliged to monitor and maintain precise records of emissions. Emission targets for industrialized Parties to the Protocol were expressed as levels of allowed emissions, or *assigned amounts*, over the 2008–2012 commitment time period. Such assigned amounts, denominated in tonnes (of CO_2 equivalent emissions), are known informally as *Kyoto units*. Parties may add to their holdings of Kyoto units through credits for clean development mechanisms (CDM) such as land use, land-use change and forestry (LULUCF) or by moving units from one country to another, for example through emissions trading.

The Kyoto Protocol was generally seen as an important first step toward a truly global emission reduction regime with intent on stabilizing GHG emissions, and providing the essential architecture for ensuing international agreements on climate change. The Protocol's stringent emission reductions were devised to align with findings and recommendations of the Intergovernmental Panel on Climate Change (IPCC).

The December 2009 United Nations Climate Change Conference in Copenhagen did not result in a new global climate protocol, however, a decision was taken to "take note" of an accord drawn up by a core group of heads of state (including the United States, China,

India, South Africa, Brazil, and the European Union). The accord (which is not legally binding) included a recognition to limit temperature rises to less than two degrees Celsius and to aid developing nations through financial support in achieving reduction in greenhouse gas emissions.[66] Discussions took place in tandem with the International Energy Agency whose proposed plan, entitled the 450 Scenario, includes an aggressive timetable of actions that would be required to limit the long-term concentration of greenhouse gases in Earth's atmosphere to 450 parts per million of carbon-dioxide equivalent; the concentration level commonly associated with a global temperature rise of around two degrees Celsius above pre-industrial levels. The plan outlined a timeline to 2030 with actions to achieve this objective including a fifty percent reduction in greenhouse gas emissions through implementation of energy efficient technologies and the use of low-carbon energy technologies to produce sixty percent of global electricity, comprised of thirty-seven percent renewables energy, eighteen percent nuclear energy, and five percent using power plants capable of CCS. The plan also calls for substantial deployment of advanced vehicle technology and a resultant shift from current combustion technology, with sixty percent of new sales attributable to hybrid, plug-in hybrid, and electric vehicles.[67] In 2013, these vehicles comprised about three percent of new car sales.

A further 2010 UN Climate Change Conference, held in Cancún, Mexico, adopted a number of proposals termed the Cancún Agreements. The Agreements acknowledged the goal of reducing emissions from industrialized countries by twenty-five to forty percent (relative to 1990) by 2020, and also supported enhanced action on climate change in the developing world.[68] At a UN climate meeting held in Durban, South Africa in November 2011, agreement was reached to begin work on a new climate deal that would have legal force and require both developed and developing countries to cut their carbon emissions. Attendees set targets to reach agreement on terms by 2015 and bring the agreement into effect in 2020.[69]

The Doha COP 18 UN Climate Change Conference served as the eighth meeting of the parties to the Kyoto Protocol in late November and early December, 2012, in Doha, Qatar.[70] Countries launched a new commitment period under the Kyoto Protocol, agreeing to a firm timetable to adopt a universal climate agreement by 2015 and a path to raise necessary ambition to respond to climate change. They also endorsed the completion of new institutions and agreed on ways and means to deliver scaled-up climate finance and technology to developing countries. The following Amendment to the Kyoto Protocol was agreed:

2.3.3.1. Amendment of the Kyoto Protocol

The Kyoto Protocol, as the only existing and binding agreement under which developed countries commit to cutting greenhouse gases, has been amended so that it would continue as of 1 January 2013.

- Governments have decided that the length of the second commitment period will be eight years.
- The legal requirements that will allow a smooth continuation of the Protocol have been agreed.

- The valuable accounting rules of the protocol have been preserved.
- Countries that are taking on further commitments under the Kyoto Protocol have agreed to review their emission reduction commitments at the latest by 2014, with a view to increasing their respective levels of ambition.
- The Kyoto Protocol's Market Mechanisms—the Clean Development Mechanism (CDM), Joint Implementation (JI) and International Emissions Trading (IET)—can continue as of 2013.
- Access to the mechanisms will be uninterrupted for all developed countries that have accepted targets for the second commitment period.
- JI will continue to operate, with the agreed technical rules allowing the issuance of credits, once a host country's emissions target has been formally established.
- Australia, the EU, Japan, Lichtenstein, Monaco, and Switzerland have declared that they will not carry over any surplus emissions trading credits (Assigned Amounts) into the second commitment period of the Kyoto Protocol.

2.3.4. EU Emissions Trading System (ETS)

The first and biggest international scheme for market-based trading of greenhouse gas emission allowances, the EU ETS works on the *cap and trade* principle. It covers approximately eleven thousand power stations and industrial plants in thirty-one countries, including the twenty-eight member states as well as Iceland, Liechtenstein, and Norway. It covers CO_2 emissions from power stations, combustion plants, oil refineries and iron and steel works, as well as factories making cement, glass, lime, bricks, ceramics, pulp, paper and board. Nitrous oxide emissions are also covered by the scheme. In 2020, emissions from sectors covered by the ETS will be twenty-one percent lower than in 2005. The aviation sector was brought into the system at the start of 2012; however, in November 2012 the European Commission deferred application of the scheme to flights operated to and from countries outside the European Union to allow more time to reach a global agreement addressing aviation emissions (please see Chapter 9 for more on EU-ETS in aviation).[71]

2.3.5. Conclusion

A sense of urgency has defined EU energy and climate policy in recent years. Key efforts are motivated by analyses of required technological and financial actions through 2050 that will help achieve required greenhouse gas reduction targets. Increased energy funding has resulted, for example through the 2007–2013 EU Framework 7 (FP7) research funding platform and is extended to Horizon 2020, with roll-out from January 2014. The Strategic Energy Technology (SET) Plan was introduced, providing a blueprint for rebalancing of supply-side energy, including a range of low carbon energy initiatives. Organizations contributing to energy policy include the European Energy Research Alliance (EERA), the International Partnership for Energy Efficiency Cooperation (IPEEC), and Electricity Liberalization consortia focused on regulation to support competition in energy generation and distribution. Energy and climate in the European Union, like many other cross-cutting policy issues, rely heavily on the cooperation of individual member states and on the close coordination of stakeholders from across public and private sectors.

A variety of tangible strides have been made across the European Union, as evidenced by creative German and Spanish policies to commercialize solar technologies, French global leadership in nuclear power generation and safety, and pioneering efforts by the Dutch in wind energy, to name just a few examples. To an even greater extent in the coming decades, Europe will be required to leverage practical technologies and effective policies in order to realize its aggressive national and international energy and climate aspirations.

2.4. China Energy and Climate Change Policy

China is an increasingly important player in twenty-first century global energy policy. China ranks second in the world in energy consumption and first in fuel-related emissions. It also has significant domestic energy resources and infrastructure and, as a permanent UN Security Council member and major world power, has elevated the role of energy and climate in its foreign and national security policies.

China's gross domestic product (GDP) grew by an estimated 9.2% in 2011 and by 7.8% in 2012. This ongoing high growth rate has, whether by cause or by effect, drastically increased Chinese energy consumption and, according to IEA, Beijing now ranks as the world's second largest oil consumer. Oil consumption growth in China accounted for a whopping fifty percent of global growth in 2011.

Chinese proven oil reserves were estimated at 20.4 billion barrels in January 2012, concentrated in China's northeast. Beijing started importing oil in 1993, and by 2009 had become the world's second largest oil importer. By 2011 its total net imports reached 5.5 million barrels per day, delivered from regions as diverse as the Persian Gulf, Sudan, Angola, and Venezuela. China is active in developing domestic oil resources in northwestern regions of the mainland as well as potential offshore energy resources in the East and South China Seas. Development and claim to resources in these regions may prove contentious, as neighboring energy-hungry powers—including Japan, Malaysia, the Philippines, and Vietnam—share similar interests. Estimates of these resources range from 28 billion barrels of oil according to the US Geological Survey (USGS) to over 105 billion barrels from Chinese sources. Natural gas resources are abundant as well, but like oil, estimates by USGS and Chinese sources vary.[72]

Chinese government energy policy is administered by the National Development and Reform Commission, which serves as the energy sector's primary policymaking and regulatory entity. A National Energy Commission established in January 2010 seeks to consolidate Chinese government energy policy, as well as formalize a more comprehensive energy agenda which incorporates new and lower carbon technologies as an element of national energy planning. Numerous national oil companies, such as the China National Petroleum Corporation (CNPC), the China Petroleum and Chemical Corporation (Sinopec), and the China National Offshore Oil Corporation (CNOOC), are major players in Chinese domestic and international production and policymaking.[73]

In 2009, coal accounted for approximately seventy percent of Chinese energy consumption.[74] That same year, China was responsible for nearly half of global coal, thus having tremendous influence on the future of the coal market. In its ambitious Twelfth

Five-Year Plan, 2011–2015, China intends to reduce energy and carbon intensity via enhanced energy efficiency and diversification of the energy mix.[75] In addition to coal, natural gas and its accompanying pipeline infrastructure is of increasing interest within China. The Chinese electricity sector is dominated by five state-owned holding companies including China Huaneng Group and China Datang Group. The Three Gorges Dam is the world's largest hydroelectric power station in terms of installed capacity (22,500 MW). Furthermore, China is the world's largest hydropower producer, and is seeking to increase its nuclear power generation from its 2010 level of two percent of net generation. Modest but growing steps are being taken to increase the use of renewable energy resources, notably with wind power. Furthermore, Beijing hopes to increase its solar production from two gigawatts in 2011 to twenty-five gigawatts by 2020.[76] Other policies are being introduced at the national and sub-national level to encourage the purchase of *new energy vehicles*, a term China uses to describe alternatives to the internal combustion engine.[77] Despite aggressive targets, such technologies are expensive and demand within China is relatively weak. In the near term, it is therefore possible that China may be more interested in innovating and manufacturing such renewable technologies for a world marketplace, as they are demonstrating in the export of solar PV panels, wind energy components, and advanced electronics comprising rare-earth metals.

Environmental pollution and climate change are significant problems confronting China. China emitted an estimated 6,666 million tons of CO_2 in 2008, an annual figure that ballooned twenty percent in just three years to an estimated 8,000 million tons in 2011, according to IEA and US DOE analysis. At such a pace, it is not difficult to understand that China became the world's largest CO_2 emitter in 2007. A 2009 conference report prepared for the US National Intelligence Council (NIC) noted that two-thirds of 338 Chinese cities for which air-quality data is available are considered polluted, that industrial pollution has occurred in more than seventy percent of Chinese rivers and lakes, and that underground water in ninety percent of Chinese cities is also polluted.[78]

Other findings from this report include the following:
1. China's average temperature has risen by 1.1°C between 1908 and 2007.
2. Sea level and sea surface temperature have increased by 90 mm and 0.9°C respectively, over the past thirty years.
3. Extreme weather events such as floods, drought, and storms have caused annual direct economic losses of between $25 and $37.5 billion per year.
4. Water resource scarcity, fast-growing urbanization and industrialization, and severe pollution may lead to a water crisis that could result in social unrest.
5. China's coastal regions are vulnerable to storms, floods, and sea-level rise due to their low and flat landscape.[79]

China is taking some steps on the domestic front to reduce its emissions and address climate change. It conducts bilateral dialogues with Australia, the European Union, Japan, and the United States while also participating in international climate change forums such as the Intergovernmental Panel on Climate Change.[80] China's participation in international forums is of critical import. It is a large developing country, with pockets of considerable wealth that have achieved a high level of development as measured by any global metrics. A primary challenge for China and the international community is to appropriately reconcile and balance the energy and climate goals associated with a manufacturing powerhouse of tremendous global influence against the urgent necessity associated with improving basic needs and quality of life for the world's most populous nation.

2.4.1. Conclusion

China's role as a major player in international energy and climate change policy must not be understated. Its growing dependence on foreign oil imports has even resulted in Beijing's use of warships to address international antipiracy patrols off the Gulf of Aden in 2008. China has also developed a chain of bases called *strings of pearls* in locales such as Chittagong, Bangladesh; Sittwe, Myanmar; Gwadar, Pakistan; Colombo, Sri Lanka; and other areas in the South China Sea and Indian Ocean to maintain sea lines of communication with its oil imports from the Middle East and to increase its geopolitical influence in these areas.[81] China's increasing reliance on imported, largely fossil, energy resources and maritime security to protect its international trade may make it an increasingly assertive power throughout the world. Sustaining recent growth rates in China will require massive new energy supplies from both sides of its borders, and have significant impacts on the global climate.

2.5. Russia Energy and Climate Change Policy

Russia possesses the world's largest conventional natural gas reserves, second largest coal reserves, and the ninth largest crude oil reserves; and not surprisingly, its economy depends heavily on energy exports. It was the world's second largest oil producer (after Saudi Arabia), and the second largest natural gas producer (after the United States) in 2011. In 2012, Russian oil production, averaging slightly over ten million barrels per day, actually surpassed Saudi oil production.[82]

Russia exports a significant portion of its energy resources to European countries, which affords it coercive leverage over these nations, such as Ukraine which receives 51.6% of its domestic natural gas supplies from Russia. Moscow has used this leverage in both January 2006 and January 2009 when it raised natural gas prices and threatened to cut off delivery to Ukraine. This led to the eventual toppling of the Kiev government and replacement by a more pro-Moscow government. Empirical evidence of Russian export leverage over European countries can be witnessed in the following statistics that show the distribution of Moscow's natural gas exports in 2010:
- Commonwealth of Independent States (CIS, made of selected former Soviet Republics): 37%
- Eastern Europe: 31%
- Germany: 27%
- Turkey: 14%
- Italy: 10%
- Other Western European Countries: 10%
- France: 8%

While Moscow allows some foreign energy companies to invest and operate in Russia, preferential treatment is given to Russian energy companies such as Gazprom (natural gas) and Transneft (oil pipeline).[83] Its oil production is heaviest in Western Siberia and occurs in areas as diverse as Sakhalin Island and the Urals-Volga. There are significant untapped oil and natural gas resources in Eastern Siberia, the Arctic, and the Northern Caspian Sea, all of which interest not only Russia but other regional neighbors, and carry the potential for increased conflict. A quick glance at the numbers make it evident why: Arctic Ocean reserves are estimated at 90 billion barrels of oil, 1,669 trillion cubic feet of natural gas, and 44 billion barrels of natural gas liquids. Estimated North Sakhalin Island oil reserves are 5.3 billion barrels, 43.8 trillion cubic feet of natural gas, and 0.8 billion barrels of natural gas liquids.[84]

Oversight for Russian energy policy is charged to the Ministry of Energy, yet multiple agencies participate in the implementation of the national energy policy. The Ministry of Natural Resources is responsible for issuing field licenses, monitoring compliance with license agreements, and levying fines for violating environmental regulations. The Finance Ministry administers energy sector tax policy, and the Ministry of Economic Development influences tariff regulation and energy sector reforms. Within these ministries the Federal Energy Commission administers oil transportation tariffs, the Commission for State Policy on the Oil Market regulates oil and oil product markets, the Commission on Protective Measures in Foreign Trade and Customs and Tariff Policy sets crude oil export tariffs, the Regional Energy Commission regulates retail gas prices, and the State Atomic Energy Corporation (Rosatom) administers Russian nuclear energy.[85]

Like China, Russia is also confronting significant climatic and environmental challenges which will impact not only its own energy and climate policies, but those of neighboring countries as well. Russia ranked fourth in the world in energy-related emissions with about 1.65 billion metric tons in 2011. A 2009 NIC report noted that a warming climate may lead to mixed impacts, including reduced energy and increased hydroelectricity production. At the same time, potential permafrost thaw could negatively affect energy infrastructure and increase river crossing hazards. While water supply may increase in Siberia, North, and Northwest Russia, water shortages are possible in southern European Russia increasing economic and social strains. Russian agricultural production may experience mixed impacts, with northern localities potentially benefitting from increased CO_2 levels and shifted rainfall, while southern Russia may grapple with reduced productivity and become more vulnerable to drought. Russia will also experience increased migration pressure from Central Asia, the Caucasus, Mongolia, and northeastern China due to water shortages in these areas. Climate change may even impact international maritime trade in the Arctic Ocean potentially resulting in positive economic and negative environmental impacts.[86]

Additional NIC report findings include the following:
1. Prediction of significant winter temperature increases in regions of the Arctic, averaging 4–5°C by 2050.
2. Projected summer temperature increases in the northern Caucasus, Volga, and southern Western Siberia projected at 2–3°C.

3. Projected snow mass decrease in European Russia of 10–15% by 2015.
4. Increased risk of fire and flooding; and outbreaks of disease carrying insects in northern Russia.
5. Concerns about government management of potential climate change induced infrastructure collapse.[87]

2.5.1. Conclusion

Russia is a critical nation in global energy and climate change policy. Its substantial energy resources generate significant export earnings, which in turn, enhance domestic economic growth and prosperity. Through these same energy resources, Russia has opportunity to gain economic and geopolitical leverage in the Arctic Ocean, Europe, the Caspian Sea region, and other former Soviet states. In one particularly acute example, Russia has protested western attempts to build a Trans-Caspian pipeline that would bypass Russia in delivering oil and natural gas to Europe. An emerging area of concern to players in the international energy arena is Russia's sparsely populated but energy rich east Asian region, and countries such as China, Japan, South Korea, and the United States would be interested in seeing its resources developed.

Of particular interest from a geopolitical perspective is the China-Russia energy relationship. Between 2000 and 2010, Russian crude oil exports to China increased nearly tenfold from 1.3 million to 12.8 million tons.[88] Moscow is particularly concerned about Chinese attempts to increase economic and energy investment in northeast Russia and about regional demographic trends.[89]

As Russia confronts the opportunities and challenges associated with energy and climate, recent history has shown that it may be tempted to leverage extensive energy resources at the expense of regional neighbors. Given the reliance and increasing inter-dependence of energy consuming states in the region, Russian leadership has a timely opportunity to balance its objectives with strategies that help optimize not only domestic but international economic, environmental, and security outcomes.

2.6. Brazil Energy and Climate Change Policy

Brazil's energy matrix is one of the cleanest in the world, owing primarily to abundant hydroelectric power, electricity generated from biomass, and ethanol derived from sugarcane. Whereas the global average for renewable energy is about 12.6%, Brazil's 45% domestic share ranks it first among the world's ten most populous nations in carbon neutral energy supply.[90] At the same time, Brazil has expertise in conventional oil development and is poised to emerge as a significant fossil energy player over the coming decades, due largely to discoveries of significant offshore oil and natural gas reserves in 2007. Meanwhile, the Amazon rainforest, sixty percent of it in Brazil, is well known for its ecological value to earth's biosphere, acting as a carbon sink with far reaching benefits for the global climate. Depletion through deforestation and loss of biodiversity since the 1960s carries significant implications for Brazil, South America, and the world. Upon the backdrop of these disparate realities, the Brazilian economy has surged in recent years, creating new challenges for a state that seeks to balance the rate of economic growth with appropriate social, industrial, and environmental policy. Brazil has been a strong voice on the global stage concerning sustainability and climate change, playing host in 2012 to the UN Rio+20 Conference on Sustainable Development, and has been a prominent figure in many of the Conferences of Parties (COP).

Authority over the issues surrounding energy and climate policy is shared by several executive branch agencies, the Casa Civil and relevant legislative committees. Brazil has developed a long range planning framework for both the public and private sectors that includes the National Energy Plan 2030 (PNE 2030) and the National Energy Matrix 2030. The stated objective of these documents is to devise strategies and develop policies that ensure the security and quality of energy supply for decades to come.[91] Key areas of focus for Brazilian energy policy include: security of energy supply, a policy of reasonable tariffs, and aspirations to expand energy services to a greater share of the population.

The lead for implementation of energy coordination and policy making is the Ministry of Mines and Energy (MME), which has oversight for energy planning and development, electricity, oil and natural gas, renewable fuels, geology, mining and materials processing. MME leads a multi-stakeholder mechanism known as the National Energy Policy Council (CNPE) which includes participation by key ministries including: Planning, Treasury, Environment, Development/Industry/Foreign Trade, and Agriculture, as well as representatives from the states, civil society, and Brazilian universities. CNPE advises the presidency of the republic for the electricity sector and is the principal forum for long term energy policy issues. The National Agency for Petroleum, Natural Gas, and Biofuels (ANP) was created in 1998 to provide regulatory authority and supervision for the production and distribution of fuels in Brazil. While an independent agency, it maintains links to MME, and has a diverse charter including the execution of geological and resource assessments, management of the tender process for exploration, development and production of oil and gas, calculation of revenues for various government entities, monitoring of prices, regulation and oversight of activities related to the supply chains for all types of fuels (including ethanol).[92]

Propelled by broad economic expansion, electricity consumption is expected to grow by five percent per year over the next ten years. Whereas hydropower currently provides about seventy-five percent of Brazil's electricity, a combination of concerns about over-reliance on a single source, risk of drought and environmental and cultural opposition to new projects will likely reduce this share as total electricity demand increases. Nuclear generation capacity is planned to grow in order to keep pace at a two percent share of a larger total base, natural gas will increasingly displace coal for thermal generation (a net one percent increase to fifteen percent overall), and a suite of alternative energy sources are planned including small hydro, wind, and biomass, doubling their composite share from eight to sixteen percent.[93]

Two significant Brazilian companies heavily influence the country's energy sector: Eletrobras and Petrobras. Eletrobras is Latin America's largest power utility with a generating capacity of forty-three gigawatts. The Brazilian federal government owns a fifty-two percent stake in the company, with the remaining shares publically traded on various international markets.[94] Petrobras is the largest company in Latin America by revenue, and again, the Brazilian federal government is the largest shareholder and maintains voting control.[95]

In oil and natural gas, a headline story not only for Brazil, but the energy world, is the 2007 discovery of extensive pre-salt layer continental shelf reserves estimated to contain over fifty billion barrels of oil equivalent (boe). Following initial resource assessment and exploration, a pilot project in the Tupi Lula fields began production in October 2010 at an output of approximately one hundred thousand barrels per day (bpd). Brazil is already a net oil exporter, and the pre-salt assets have the potential to increase domestic production from about 2.7 million bpd nearly twofold over the coming decades. Formidable technical challenges surround the extraction of these reserves, given the depth and pressures involved, associated freezing temperatures, and distance offshore. In addition, the fields have the potential to include large volumes of associated natural gas for which major infrastructure would be required to either transport it to markets via pipeline or liquefy it at sea, both of which introduce additional technical, logistical, economic, and safety considerations. Pre-salt resources have raised legislative questions at the federal level to ensure the distribution of royalty income is equitable and undergirds socio-economic development for all Brazilian states, including many not endowed with fossil fuels.[96]

Brazil is renowned for its ethanol industry, ranking second to the United States in total production. Sugarcane juice is fermented to produce the ethanol, while the residual biomass, or *bagasse*, is burned to produce process steam for distillation and the generation of surplus green energy for the local electrical grid. In all, the process is extremely efficient with an energy ratio that yields between five and nine parts energy output for every one part of energy input (please see chapter 6). When combusted in motor vehicle engines, sugarcane ethanol has lifecycle greenhouse gas emissions that are approximately sixty percent lower than standard gasoline, according to US EPA estimates. This qualifies it as a so-called *advanced biofuel*. Sugarcane ethanol is a very sustainable form of bioenergy, and has provided opportunities to de-carbonize the Brazilian vehicle fleet, in which more than ninety percent of vehicles can operate as *flex-fuel*. Brazilian statute mandates a minimum ethanol blend level on a quarterly basis, which typically varies between eighteen and twenty-five percent. Though the commodity is freely traded, the sugarcane industry has received indirect government support in the past, and recent policies are aimed at providing favorable loan terms for upgrading infrastructure, optimizing agricultural practices, and incentivizing research and development for advanced biofuels. Gasoline as a commodity has been kept artificially low in recent years in efforts to stem inflation. Though prices are adjusted periodically, this policy has been known to make ethanol less competitive and has unintentionally created supply shortages for gasoline and inventory imbalances for ethanol. Ethanol as a share of Brazilian transport fuel declined from fifty-five to thirty-five percent between 2008 and 2012.[97] Ironically, during part of 2011, Brazil increased imports of US corn ethanol to help meet statutory blend levels, while exporting sugarcane ethanol to several US states, including California, where local policies favor advanced biofuels. In the long term, Brazil is coordinating its energy, agricultural, and environmental policies to optimize the value that sugarcane delivers across the sectors of food, fiber, and energy.

Connecting energy policy with sustainable development and environmental policy has been a priority for Brazil, both domestically and internationally. While Brazil ranks fourth in the world in total greenhouse gas emissions, a small percentage of these emissions are attributable to the conversion of energy (please refer to Table 2.1). This is principally because fifty-five percent of total Brazilian emissions derive from land use, land-use change, and forestry, twenty-five percent from agriculture and livestock; and the industrial and energy sectors account for only twenty percent. Brazil's clean energy matrix and modest per capita energy intensity result in very low energy-specific emissions for the nation as a whole.

Brazil's position in the context of international climate change negotiations has generally focused on three issues, namely:
1. Insistence upon each country's individual sovereign right to national development;
2. Strong opposition to any suggestion that the Amazon rainforest be put under international control for its protection;
3. Insistence of acceptance of obligations by industrialized countries for their emissions to date.[98]

Since 2008, Brazil has made some policy adjustments with regard to international agreements about forests, and importantly, introduced policies aimed at identifying CO_2 reduction targets by sector. For example, under a 2009 law, Brazil would target a thirty-six to thirty-nine percent reduction in CO_2 by 2020; which, if implemented, will keep total emissions at the 2005 level of 2.0 Gt CO_2e in lieu of the business as usual projection level of 3.2 Gt CO_2e. Though energy would be a minor component of these strategies, given the dominant share of non-energy emissions, the Brazilian energy sector has suggested ways to assess and monitor energy performance and emissions, fiscal incentives, and energy efficiency projects. Brazil established, in 2009, a National Climate Change Fund, in order to assist in formalizing emission reductions across all sectors.[99]

2.6.1. Conclusion

Brazil has been blessed with considerable natural resources with respect to both energy and the environment. The country has postured itself well by responsibly leveraging both renewable and fossil energy sources to promote social development and economic growth in a global market. In the coming decade, Brazil will confront new challenges associated with rapid growth in electricity demand and delicate decisions about benefits and risks associated with the development of large-scale hydropower and, significantly, deepwater oil and natural gas. Land use change and market optimization questions surrounding sugarcane will continue to require careful policy vision. With respect to the environment and climate change, the international community can benefit from the contributions of Brazil, recognizing that large segments of developing countries will increasingly enjoy the benefits of industrialization and energy access in the coming decades. As that occurs, it will remain imperative that Brazil balance the complicated forces between increased economic growth, supply of services, social reform, increased output of fossil and renewable energy, and environmental impacts. It is clear that whether the context is demand growth, renewables, oil, environmental impacts, or global collaboration, Brazilian energy and climate policy will be of critical global importance.

2.7. India Energy and Climate Change Policy

Though India is the world's fourth largest energy consumer, it has the lowest per capita energy consumption of the countries discussed in this section. Yet India's energy and climate footprint is bifurcated—large urban centers are responsible for high levels of coal-based emissions; and rural areas have little or no energy access. Economic growth in India has been steady over the past decade and more than half of India's economic output is attributable to the service industry.[100] Annual GDP growth based upon purchasing power parity (PPP) for the period 2015–2030 is estimated to be 5.9% in India. This rate is notably greater than other major economies, including China. Should these projections prove accurate, Indian economic growth would likely be less energy intensive than in neighboring China, where manufacturing and construction constitute a larger share.

First and foremost, Indian energy policy is focused on securing energy sources to sustain economic development. Much of this has come from coal which accounts for about fifty-five percent of the country's commercial energy supply. Though India ranks fourth in coal reserves with about seven percent of proven global totals, demand growth has outstripped domestic supply for coal, forcing India to increase coal imports by more than thirteen percent per year since 2001.[101]

Similar to China, India's energy planning is coordinated within the central government's revolving five year plans. The current Twelfth Five Year Plan (2012–2016) emphasizes energy and climate initiatives, yet conveys a candid outlook of the critical realities:

> A GDP growth rate of about 8 per cent requires growth rate of about 6 per cent in total energy use from all sources. Unfortunately, the capacity of the economy to expand domestic energy supplies to meet this demand is severely limited. The country is not well-endowed with energy resources, except coal, and the existence of policy distortions makes management of demand and supply more difficult.[102]

India has recently made significant strides to re-align energy prices notably for coal and liquid petroleum products. Despite near term price increases, these policy adjustments are nevertheless expected to have positive long term impacts on conservation and efficient use.[103]

A serious issue confronting India in the midst of its emergence as a major global economy is poverty. Though many have entered a rising middle class, the rural poor have been largely bypassed. A 2012 IEA report estimated that nearly twenty-five percent of the Indian population lacks basic access to electricity, while electrified areas suffer from intermittent service.[104] The government of India launched a rural electrification initiative in 2005 known as the Rajiv Gandhi Grameen Vidyutikaran Yojana (RGGVY) aimed at increasing rural household access by creating additional electricity infrastructure. Capital subsidies and preferential policies specifically target below poverty line households in un-electrified villages and rural communities.[105]

Oil reserves in India are extremely limited, and reliance on foreign petroleum is perhaps the country's weakest energy link. In its twelfth and provisional thirteenth Five Year Plans, India anticipates meeting approximately seventy percent of its expected energy consumption with domestic resources,[106] however growing imported oil demands will preclude greater levels of near term energy self-sufficiency. In 2013, India's petroleum minister announced that the ministry would work toward energy independence by 2030 through a series of steps aimed at increasing supply or reducing demand. Among them are the following: Increased hydrocarbon production; unconventional resources such as coalbed methane and shale gas; foreign acquisitions by domestic Indian companies; and reduced subsidies on motor fuels.[107]

India has increased its development offshore, where about half its oil and three-fourths of its natural gas is known to reside. India's New Exploration Licensing Policy (1999) was successful in attracting largely domestic private investment and in identifying new oil and natural gas finds. Despite this, India is a net importer of all fossil fuels. Over the last four decades, energy supply from imported sources has increased from ten to more than thirty percent, a sobering statistic for a country poised for sustained economic expansion.

Traditional biomass and waste are thermally converted to provide nearly a quarter of India's energy supply. Much of this is used to provide for the energy requirements of buildings, heating and cooling, and other industrial needs. In rural areas, firewood, animal dung and agricultural residue are used as fuel for cooking, heating and lighting due to a lack of grid connectivity or access to alternate energy services. According to the Ministry of New and Renewable Energy (MNRE), India has 288 biomass power and co-generation plants that generate 2.7 GW of installed capacity with the potential to reach 18 GW in total generating capacity.[108]

India generates about three percent of the world's hydropower with 113 billion kilowatt-hours (kWh) generated in 2010 (ranked seventh in the world).[109] Due to the tropical climate, India has identified opportunities to increase its hydroelectric generating capacity from 39.3 GW to more than 100 GW if all projects currently under survey and investigation are approved and constructed.[110] India currently has six nuclear power plants in total with a combined 4.4 GW of generating capacity. As part of its energy growth strategy, the government has indicated that it plans to increase the share of nuclear power from four percent in 2011 to twenty-five percent over the long term.[111]

The country's energy sector is administered and managed via a multi-ministerial structure that includes the Ministries of Power, Coal, Petroleum and Natural Gas, New and Renewable Energy, Environments and Forests, the Department of Atomic Energy, and the Planning Commission, among others.

Several national policies have been implemented to ensure a smooth functioning framework for the power sector, open access to transmission and distribution networks, regulation of tariffs, and to improve rural electrification. The government established the Power Grid Corporation of India (POWERGRID) to operate five regional electricity grids, while states and private companies operate transmission/distribution segments. Other policies have liberalized the hydrocarbon market and encouraged private sector investment throughout the energy supply chain. This advocacy applies both to investment within India, as well as to Indian investment in foreign energy projects that will accrue value in India, such as from imports of oil or natural gas.[112] One such example is in liquefied natural gas (LNG), in which the government of India began an import arrangement with Qatar in 2004. Indian firms, such as Petronet, have established trading relationships with foreign and domestic partners to ensure a stable supply to Indian markets. Due to increasing demand, India is an attractive trading partner, and LNG is now being acquired on the spot market from the Middle East and Africa.

IEA has noted that strategic technologies may enable India to sustain social-economic growth while developing increased energy resources including: clean coal technologies; nuclear power through a three stage nuclear program; energy efficiency in industry and buildings through such approaches as audits, trading schemes and labeling; increased use of biodiesel and ethanol in transportation fuels; and improved transmission and distribution networks.[113]

India has taken some critical steps domestically to reduce the environmental footprint associated with energy production and use. For example, the Ministry of New and Renewable Energy aims to increase the share of renewable energy to six percent of India's total energy matrix and to ten percent of the electricity mix by 2022.

India's National Action Plan on Climate Change (NAPCC) identified eight priority national missions to address climate change mitigation and adaptation. Among these, the National Mission on Enhanced Energy Efficiency (NMEEE) seeks to create a regulatory and policy framework that is conducive to sustainable business models and innovation. Fuel savings, avoided capacity additions, and emissions reductions are among the benefits. Other campaigns are directed at increasing the contribution of solar energy and accelerating the adoption and use of sustainable biofuels, setting an indicative target of 20% blending of biodiesel or bioethanol by 2017.[114]

In December 2009, India voluntarily agreed to a twenty to twenty-five percent reduction in emission intensity by 2020 from 2005 levels, exclusive of agricultural emissions. The government projects that its per capita emission in 2030 (< 4 t CO_2e) will remain lower than the 2005 global average (4.22 t CO_2e).[115] As revealed in the details of its recent policy plans, India has made significant progress in outlining steps required to align and implement its energy and climate agendas to continue strong economic growth in a sustainable manner.

2.7.1. Conclusion

India's economic growth is projected to outpace other developing countries as well as its capability to procure domestic energy resources. Going forward, it must grapple with major increases in energy demand while balancing imports against the development of domestic supplies. Energy analysts warn than India must manage its energy growth "without locking in high emissions."[116] Thus, environmentally sound solutions will be imperative from more efficient use of coal to increasing shares of natural gas, nuclear and biomass. The Indian energy and climate agenda includes some familiar essentials: improving access and reliability, increasing exploration and capacity, implementing regulatory and pricing reform, reducing petroleum reliance, increasing energy diversity with lower carbon sources, and increasing efficiency. Effective implementation will raise countless millions more to a higher quality of life while controlling the modest per capita energy intensities that typify the country today. India has made significant investments in foreign partnerships, and encouraged private interests, both foreign and domestic, to participate in Indian projects. This inclusive and global vision may come by necessity, but it may bring creative new solutions to the global energy and climate dialogue.

Notes

1. Vaclav Smil, *Energy Transitions: History, Requirements, Prospects* (Santa Barbara, CA: ABC-CLIO, 2010), 25, 105.
2. "Key Figures," European Commission Market Observatory for Energy, Directorate-General for Energy, last modified June 2011.
3. International Energy Agency (IEA), *2013 Key World Energy Statistics* (Paris: IEA, 2013), 48–57.
4. Ibid.; IEA, *CO2 Emissions from Fuel Combustion: Highlights,* 2012 ed. (Paris, IEA: 2012), 99–101.
5. IEA, *World Energy Outlook 2010* (Paris, IEA: 2010).
6. Smil, *Energy Transitions*; British Petroleum, *BP Statistical Review of World Energy: June 2012* (London: British Petroleum, 2012).
7. Hillary Rodham Clinton, *Remarks on American Leadership at the Council on Foreign Relations,* 31 January 2013.
8. IEA, *2013 Key World Energy Statistics*; Council on Foreign Relations, *The Global Climate Change Regime: Issue Brief* (Washington, D.C.: Council on Foreign Relations, 2013).
9. IEA, *2013 Key World Energy Statistics*.
10. Ibid.
11. See Linda Qaimmaqami, ed. *Foreign Relations of the United States 1969–1976, Volume XXXVI: Energy Crisis, 1969–1974* (Washington, D.C.: Government Printing Office, 2011), 574–953, for the official US foreign policy response. See also Daniel Yergin, *The Prize: The Epic Quest for Oil, Money, & Power* (New York: Simon and Schuster, 1990), 606–32; Energy Information Administration (EIA), *Annual Energy Review 2010 (DOE/EIA-0384(2010))* (Washington, D.C.: Energy Information Administration, 2011), 175, 187.
12. Qaimmaqami, *Foreign Relations, Volume XXXVI*.
13. Department of Energy Organization Act, 42 U.S.C. § 7101 (1977); Terrence R. Fehner, *Department of Energy, 1977–1994: A Summary History* (Washington, D.C.: United States Department of Energy, 1994).
14. Energy Policy and Conservation Act of 1975, 42 U.S.C. § 6201 (1975); *Immediate Relief from High Oil Prices: Deploying the Strategic Petroleum Reserves, Hearing before the Select Committee on Energy Independence and Global Warming,* 110th Cong. (2008); Strategic Petroleum Reserve, "Strategic Petroleum Reserve Inventory," US Department of Energy, last modified November 1, 2013. http://www.spr.doe.gov/dir/dir.html.
15. International Energy Agency, "FAQs: Organisation and Structure," *IEA.org*, last modified 2013. http://www.iea.org/aboutus/faqs/organisationandstructure/.

16. EIA, *Annual Energy Review 2010*, 175–187; Council of Economic Advisors, *Economic Report of the President 1981* (Washington, D.C.: Government Printing Office, 1981): 7–8.
17. Ayesha Rascoe, "NRC Approves First New Nuclear Power Plant in a Generation," *Reuters.com*, February 9, 2012, http://www.reuters.com/article/2012/02/09/us-usa-nuclear-license-idUSTRE8181T420120209.
18. Southern Company, "Southern Nuclear," *Southerncompany.com*, last modified 2013, http://www.southerncompany.com/about-us/our-business/southern-nuclear/home.cshtml; SCANA, "SCE&G Completes First Nuclear Concrete Placement," *Sceg.com*, March 13, 2013, http://www.sceg.com/NR/rdonlyres/3CCBEA1-8009-44A5-95F9-3CDB98C6C061/0/03112013SCEGFirstNuclearConcrete.pdf.
19. Nuclear Regulatory Commission, "Combined License Applications for New Reactors," *NRC.gov*, last modified September 23, 2013, http://www.nrc.gov/reactors/new-reactors/col.html.
20. *Statement of Commissioner William D. Magwood, IV*, Nuclear Regulatory Commission, February 9, 2012, http://www.nrc.gov/about-nrc/organization/commission/comm-william-magwood/comm-magwood-statement-02-09-12.pdf; EPA, "Nuclear Incidents: Three Mile Island Nuclear Plant," *EPA Radiation Protection*, last modified February 15, 2012, http://www.epa.gov/rpdweb00/rert/tmi.html; Subcommittee on Nuclear Regulation, *Nuclear Accident and Recovery at Three Mile Island: A Report* (Washington, D.C.: Government Printing Office, 1980); and EIA, *Annual Energy Review 2010*, 283.
21. Natural Gas Policy Act of 1978, 15 U.S.C. § 3301 (1978); Exec. Order No. 12,287, 46 Fed. Reg. 9909 (Jan. 30, 1981); Mack Ott and John A. Taton, "A Perspective on the Economics of Natural Gas Decontrol," *Federal Reserve Bank of St. Louis Review*, 64 (November 1982): 19–31; EIA, *Annual Energy Review 2010*, 165, 207; National Highway Traffic Safety Administration, "CAFE: Fuel Economy," US Department of Transportation, last modified 2012, http://www.nhtsa.gov/fuel-economy.
22. EIA, *Annual Energy Outlook 2013* (Washington, D.C.: US Energy Information Administration, 2013). http://www.eia.gov/forecasts/aeo/.
23. Adam Sieminksi, "U.S. Energy Outlook," presentation given at the IEA Bilateral Meetings, Paris, France, March 14, 2013, http://www.eia.gov/pressroom/presentations/sieminski_03142013_iea.pdf. See especially slides 2 and 8.
24. Stratospheric Ozone Protection Plan, 51 Fed. Reg. 1257 (1986); Bjorn Lomborg, *Cool It: The Skeptical Environmentalist's Guide to Global Warming* (New York: Random House, 2007).
25. Stephen Seidel and Dale Keyes, *Can We Delay a Greenhouse Warming? The Effectiveness and Feasibility of Options to Slow a Build-Up of Carbon Dioxide in the Atmosphere*, (Washington, D.C.: EPA, 1983); U.S. Congress, Senate Committee on Environment and Public Works, Subcommittee on Toxic Substances and Environmental Oversight, *Global Warming, Hearing Before the Subcommittee on Toxic Substances and Environmental Oversight of the Committee on Environment and Public Works, United States Senate*, 96th Cong. (1986); National Aeronautics and Space Administration, *Report of the International Ozone Trends Panel, 1988* (Washington, D.C.: NASA, 1988); Subcommittee on Energy and Power, *Energy Policy Implications of Global Warming* (Washington, D.C.: Government Printing Office, 1989); Congressional Budget Office, *Carbon Charges as a Response to Global Warming: The Effects of Taxing Fossil Fuels*, (Washington, D.C.: CBO, 1990); Department of Energy, Assistant Secretary for Fossil Energy, Office of Planning and Environment, *A Fossil Energy Perspective on Global Climate Change* (Washington, D.C.: DOE, 1990).
26. United Nations (UN), "Earth Summit: UN Conference on Environment and Development," UN Department of Public Information, last modified May 23, 1997. http://www.un.org/geninfo/bp/enviro.html.
27. Council on Environmental Quality, *United States of America National Report: United Nations Conference on Environment and Development*, (Washington, D.C.: CEQ, 1992); Susan R. Fletcher, *Financing New International Environmental Commitments*, (Washington, D.C.: GPO, 1992); House Committee on Foreign Affairs, Subcommittee on Western Hemisphere Affairs, *The United Nations Conference on Environment and Development*, (Washington, D.C.: GPO, 1993).

28. Energy Information Administration, *Impacts of the Kyoto Protocol on U.S. Energy Markets and Economic Activity*, (Washington, D.C.: EIA, 1998); U.S. Congress, House Committee on International Relations, *The Kyoto Protocol: Problems With U.S. Sovereignty and the Lack of Developing Country Participation*, (Washington, D.C.: Government Printing Office, 1998); House Committee on Government Reform and Oversight, Subcommittee on National Economic Growth, Natural Resources, and Regulatory Affairs, *The Kyoto Protocol: Is the Clinton-Gore Administration Selling Out Americans?: Parts 1–VI*, (Washington, D.C.: Government Printing Office, 1999); Charli E. Coon, "Why President Bush is Right to Abandon the Kyoto Protocol," Heritage Foundation, March 18, 2001, http://www.heritage.org/research/reports/2001/05/president-bush-right-to-abandon-kyoto-protocol; Statement by Minister Kent, *Environment Canada*, December 12, 2011, http://www.ec.gc.ca/default.asp?lang=En&n=FFE36B6D-1&news=6B04014B-54FC-4739-B22C-F9CD9A840800.

29. Energy Policy Act of 2005, 119 Stat. 594; Robert L. Bamberger and Carl E. Behrens, *Energy Policy: Comprehensive Energy Legislation (H.R. 6, S. 10) in the 109th Congress* (Washington, D.C.: Library of Congress, Congressional Research Service, 2005); Task Force on Strategic Unconventional Fuels, *Development of America's Strategic Unconventional Fuels Resources* (Washington, D.C.: Task Force on Strategic Unconventional Fuels, 2006).

30. Solar Trust of America LLC, et al., No. 12-11136 (United States Bankruptcy Court, 2012); Steven Mufson and Juliet Elperin, "Loan-Guarantee Recipient Founders," *Washington Post*, November 1, 2011; House Committee on Energy and Commerce, Subcommittee on Oversight and Investigations, *The Solyndra Failure: Majority Staff Report*, http://energycommerce.house.gov/sites/republicans.energycommerce.house.gov/files/analysis/20120802solyndra.pdf.

31. Scott J. Wallsten, "The Effects of Government-Industry R&D Programs on Private R&D: The Case of the Small Business Innovation Research Program," *Rand Journal of Economics* 31, no. 1 (2000): 82–100; American Recovery and Reinvestment Act of 2009, Public Law 111-5 (2009); Department of Labor, Office of Inspector General, *Recovery Act: Slow Pace Placing Workers Into Jobs Jeopardizes Employment Goals of the Green Jobs Program* (Washington, D.C.: Dept. of Labor Office of Inspector General, 2011), 1–4; Daniel Steinberg, Gian Porro, and Marshall Goldberg, *Preliminary Analysis of the Jobs and Economic Impacts of Renewable Enegy Projects Supported by the §1603 Treasury Grant Program* (Golden, CO: National Renewable Energy Laboratory, 2012), iv–v, www.nrel.gov/docs/fy12osti/52739.pdf; U.S. Congress, House Committee on Government Oversight and Reform, Subcommittee on Regulatory Affairs, Stimulus Oversight, and Government Spending, *The Green Energy Debacle: Where Has All the Taxpayer Money Gone?* (Washington, D.C.: Government Printing Office, 2012).

32. Environmental Protection Agency, "Carbon Dioxide Capture and Sequestration," *EPA: Climate Change*, last modified 2012, www.epa.gov/climatechange/ccs/; *Massachusetts v. EPA*, 549 U.S. 497; *American Electric Power v. Connecticut*, 131 S. Ct. 2527; Environmental Protection Agency, "Standards of Performance for Greenhouse Gas Emissions for New Stationary Sources: Electric Utility Generating Units," 77 Fed. Reg. 72 (2012): 22393–441.

33. Jason Scott Johnston, "Climate Change Hysteria and the Supreme Court: The Economic Impact of Global Warming on the U.S. and the Misguided Regulation of Greenhouse Gas Emissions Under the Clean Air Act," *Scholarship at Penn Law*, Paper 209 (2008), http://lsr.nellco.org/upenn_wps/209; Marlo Lewis, Jr. "Carbon Pollution Standard: 4 Ways Weird," *National Journal*, April 2, 2012, http://www.nationaljournal.com/2012/04/what-will-be-upshots-of-epas-c.php.

34. "United States 09 Copenhagen," *State Department*, last modified 2009, http://cop15.state.gov/; House Committee on Foreign Affairs, Subcommittee on Asia, Pacific, and the Global Environment, *Global Change Finance: Providing Assistance for Vulnerable Countries*, (Washington, D.C.: Government Printing Office, 2010).

35. National Security Presidential Directive 66 and Homeland Security Presidential Directive 25, *White House Office of the Press Secretary*, January 12, 2009, http://georgewbush-whitehouse.archives.gov/news/releases/2009/01/20090112-3.html.

36. *National Security Strategy* (Washington, D.C.: The White House, 2010), www.whitehouse.gov/sites/default/files/rss_viewer/national_security_strategy.pdf.
37. Energy Information Administration, "U.S. Crude Oil Production On Track to Surpass Imports for First Time Since 1995," *EIA.gov*, March 20, 2013, http://www.eia.gov/todayinenergy/detail.cfm?id=10451.
38. "CIA Opens Center on Climate Change and National Security," *CIA.gov*, 2009, https://www.cia.gov/news-information/press-releases-statements/center-on-climate-change-and-national-security.html.
39. See for example Scott Thomas and David Kerner, Defense Energy Resilience: Lessons from Ecology (Carlisle, PA: U.S. Army War College, 2010); "The DOD Energy Blog: Rethinking Military Power," U.S. Department of Defense, http://dodenergy.blogspot.com/; Defense Science Board, Trends and Implications of Climate Change for National and International Security (Washington, D.C.: Defense Science Board, 2011).
40. The White House, *Blueprint for a Secure Energy Future* (Washington, D.C.: The White House, 2011): 4–7.
41. The White House, *The Blueprint for a Secure Energy Future: Progress Report* (Washington, D.C.: The White House, 2012): 2–8.
42. Energy Information Administration, "What is the Status of the U.S. Nuclear Industry?," *EIA Energy in Brief*, last modified December 14, 2012, http://www.eia.gov/energy_in_brief/article/nuclear_industry.cfm; Frank Newport, "Americans Still Favor Nuclear Power a Year After Fukushima," *Gallup Politics*, March 26, 2012, http://www.gallup.com/poll/153452/Americans-Favor-Nuclear-Power-Hear-Fukushima.aspx; Wendy Kiska and Deborah Lucas, *Federal Loan Guarantees for the Construction of Nuclear Power Plants*, (Washington, D.C.: Congressional Budget Office, 2011); A Review of Nuclear Safety in Light of the Impact of Natural Disasters on Japanese Nuclear Facilities, Hearing before a Subcomm. of the Comm. On Appropriations, S. Hrg. 112-193 (2011).
43. International Atomic Energy Agency (IAEA), "PRIS: The Database on Nuclear Power Reactors," *IAEA.org*, last modified November 10, 2013, http://www.iaea.org/PRIS/.
44. Newport, "Americans Still Favor Nuclear Power."
45. "Sustainable Energy for All," United Nations, last modified 2013, http://www.sustainableenergyforall.org/.
46. Myron Ebell, *Increase Access to Energy* (Washington, D.C.: Competitive Enterprise Institute, 2009); Energy Information Administration, "Impact of Limitations on Access to Oil and Natural Gas Resources in the Federal Outer Continental Shelf," *EIA Issues in Focus*, last modified 2009, http://www.eia.gov/oiaf/aeo/otheranalysis/aeo_2009analysispapers/aongr.html; EIA, "U.S. Crude Oil and Natural Gas Proved Reserves," *EIA Natural Gas*, last modified August 1, 2013, http://www.eia.gov/naturalgas/crudeoilreserves/; "National Oil and Gas Assessment," *U.S. Geological Survey*, last updated August 6, 2013, http://energy.usgs.gov/OilGas/AssessmentsData/NationalOilGasAssessment.aspx; Nicholas Loris, "Ten Actions Congress Can Take to Lower Gas Prices," *Heritage Foundation Backgrounder*, May 31, 2012, http://www.heritage.org/research/reports/2012/05/ten-actions-congress-can-take-to-lower-gas-prices; Office of the Spokesperson, "U.S. Support for the Sustainable Energy for All Global Action Agenda," *U.S. Department of State* (June 20, 2012), http://www.state.gov/r/pa/prs/ps/2012/06/193500.htm; David Rothbard and Craig Rucker, "The U.N.'s Rio+20 Agenda," *National Review Online* (June 20, 2012), http://www.nationalreview.com/articles/303268/un-s-rio20-agenda-david-rothbard.
47. Minority Staff, Senate Committee on Environment and Public Works, *'Consensus' Exposed: The CRU Controversy*, (Washington, D.C.: U.S. Senate, 2010); http://www.epw.senate.gov/public/index.cfm?FuseAction=Files.View&FileStore_id=7db3fbd8-f1b4-4fdf-bd15-12b7df1a0b63; Majority Staff, House Committee on Energy and Commerce, Subcommittee on Oversight and Investigations, *The Solyndra Failure*, (Washington, D.C.: U.S. House of Representatives, 2012), http://energycommerce.house.gov/sites/republicans.energycommerce.house.gov/files/analysis/20120802solyndra.pdf; Mufson and Elperin, "Loan-Guarantee Recipient Founders."
48. Gal Luft, "United States: A Shackled Superpower," in *Energy Security Challenges for the 21st Century: A Reference Handbook*, ed. Gal Luft and Anne Korin (Santa Barbara: Praeger Security International, 2009), 143–59; Daniel Yergin, *The Quest: Energy, Security, and the Remaking of the Modern World*

(New York: Penguin Press, 2011); Patrick J. Michaels, ed., *Climate Coup: Global Warming's Invasion of Our Government and Our Lives* (Washington, D.C.: CATO Institute, 2011); Woodrow W. Clarke II and Grant Cooke, *Global Energy Innovation: Why America Must Lead* (Santa Barbara: Praeger, 2012); *American Energy Initiative: Identifying Roadblocks to Wind and Solar Energy on Public Lands and Waters, Oversight Hearing before the House Natural Resources Comm.* (Washington, D.C.: GPO, 2012).

49. Center for European Studies, "The Congress of Europe in The Hague (7–10 May, 1948)," *CVCE.eu*, last modified 2013, http://www.cvce.eu/recherche/unit-content/-/unit/04bfa990-86bc-402f-a633-11f39c9247c4.
50. The full text of Robert Schuman's May 9, 1950 speech can be accessed on the Schuman Project website at http://www.schuman.info/9May1950.htm.
51. Paul-Henri Spaak and the Intergovernmental Committee on European Integration, *The Brussels Report on the General Common Market* (1956). The English translation of this report is commonly referred to as the Spaak Report.
52. "Brief History," *OPEC*, 2013, http://www.opec.org/opec_web/en/about_us/24.htm.
53. Barry Turner, *Suez 1956: The Inside Story of the First Oil War* (London: Hodder & Stoughton, 2006).
54. Roger Owen, "Suez Crisis," in *The Oxford Companion to the Politics of the World* (Second ed.), ed. Joel Krieger (London: Oxford University Press, 2001).
55. Rose McDermott, *Risk-Taking in International Politics: Prospect Theory in American Foreign Policy* (Ann Arbor: University of Michigan Press, 2001).
56. Nicholas Moussis, "The 1973 Energy Crisis," *Europedia*, 2011, http://www.europedia.moussis.eu/books/Book_2/6/19/01/01/index.tkl?all=1&pos=267.
57. European Commission, *EURATOM Supply Agency: Annual Report 2011* (Luxembourg: Publications Office of the European Union, 2012). http://dx.doi.org/10.2833/11309.
58. European Commission, *A European Strategy for Sustainable, Competitive and Secure Energy* {SEC(2006) 317} (Brussels, Belgium: European Commission, 2006).
59. Commission to the Council, the European Parliament, the European Economic and Social Committee, and the Committee of the Regions, *A European Strategic Energy Technology Plan (SET-Plan): Towards a Low Carbon Future* {COM(2007) 723} (Brussels, Belgium: European Commission, 2007).
60. Strategic Energy Technologies Information System, "Overview," *European Commission*, last modified August 11, 2013, http://setis.ec.europa.eu/about-setis/overview.
61. "Home," *United Nations Framework Convention on Climate Change*, last modified 2013, http://unfccc.int/2860.php.
62. "Kyoto Protocol," *United Nations Framework Convention on Climate Change*, last modified 2013, http://unfccc.int/kyoto_protocol/items/2830.php.
63. UNFCCC Conference of the Parties, *Report of the Conference of the Parties on its Seventh Session* (FCCP/CP/2001/13), Marrakech, Morocco, October–November 2001.
64. Decision No 280/2004/EC of the European Parliament and of the Council of 11 February 2004 concerning a mechanism for monitoring Community greenhouse emissions and for implementing the Kyoto Protocol.
65. "The Mechanisms under the Kyoto Protocol: Emissions Trading, the Clean Development Mechanism and Joint Implementation," *United Nations Framework Convention on Climate Change*, last modified 2013, http://unfccc.int/kyoto_protocol/mechanisms/items/1673.php
66. "Copenhagen Climate Change Conference 2010," *United Nations Framework Convention on Climate Change*, last modified 2013, http://unfccc.int/meetings/copenhagen_dec_2009/meeting/6295.php.
67. IEA, *World Energy Outlook 2010*.
68. "The Cancun Agreements," *United Nations Framework Convention on Climate Change*, last modified 2010. http://cancun.unfccc.int/.
69. "Durban Climate Change Conference: November/December 2011," *United Nations Framework Convention on Climate Change*, last modified 2013, http://unfccc.int/meetings/durban_nov_2011/meeting/6245.php.

70. "Doha Climate Change Conference: November 2012" *United Nations Framework Convention on Climate Change*, last modified 2013, http://unfccc.int/meetings/doha_nov_2012/meeting/6815.php.
71. European Commission, "The EU Emissions Trading System (EU ETS)," *EU Climate Action*, last modified 2013, http://ec.europa.eu/clima/policies/ets/index_en.htm.
72. Energy Information Administration, *Country Analysis Brief: China* (Washington, D.C.: EIA, 2012), 1–2; http://www.eia.gov/countries/analysisbriefs/China/china.pdf.
73. Bernard D. Cole, *Oil for the Lamps of China: Beijing's 21st Century Search for Energy* (Washington, D.C.: National Defense University Press, 2003): 3, 15–16, http://permanent.access.gpo.gov/websites/nduedu/www.ndu.edu/inss/mcnair/mcnair67/198_428.McNair.pdf; Energy Information Administration, *Country Analysis Brief: South China Sea* (Washington, D.C.: EIA, 2008).
74. EIA, *China*, 4–9; Sabrina Howell, "Jia You! (Add Oil!): Chinese Energy Security Strategy," in *Energy Security Challenges for the 21st Century: A Reference Handbook*, ed. Gal Luft and Anne Korin (Santa Barbara, CA: Praeger Security International, 2009), 191–219; and Bo Kong, *China's International Petroleum Policy* (Santa Barbara, CA: Praeger Security International, 2010).
75. International Energy Agency, *World Energy Outlook 2011* (Paris: IEA, 2010), http://www.iea.org/publications/freepublications/publication/WEO2011_WEB.pdf.
76. EIA, *China*, 20–33.
77. David Fridley, Nina Zheng, and Yining Qin, *Inventory of China's Energy-Related CO_2 Emissions in 2008* (Berkeley, CA: Lawrence Berkeley National Laboratory, 2011): 1, http://www.osti.gov/scitech/servlets/purl/1016716.
78. IEA, *2013 Key World Energy Statistics*; EIA, *China*; and Joint Global Change Research Institute and Battelle Memorial Institute, Pacific Northwest Division, *China: The Impact of Climate Change to 2030; A Commissioned Research Report* (Washington, D.C.: National Intelligence Council, 2009): 7–8, http://permanent.access.gpo.gov/lps122078/LPS122078.pdf.
79. Ibid., 33.
80. Christopher J. Pehrson, *String of Pearls: Meeting the Challenge of China's Rising Power Across the Asian Littoral* (Carlisle, PA: U.S. Army War College Strategic Studies Institute, 2006), http://permanent.access.gpo.gov/lps77746/PUB721.pdf; U.S. Joint Forces Command, *The Joint Operating Environment (JOE) 2008: Challenges and Implications for the Future Joint Force* (Norfolk, VA: Joint Forces Command, 2008): 21; and Daniel J. Kostecka, "Places and Bases: The Chinese's Navy's Emerging Support Network in the Indian Ocean," *Naval War College Review* 64, no. 1 (2011): 59–78.
81. Pehrson, *String of Pearls*.
82. Energy Information Administration, *Country Analysis Brief: Russia* (Washington, D.C.: EIA, 2012), 1–2, http://www.eia.gov/countries/analysisbriefs/Russia/russia.pdf; Vladimir Soldatkin, "Rosneft Leads Russian Oil Output to New High," *Reuters.com*, January 2, 2013, http://www.reuters.com/article/2013/01/02/russia-oil-idUSL5E9C215X20130102.
83. Ariel Cohen, "Russia: The Flawed Energy Superpower," *Energy Security Challenges for the 21st Century: A Reference Handbook*, ed. Gal Luft and Anne Korin (Santa Barbara, CA: Praeger Security International, 2009), 91–108; Alexander Ghaleb, *Natural Gas as an Instrument of Russian State Power* (Carlisle, PA: U.S. Army War College Strategic Studies Institute, 2011), 88–91; and IEA, *Russia*, 17.
84. U.S. Geological Survey (USGS), *Circum-Arctic Resource Appraisal: Estimates of Undiscovered Oil and Gas North of the Arctic Circle* (Reston, VA: USGS, 2008), http://pubs.usgs.gov/fs/2008/3049/fs2008-3049.pdf; Barry S. Zellen, *Arctic Doom, Arctic Boom: The Geopolitics of Climate Change in the Arctic* (Santa Barbara, CA: Praeger Security International, 2009); USGS, *Assessment of Undiscovered Oil and Gas Resources of the North Sakhalin Basin Province, Russia, 2011* (Reston, VA: USGS, 2012), http://permanent.access.gpo.gov/gpo26597/FS11-3149-508.pdf; and IEA, *Russia*, 3–7.
85. IEA, *Russia*, 8, 14.
86. Joint Global Change Research Institute and Battelle Memorial Institute, Pacific Northwest Division, *Russia: The Impact of Climate Change to 2030; A Commissioned Research Report* (Washington, D.C.: National Intelligence Council, 2009), 3–4, http://permanent.access.gpo.gov/lps122080/climate2030_

russia.pdf; Arctic Research Commission, *Arctic Marine Shipping Assessment: Scenarios, Futures and Regional Futures to 2020* (Washington, D.C.: Arctic Council, 2009), 92–121; http://www.arctic.gov/publications/AMSA/scenarios.pdf; Hsiao-Tien Pao, Hsiao-Cheng Yu, and Yeou-Herng Yang, "Modeling the CO_2 Emissions, Energy Use, and Economic Growth in Russia," *Energy* 36 no. 8 (2011): 5094–100; and Energy Information Administration, "Total Carbon Dioxide Emissions from Consumption of Energy (Million Metric Tons)"; *EIA.gov*, last modified 2013. http://www.eia.gov/cfapps/ipdbproject/iedindex3.cfm?tid=90&pid=44&aid=8&cid=RS,&syid=2006&eyid=2010&unit=MMTCD.

87. Joint Global Change Research Institute and Battelle Memorial Institute, Pacific Northwest Division, *Russia*, 16–18.
88. Jonathan Haslam, "A Pipeline Runs Through It," *The National Interest* 92 (November/December 2007): 73–79; Stephen J. Blank, ed., *Russia and the Arctic* (Carlisle, PA: U.S. Army War College Strategic Studies Institute, 2011), http://permanent.access.gpo.gov/gpo10677/PUB1073.pdf; and Shoichi Itoh, *Russia Looks East: Energy Markets and Geopolitics in Northeast Asia* (Washington, D.C.: Center for Strategic and International Studies, 2011): 31; http://csis.org/files/publication/110721_Itoh_RussiaLooksEast_Web.pdf.
89. Christopher Marsh, "Russia Plays the China Card," *The National Interest* 92 (November/December 2007): 68–73; and Itoh, *Russia Looks East*, 38–39.
90. Ministério de Minas e Energia, *Balanço Energético Nacional 2011: Resultados Preliminares* [National Energy Balance 2011: Preliminary Results] (Rio de Janeiro: Empresa de Pesquisa Energética, 2011), https://ben.epe.gov.br/downloads/resultados_pre_ben_2011.pdf.
91. Ministério de Minas e Energia (MME), *Plano Nacional de Energia 2030* [National Energy Plan] (Rio de Janeiro, Brazil: MME, 2007), http://www.mme.gov.br/mme/galerias/arquivos/publicacoes/pne_2030/PlanoNacionalDeEnergia2030.pdf.
92. Ibid.
93. Ministério de Minas e Energia (MME), *Plano Decenal de Expansão de Energia 2020: Sumário* (Rio de Janiero, Brazil: MME, 2010), http://www.epe.gov.br/PDEE/20120302_2.pdf.
94. See the Eletrobras corporate website at http://www.eletrobras.com/elb/data/Pages/LUMIS293E16C4ENIE.htm.
95. See the Petrobras corporate website at http://www.petrobras.com/en/home.htm.
96. Energy Information Administration (EIA), *Country Analysis Brief: Brazil* (Washington, D.C.: EIA, 2013), http://www.eia.gov/countries/analysisbriefs/Brazil/brazil.pdf.
97. MME, *Plano Nacional de Expansão de Energia*.
98. Karl Hallding et al., *Together Alone: Brazil, South Africa, India, China (BASIC) and the Climate Change Conundrum (Report Preview)* (Stockholm, Sweden: Stockholm Environment Institute, 2011), 3.
99. Brazilian Laws no. 12.114 and 12.187 (http://www.planalto.gov.br/ccivil_03/_Ato2007-2010/2009/Lei/L12114.htm and http://www.planalto.gov.br/ccivil_03/_Ato2007-2010/2009/Lei/L12187.htm); Brazilian Decree no. 7.390 (http://www.planalto.gov.br/ccivil_03/_Ato2007-2010/2010/Decreto/D7390.htm); Ministério de Minas Energia (MME), *Plano Nacional de Eficiencia Energetica: Premissas e Directrizes Basicas* [National Energy Plan for Energy Efficiency] (Rio de Janiero, Brazil: MME, 2011); Eduardo Fernandez Silva, "Rio+20 and Brazil's Policy on Climate Change," *Nature Climate Change* 2 (2012): 379–80, http://dx.doi.org/10.1038/nclimate1525.
100. Energy Information Administration (EIA), *Country Analysis Brief: India* (Washington, D.C.: EIA, 2013), http://www.eia.gov/countries/analysisbriefs/india/india.pdf.
101. Ibid.
102. Planning Commission, Government of India, *Twelfth Five Year Plan (2012–2017): Volume II, Economic Sectors* (New Delhi, India: Sage, 2013).
103. Planning Commission, *Twelfth Five Year Plan*.
104. EIA, *India*.
105. Planning Commission, *Twelfth Five Year Plan*.
106. Ibid.

107. EIA, *India*.
108. International Energy Agency, *Energy Technology Perspectives 2010* (Paris: IEA, 2010), http://www.iea.org/publications/freepublications/publication/etp2010.pdf.
109. EIA, *India*.
110. Planning Commission, *Twelfth Five Year Plan*.
111. EIA, *India*.
112. Ibid.; 100.
113. 100.
114. 100.
115. Planning Commission, *Twelfth Five Year Plan*.
116. Nathalie Trudeau, Cecilia Tam, Dagmar Graczyk, and Peter Taylor, *Energy Transition for Industry: India and the Global Context* (Paris: International Energy Agency, 2011), 15.

Bibliography

Arctic Research Commission. *Arctic Marine Shipping Assessment: Scenarios, Futures and Regional Futures to 2020.* Washington, D.C.: Arctic Council, 2009. http://www.arctic.gov/publications/AMSA/scenarios.pdf.

Blank, Stephen J., ed. *Russia and the Arctic.* Carlisle, PA: U.S. Army War College Strategic Studies Institute, 2011. http://permanent.access.gpo.gov/gpo10677/PUB1073.pdf.

British Petroleum. *BP Statistical Review of World Energy: June 2012.* London: British Petroleum, 2012. http://www.bp.com/statisticalreview.

Clarke II, Woodrow W., and Grant Cooke. *Global Energy Innovation: Why America Must Lead.* Santa Barbara: Praeger, 2012.

Clinton, Hillary Rodham. *Remarks on American Leadership at the Council on Foreign Relations.* 31 January 2013. http://www.state.gov/secretary/rm/2013/01/203608.htm.

Cohen, Ariel. "Russia: The Flawed Energy Superpower." *Energy Security Challenges for the 21st Century: A Reference Handbook*, edited by Gal Luft and Anne Korin, 91–108. Santa Barbara, CA: Praeger Security International, 2009.

Cole, Bernard D. *Oil for the Lamps of China: Beijing's 21st Century Search for Energy.* Washington, D.C.: National Defense University Press, 2003. http://permanent.access.gpo.gov/websites/nduedu/www.ndu.edu/inss/mcnair/mcnair67/198_428.McNair.pdf.

Commission to the Council. the European Parliament, the European Economic and Social Committee, and the Committee of the Regions. *A European Strategic Energy Technology Plan (SET-Plan): Towards a Low Carbon Future* {COM(2007) 723}. Brussels, Belgium: European Commission, 2007.

Coon, Charli E. "Why President Bush is Right to Abandon the Kyoto Protocol." *Heritage Foundation*, March 18, 2001. http://www.heritage.org/research/reports/2001/05/president-bush-right-to-abandon-kyoto-protocol.

Council of Economic Advisors. *Economic Report of the President 1981.* Washington, D.C.: Government Printing Office, 1981. http://fraser.stlouisfed.org/docs/publications/ERP/1981/ERP_1981.pdf.

Council on Foreign Relations. *The Global Climate Change Regime: Issue Brief.* Washington, D.C.: Council on Foreign Relations, 2013. http://www.cfr.org/climate-change/global-climate-change-regime/p21831.

Defense Science Board. *Trends and Implications of Climate Change for National and International Security.* Washington, D.C.: Defense Science Board, 2011.

Ebell, Myron. *Increase Access to Energy.* Washington, D.C.: Competitive Enterprise Institute, 2009. http://cei.org/sites/default/files/Myron%20Ebell%20-%20Increase%20Access%20to%20Energy.pdf.

Energy Information Administration. *Annual Energy Outlook 2013.* Washington, D.C.: US Energy Information Administration, 2013. http://www.eia.gov/forecasts/aeo/.

Energy Information Administration. *Annual Energy Review 2010 (DOE/EIA-0384(2010)*. Washington, D.C.: US Energy Information Administration, 2011. http://www.eia.gov/totalenergy/data/annual/archive/038410.pdf.

Energy Information Administration. *Country Analysis Brief: Brazil.* Washington, D.C.: US Energy Information Administration, 2013. http://www.eia.gov/countries/analysisbriefs/Brazil/brazil.pdf.

Energy Information Administration. *Country Analysis Brief: China.* Washington, D.C.: US Energy Information Administration, 2012. http://www.eia.gov/countries/analysisbriefs/China/china.pdf.

Energy Information Administration. *Country Analysis Brief: India.* Washington, D.C.: US Energy Information Administration, 2013. http://www.eia.gov/countries/analysisbriefs/india/india.pdf.

Energy Information Administration. *Country Analysis Brief: Russia.* Washington, D.C.: US Energy Information Administration, 2012. http://www.eia.gov/countries/analysisbriefs/Russia/russia.pdf.

Energy Information Administration. *Country Analysis Brief: South China Sea.* Washington, D.C.: US Energy Information Administration, 2008.

Energy Information Administration. "Impact of Limitations on Access to Oil and Natural Gas Resources in the Federal Outer Continental Shelf." *EIA Issues in Focus*, last modified 2009. http://www.eia.gov/oiaf/aeo/otheranalysis/aeo_2009analysispapers/aongr.html.

Energy Information Administration. *Impacts of the Kyoto Protocol on U.S. Energy Markets and Economic Activity.* Washington, D.C.: US Energy Information Administration, 1998. http://www.eia.gov/oiaf/kyoto/pdf/sroiaf9803.pdf.

Energy Information Administration. "What is the Status of the U.S. Nuclear Industry?" *EIA Energy in Brief.* Last updated December 14, 2012. http://www.eia.gov/energy_in_brief/article/nuclear_industry.cfm.

Environmental Protection Agency (EPA). "Nuclear Incidents: Three Mile Island Nuclear Plant." *EPA Radiation Protection.* Last modified February 15, 2012. http://www.epa.gov/rpdweb00/rert/tmi.html.

Environmental Protection Agency. "Standards of Performance for Greenhouse Gas Emissions for New Stationary Sources: Electric Utility Generating Units." 77 *Fed. Reg.* 72 (2012): 22393–441.

European Commission. *A European Strategy for Sustainable, Competitive and Secure Energy* {SEC(2006) 317}. Brussels, Belgium: European Commission, 2006. http://ec.europa.eu/energy/green-paper-energy/doc/2006_03_08_gp_document_en.pdf.

European Commission. *EURATOM Supply Agency: Annual Report 2011.* Luxembourg: Publications Office of the European Union, 2012. http://dx.doi.org/10.2833/11309.

European Commission. "Key Figures." Market Observatory for Energy, Directorate-General for Energy. Last modified June 2011. http://ec.europa.eu/energy/observatory/countries/doc/key_figures.pdf.

Fehner, Terrence R. *Department of Energy, 1977–1994: A Summary History.* Washington, D.C.: United States Department of Energy, 1994.

Fridley, David, Nina Zheng, and Yining Qin. *Inventory of China's Energy-Related CO_2 Emissions in 2008.* Berkeley, CA: Lawrence Berkeley National Laboratory, 2011. http://www.osti.gov/bridge/servlets/purl/1016716-Hv9lcU/1016716.pdf.

Ghaleb, Alexander. *Natural Gas as an Instrument of Russian State Power.* Carlisle, PA: U.S. Army War College Strategic Studies Institute, 2011.

Hallding, Karl, Marie Olsson, Aaron Atteridge, Marcus Carson, Antto Vihma, and Mikael Roman. *Together Alone: Brazil, South Africa, India, China (BASIC) and the Climate Change Conundrum (Report Preview).* Stockholm, Sweden: Stockholm Environment Institute, 2011.

Haslam, Jonathan. "A Pipeline Runs Through It." *The National Interest* 92 (November/December 2007): 73–79.

Howell, Sabrina. "Jia You! (Add Oil!): Chinese Energy Security Strategy." In *Energy Security Challenges for the 21st Century: A Reference Handbook*, edited by Gal Luft and Anne Korin, 191–219. Santa Barbara, CA: Praeger Security International, 2009.

International Atomic Energy Agency (IAEA). "PRIS: The Database on Nuclear Power Reactors." *IAEA.org.* Last modified November 10, 2013. http://www.iaea.org/PRIS/.

International Energy Agency. *2013 Key World Energy Statistics*. Paris: IEA, 2013. http://www.iea.org/publications/freepublications/publication/KeyWorld2013_FINAL_WEB.pdf.

International Energy Agency. *CO_2 Emissions from Fuel Combustion: Highlights*. 2012 ed. Paris: IEA, 2012. http://www.iea.org/co2highlights/CO2highlights.pdf.

International Energy Agency. *Energy Technology Perspectives 2010*. Paris: IEA, 2010. http://www.iea.org/publications/freepublications/publication/etp2010.pdf.

International Energy Agency. "FAQs: Organisation and Structure." *IEA.org*. Last modified 2013. http://www.iea.org/aboutus/faqs/organisationandstructure/.

International Energy Agency. *World Energy Outlook 2010*. Paris: IEA, 2010. http://www.iea.org/publications/freepublications/publication/weo2010.pdf.

International Energy Agency. *World Energy Outlook 2011*. Paris: IEA, 2010. http://www.iea.org/publications/freepublications/publication/WEO2011_WEB.pdf.

Itoh, Shoichi. *Russia Looks East: Energy Markets and Geopolitics in Northeast Asia*. Washington, D.C.: Center for Strategic and International Studies, 2011. http://csis.org/files/publication/110721_Itoh_RussiaLooksEast_Web.pdf.

Johnston, Jason Scott. "Climate Change Hysteria and the Supreme Court: The Economic Impact of Global Warming on the U.S. and the Misguided Regulation of Greenhouse Gas Emissions Under the Clean Air Act." *Scholarship at Penn Law*, Paper 209 (2008). http://lsr.nellco.org/upenn_wps/209.

Joint Global Change Research Institute and Battelle Memorial Institute, Pacific Northwest Division. *China: The Impact of Climate Change to 2030; A Commissioned Research Report*. Washington, D.C.: National Intelligence Council, 2009. http://permanent.access.gpo.gov/lps122078/LPS122078.pdf.

Joint Global Change Research Institute and Battelle Memorial Institute, Pacific Northwest Division. *Russia: The Impact of Climate Change to 2030; A Commissioned Research Report*. Washington, D.C.: National Intelligence Council, 2009. http://permanent.access.gpo.gov/lps122080/climate2030_russia.pdf.

Kiska, Wendy, and Deborah Lucas. *Federal Loan Guarantees for the Construction of Nuclear Power Plants*. Washington, D.C.: Congressional Budget Office, 2011. http://permanent.access.gpo.gov/gpo10958/08-03-NuclearLoans.pdf.

Kong, Bo. *China's International Petroleum Policy*. Santa Barbara, CA: Praeger Security International, 2010.

Kostecka, Daniel J. "Places and Bases: The Chinese's Navy's Emerging Support Network in the Indian Ocean." *Naval War College Review* 64, no. 1 (2011): 59–78.

Lewis, Marlo Jr. "Carbon Pollution Standard: 4 Ways Weird." National Journal, April 2, 2012. http://www.nationaljournal.com/2012/04/what-will-be-upshots-of-epas-c.php.

Lomborg, Bjorn. *Cool It: The Skeptical Environmentalist's Guide to Global Warming*. New York: Random House, 2007.

Loris, Nicholas. "Ten Actions Congress Can Take to Lower Gas Prices." *Heritage Foundation Backgrounder*, May 31, 2012. http://www.heritage.org/research/reports/2012/05/ten-actions-congress-can-take-to-lower-gas-prices.

Luft, Gal. "United States: A Shackled Superpower." In *Energy Security Challenges for the 21st Century: A Reference Handbook*, edited by Gal Luft and Anne Korin, 143–59. Santa Barbara: Praeger Security International, 2009.

Magwood IV, William D. *Statement of Commissioner William D. Magwood, IV*. Nuclear Regulatory Commission. February 9, 2012. http://www.nrc.gov/about-nrc/organization/commission/comm-william-magwood/comm-magwood-statement-02-09-12.pdf.

Marsh, Christopher. "Russia Plays the China Card." *The National Interest* 92 (November/December 2007): 68–73.

McDermott, Rose. *Risk-Taking in International Politics: Prospect Theory in American Foreign Policy*. Ann Arbor: University of Michigan Press, 2001.

Michaels, Patrick J., ed. *Climate Coup: Global Warming's Invasion of Our Government and Our Lives*. Washington, D.C.: CATO Institute, 2011.

Ministério de Minas e Energia (MME). *Balanço Energético Nacional 2011: Resultados Preliminares* [National Energy Balance 2011: Preliminary Results]. Rio de Janeiro, Brazil: Empresa de Pesquisa Energética, 2011. https://ben.epe.gov.br/downloads/resultados_pre_ben_2011.pdf.

Ministério de Minas e Energia (MME). *Plano Decenal de Expansão de Energia 2020: Sumário.* Rio de Janeiro, Brazil: MME, 2010. http://www.epe.gov.br/PDEE/20120302_2.pdf.

Ministério de Minas e Energia (MME). Plano Nacional de Eficiencia Energetica: Premissas e Directrizes Basicas. [National Energy Plan for Energy Efficiency: Basic Premises and Directives] Rio de Janiero, Brazil: MME, 2011.

Ministério de Minas e Energia (MME). *Plano Nacional de Energia 2030* [National Energy Plan]. Rio de Janiero, Brazil: MME, 2007. http://www.mme.gov.br/mme/galerias/arquivos/publicacoes/pne_2030/PlanoNacionalDeEnergia2030.pdf.

Mufson, Steven, and Juliet Elperin. "Loan-Guarantee Recipient Founders." *Washington Post*, November 1, 2011.

National Highway Traffic Safety Administration. "CAFE: Fuel Economy." *US Department of Transportation*. Last modified 2012. http://www.nhtsa.gov/fuel-economy.

Newport, Frank. "Americans Still Favor Nuclear Power a Year After Fukushima." *Gallup Politics*, March 26, 2012. http://www.gallup.com/poll/153452/Americans-Favor-Nuclear-Power-Hear-Fukushima.aspx.

Nuclear Regulatory Commission. "Combined License Applications for New Reactors." *NRC.gov*. Last modified September 23, 2013. http://www.nrc.gov/reactors/new-reactors/col.html.

Ott, Mack, and John A. Taton. "A Perspective on the Economics of Natural Gas Decontrol." *Federal Reserve Bank of St. Louis Review*, no. 64 (November 1982): 19–31.

Owen, Roger. "Suez Crisis." In *The Oxford Companion to the Politics of the World*. 2nd ed., edited by Joel Krieger. London: Oxford University Press, 2001.

Pao, Hsiao-Tien, Hsiao-Cheng Yu, and Yeou-Herng Yang. "Modeling the CO_2 Emissions, Energy Use, and Economic Growth in Russia." *Energy* 36, no. 8 (2011): 5094–100.

Pehrson, Christopher J. *String of Pearls: Meeting the Challenge of China's Rising Power Across the Asian Littoral.* Carlisle, PA: U.S. Army War College Strategic Studies Institute, 2006. http://permanent.access.gpo.gov/lps77746/PUB721.pdf.

Planning Commission, Government of India, Twelfth Five Year Plan (2012–2017): Volume II, Economic Sectors (New Delhi, India: Sage, 2013).

Qaimmaqami, Linda, ed. *Foreign Relations of the United States 1969–1976, Volume XXXVI: Energy Crisis, 1969–1974.* Washington, D.C.: Government Printing Office, 2011.

Rascoe, Ayesha. "NRC Approves First New Nuclear Power Plant in a Generation." *Reuters.com*, February 9, 2012. http://www.reuters.com/article/2012/02/09/us-usa-nuclear-license-idUSTRE8181T420120209.

Rothbard, David, and Craig Rucker. "The U.N.'s Rio+20 Agenda." *National Review Online* (June 20, 2012). http://www.nationalreview.com/articles/303268/un-s-rio20-agenda-david-rothbard.

SCANA. "SCE&G Completes First Nuclear Concrete Placement." Sceg.com, March 11, 2013. http://www.sceg.com/NR/rdonlyres/A3CCBEA1-8009-44A5-95F9-3CDB98C6C061/0/03112013SCEGFirstNuclearConcrete.pdf.

Sieminksi, Adam. "U.S. Energy Outlook." Presentation given at the IEA Bilateral Meetings, Paris, France, March 14, 2013. http://www.eia.gov/pressroom/presentations/sieminski_03142013_iea.pdf.

Silva, Eduardo Fernandez. "Rio+20 and Brazil's Policy on Climate Change." *Nature Climate Change* 2 (2012): 379–80. http://dx.doi.org/10.1038/nclimate1525.

Smil, Vaclav. *Energy Transitions: History, Requirements, Prospects.* Santa Barbara, CA: ABC-CLIO, 2010.

Soldatkin, Vladimir. "Rosneft Leads Russian Oil Output to New High." *Reuters.com*, January 2, 2013. http://www.reuters.com/article/2013/01/02/russia-oil-idUSL5E9C215X20130102.

Southern Company. "Southern Nuclear." *Southerncompany.com*. Last modified 2013. http://www.southerncompany.com/about-us/our-business/southern-nuclear/home.cshtml.

Steinberg, Daniel, Gian Porro, and Marshall Goldberg. *Preliminary Analysis of the Jobs and Economic Impacts of Renewable Enegy Projects Supported by the §1603 Treasury Grant Program.* Golden, CO: National Renewable Energy Laboratory, 2012. www.nrel.gov/docs/fy12osti/52739.pdf.

Strategic Energy Technologies Information System. "Overview." *European Commission*. Last modified August 11, 2013. http://setis.ec.europa.eu/about-setis/overview.

Strategic Petroleum Reserve. "Strategic Petroleum Reserve Inventory." *US Department of Energy*. Last modified November 1, 2013. http://www.spr.doe.gov/dir/dir.html.

Subcommittee on Nuclear Regulation. *Nuclear Accident and Recovery at Three Mile Island: A Report*. Washington, D.C.: Government Printing Office, 1980.

Task Force on Strategic Unconventional Fuels. *Development of America's Strategic Unconventional Fuels Resources*. Washington, D.C.: Task Force on Strategic Unconventional Fuels, 2006.

Thomas, Scott, and David Kerner. *Defense Energy Resilience: Lessons from Ecology*. Carlisle, PA: U.S. Army War College, 2010.

Trudeau, Nathalie, Cecilia Tam, Dagmar Graczyk, and Peter Taylor. *Energy Transition for Industry: India and the Global Context*. Paris: International Energy Agency, 2011.

Turner, Barry. *Suez 1956: The Inside Story of the First Oil War*. London: Hodder & Stoughton, 2006.

United Nations. (UN). "Earth Summit: UN Conference on Environment and Development." UN Department of Public Information, Last modified May 23, 1997. http://www.un.org/geninfo/bp/enviro.html.

U.S. Geological Survey (USGS). *Assessment of Undiscovered Oil and Gas Resources of the North Sakhalin Basin Province, Russia, 2011*. Reston, VA: USGS, 2012. http://permanent.access.gpo.gov/gpo26597/FS11-3149-508.pdf.

U.S. Geological Survey (USGS). *Circum-Arctic Resource Appraisal: Estimates of Undiscovered Oil and Gas North of the Arctic Circle*. Reston, VA: USGS, 2008. http://pubs.usgs.gov/fs/2008/3049/fs2008-3049.pdf.

U.S. Joint Forces Command. *The Joint Operating Environment (JOE) 2008: Challenges and Implications for the Future Joint Force*. Norfolk, VA: Joint Forces Command, 2008.

Wallsten, Scott J. "The Effects of Government-Industry R&D Programs on Private R&D: The Case of the Small Business Innovation Research Program." *Rand Journal of Economics* 31, no. 1 (2000): 82–100.

White House. The. *Blueprint for a Secure Energy Future*. Washington, D.C.: The White House, 2011.

White House. The. *The Blueprint for a Secure Energy Future: Progress Report*. Washington, D.C.: The White House, 2012.

Yergin, Daniel. *The Prize: The Epic Quest for Oil, Money, and Power*. New York: Simon and Schuster, 1990.

Yergin, Daniel. *The Quest: Energy, Security, and the Remaking of the Modern World*. New York: Penguin, 2011.

Zellen, Barry S. *Arctic Doom, Arctic Boom: The Geopolitics of Climate Change in the Arctic*. Santa Barbara, CA: Praeger Security International, 2009.

Chapter 3
Social Engagement by the Engineer

MELISSA DARK, IDA NGAMBEKI,
DENNIS DEPEW, AND RYLAN CHONG

Abstract

The American Engineers' Council for Professional Development, the precursor institution of the Accreditation Board for Engineering and Technology (ABET), defined engineering as the "creative application of scientific principles to design or develop structures, machines, apparatus, or manufacturing processes, or works utilizing them singly or in combination; or to construct or operate the same with full cognizance of their design; or to forecast their behavior under specific operating conditions; all as respects an intended function, economics of operation or safety to life and property."[1] In light of this definition, engineers must have a working familiarity beyond the scope of their technical work. They must be able to identify and understand the social environments and its interactions in order to develop solutions to global engineering challenges. Chapter 3 endeavors to provide a necessary social and global framework for the more detailed examination of specific energy topics undertaken in later chapters. The chapter describes systems within the social environment; introduces theories, concepts, and ideas to help students understand the social context and engineering's place within it; and addresses the necessity for social engagement among engineers. The chapter also provides two energy themed case studies as examples of how the social environment affects engineering practice. Case Study 1 is designed to complement chapters 4 and 7 and Case Study 2 is designed to demonstrate how political, social, and economic forces may emerge in the energy sector.

3.1. Introduction

Questions of energy use and distribution and the global energy crisis must be considered in the context of energy producers, users, and the social environment that shapes energy use. To do so it is necessary to understand the social environment. A social environment is a context; it is a set of circumstances in which an event occurs; a time or place in which people live or in which something happens or develops. In the largest sense, the social environment is the milieu developed by humans (as opposed to the natural environment): it is society as a whole. Social environments are dynamic and multidimensional with many different aims, qualities, and aspects.

Social environments are comprised of economic, political, and social systems. Within each of these systems there are other systems; for example, the political system contains executive, legislative, and judicial systems. The economic system contains systems of production, distribution, property rights, and labor, among others. We can think of these systems as "nested" (Figure 3.1); each of these systems is an integrated whole and at the same time is part of larger systems. Changes within a system can affect the systems that are nested within it as well as the larger system or systems within

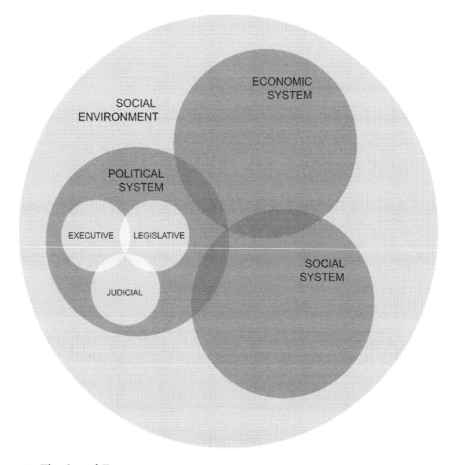

Figure 3.1. The Social Environment.

which it exists. Energy production, distribution and consumption exists within this social environment and therefore is fully integrated within, affected by, and in fact comprises these systems. For example, energy production, distribution, and consumption is an economic system with considerations of property rights and labor.

As individuals, we all experience social environments. We are shaped by the culture that we are educated or live in, or the people and institutions with whom we interact. Persons within the same social environment often develop a sense of social cohesion based on shared experiences. Solidarity produces trust, reciprocity, and a sense of belonging, which can be desirable characteristics in a society. Taken too far though, cohesion can become collectivism and unquestioning allegiance—undesirable characteristics in a society. The opposite extreme, rational self-interest, can be equally undesirable.

The global scope and temporal magnitude of the energy challenge requires that those whose work and calling is energy-related understand social environments in ways that helps them see beyond their own lived experiences, customs, and self-interest. The social environment of today's global energy crisis is a complex and dynamic environment where the political, economic, and social systems are both intra- and interacting; the same issue is different in different places not only because of geographic differences, but as a consequence of political, economic and social factors constituting the context. Consider for example nuclear power generation. As described in Case Study 1, nuclear power generation in Germany, which accounts for just over fifteen percent of that nation's power generation, is being phased out as a result of public fears over safety. On the other hand, in neighboring France, where nuclear power accounts for over seventy-five percent of power generation, nuclear power generation is increasing. The drivers for these different directions in energy policy differ in the two contexts and include social factors such as public opposition in Germany, economic considerations such as the revenue from energy export in France, and political will. The safety of nuclear energy is discussed in greater detail in Chapter 7.

This chapter begins by briefly describing foundational concepts of social systems, then focuses more specifically on political systems as an important type of social system, including interactions and dynamics within and among these systems and variations in political systems across the globe, discussing the implications of these interactions and variations on the issue of energy. Then, the chapter addresses common authentic values, examining how values and belief systems result in behavior, how common values impact the common good, and the relationship between the common good and individual rights. Finally, we review the effect of globalization on social systems and discuss the implications for energy policy.

3.2. Social Systems

Social systems are all around us; neighborhoods, workplaces, class systems, traffic systems, families, marriage traditions, cities, and political systems are all examples of social systems. Social systems are comprised of individuals and groups. The most essential characteristic of any social system is interaction. Social systems are enduring, patterned interrelationships between individuals, groups, and institutions. These enduring behavior patterns become embedded and shape the choices, opportunities, and actions of agents within the social system. Social systems supervene on individuals, influencing us through a variety of mechanisms. Social systems organize the behavior of large numbers of actors, coerce individual and group behavior, assign roles and power to individual agents, and have distributive consequences for individuals and groups. They are embodied in the actions, thoughts, beliefs, attitudes, and durable dispositions of individual human beings. Social systems are *autopoietic*; they exist only by reproducing the events that serve as components of the system. The characteristic patterns of interaction within a social system define it and differentiate it from other social systems. The mechanisms for control and influence are context specific; the mechanisms that one particular social entity uses for influence may fail in another context.

3.2.1. Political Systems

One of the most influential social systems is a political system. A political system is a manmade structure that regulates the processes and activities of human co-existence; specifically it is the social structure and methods used to manage a community, government, or state. The outputs from the political system can be broadly categorized as policies: the rules, understandings, and institutions that organize and direct human action. A political system can refer to either a particular form/system of government, for example, democracy, totalitarianism, or authoritarianism, or to a singular state or one of its subordinate authorities, such as a district or province, county, or city. A state is generally understood to be a sovereign entity with a government, defined territory, permanent population, and the capacity to enter into relations with other sovereign states. A political system is comprised of various political entities and their functions. This includes the political and legal structures manifested in governmental, civil society, and private sector domains. A democracy has different political entities than a totalitarian or authoritarian political system. Not only do the entities and functions of the different political systems vary, so do the interactions among the entities, and this has implications for how decisions are made regarding energy generation and distribution. For example, the Three Gorges Dam built on the Yangtze River in China was completed in 2012 despite widespread opposition from the local population because of the environmental impact and the necessity of relocating over one million residents of the flood plain. The political system in China allowed these decisions to be made, while in a different political context the project may have been stalled.

A political system is nested in a larger social environment that includes other social systems, such as the economic systems, and the values system. A political system interacts with these other systems. The legal structures and laws of a political system are made by the individuals and groups in the society or state, and they are influenced by the values and relationships of those actors. In turn, the political system influences the economic and value systems and, therefore, the actors—both group and individual. The process by which this occurs is manifest and perceivable in specific contexts. Later chapters explore specific contexts looking at different energy technologies and policies in different geographical locations and in so doing, provide snapshots that capture, compare and contrast particular social environments and their effects on energy production, distribution and use. There is value in understanding a particular energy technology in a particular locale at a particular point in time. There is even more value in understanding various technologies in various locales across various times, as this helps us to better understand the global energy crisis. Any movement toward shared solutions requires more robust understanding of the complex and emergent nature of the systems responsible for and impacted by the global energy crisis. We present an overview of structuration theory as a tool for thinking beyond surface features to the more deeply embedded understanding of the roles of policy in problems of human action, such as energy.

3.2.2. Structuration

Structuration is useful for thinking about "problems of human action," such as the global energy crisis, and policy solutions to these problems. A basic tenet of structuration is that public policy is a multi-layered system, which means that we cannot look simply at policy actions (laws, regulations), we need to look more deeply into the systems that produce the policy actions. According to structuration theory, systems are made up of both agents and structures.[2] Agents are individuals and collectives who act with purpose, intention, and motive. Structures are social properties that make it possible for social practices to exist. Structures can be thought of as the "rules of social life" that are applied in the enactment and reproduction of social life. All social states need agents and social practices to help them survive in a recognizably similar form. Structuration looks at how humans exert agency within structural contexts. More specifically, structuration looks at how social practices are ordered across time and space; how agents continuously reproduce these practices in the process of expressing themselves as actors; how actors reproduce the conditions that make these activities possible; and patterns of social outcomes resulting from these enacted social practices. Figure 3.2 presents these ideas more concretely.

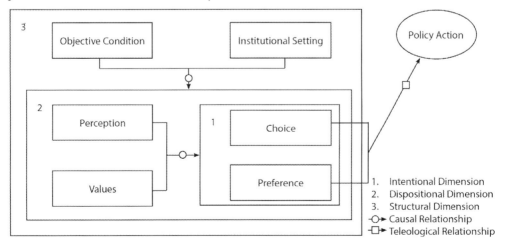

Figure 3.2. Theory of Structuration.[3]

Policy action can be, and often is, explained by choice and preference. The relationship between choice and preference, and policy action is labeled as teleological, which means that there is perception of purposeful development toward an end; that is, the choice or preference set logically leads to the given policy action. Explanations solely from the intentional dimension are often used in instrumental and rational theories of policy making. However, the analyst can go deeper by chaining back to look at perceptions and values making up the dispositional dimension, suggesting a causal determination for given choice or preference sets. The distinction between levels one and two can be described as an *in order to* rather than a *because of* relationshiip, with the former referring to the intentional sphere and the latter constituting the link between this intention and the basis for it (the disposition): how a particular actor's intention has come to be.[4] The two constructs, perception and values, which Carlsnaes suggests as belonging to the dispositional dimension, fit Giddens' model of structuration as the motives or purposes for action. Lastly, Figure 3.2 depicts a third level, the structural dimension, which represents the interaction between the dispositional and intentional dimensions. Structural factors include the contextual structures and institutions and their functions, and the manner and extent to which they enable and constrain conditions under which contingent actors (the only causal entities) necessarily have to operate. Differences at all three levels result in differences in policy action.

Take the case of the renewable energies discussed in Chapter 4. The European Union (EU) is an example of a nested system. It comprises twenty-seven member states, each with their own political, social, and economic systems. However, the EU as a whole also shares common economic and governing political systems. The European Union set a goal that twenty percent of energy consumption should come from renewable sources by 2020. However, member states have the discretion to decide how this goal will be attained. There are actors in each of these states—such as companies, politicians, and individuals—all acting to influence the choices made in the national political, social, and economic systems about what technologies are used to generate power. Therefore, these systems are influenced by the agents within them. However, the agents are also influenced by the systems; they choose their actions and responses based on recognitions of how the systems work and how they can be influenced. Case Study 1 gives an example of how political and social issues can have an impact on energy generation.

While understanding differences is important, it is not the entire story. As important as it is that students understand differences in social environments, we also believe it essential that students understand commonalities that all humans share. The next section discusses how common beliefs and values are formed and how goals are negotiated.

3.3. Common Authentic Values and Principles

Societies regularly develop a set of commonly accepted values arising from a common belief system. These values arise from common needs and desires and include those things seen as being good for the society as a whole, or for the common good. In a trea-

Case Study 1

Energy is a global necessity. However, societies struggle to find answers to address the wicked problems of energy sustainability, environmental/climate concerns, and health. A potential alternative would be the increased usage of nuclear power. It provides significant benefits, such as clean emissions, cost effectiveness (that is, there is not a high demand to develop newer technologies or purchase additional technologies to reduce the carbon footprint), and it is reliable and provides a predictable base-load. Moreover, nuclear power plants do not have to refuel often. Although nuclear power plants present benefits, Chancellor Merkel of Germany announced a plan to phase out their nuclear power plants by the year 2022.

To understand the reasoning for phasing out nuclear power plants, it is essential to understand the history that led up to the decision. Germany's first nuclear plant went online in 1957, which represented the political and economic movements of the time.[5] In 1960, Germany introduced the Atomic Energy Act with the primary aim of encouraging nuclear energy. However, not every actor was on board with the nuclear program. An anti-nuclear interest group made up of environmentalists and peace activists believed there should be policy changes regarding the military usage of nuclear power and nuclear waste disposal. The oil crisis in the 1970s promoted the idea that energy diversity and therefore nuclear energy was good for the state, but the 1986 Chernobyl incident revitalized the anti-nuclear movement and discouraged public support for nuclear power plants.[6] However, it was not until 2002 that Chancellor Schroeder, with the support of the Social Democratic Party and the Green Party, passed the first nuclear phase out deal to end all nuclear energy production by 2021.[7] The phase-out deal modified the Atomic Energy Act of 1959 by including a section that no new licensing would be distributed, which prohibited the building and operation of new nuclear power plants and other nuclear facilities.[8] Schroeder provided three reasons for the phase-out: a) there was growing concern about how to handle nuclear waste, b) nuclear power was a social problem, and c) nuclear power had no economic purpose.[9]

In 2010 Chancellor Merkel, the Christian Democratic Union, Christian Social Union, and the Free Democratic Party expressed their disagreement with Chancellor Schroeder and his supporting parties' phase out deal, because of Germany's commitment toward the European Union's 2020 energy strategy requiring twenty percent use of renewable energy, energy supply, and steady energy prices.[10] Merkel developed a new strategy that gave all nuclear plants that were constructed before 1980 an eight year license extension and those plants constructed after 1980 a fourteen year license extension.[11] In 2011, twenty-three percent of Germany's electricity was generated through nuclear power.[12] On March 11, 2011, an earthquake and tsunami struck Japan, resulting in a meltdown at the Fukushima Nuclear Plant. Days after the event, Merkel formed the Ethics Commission on Safe Energy Supply that evaluated Germany's seventeen nuclear power plants. Months later, in light of Japan's nuclear disaster and public opposition, the Ethics Commission on Safe Energy Supply took a deontological position and recommended Merkel to shut down eight of the seventeen nuclear plants and return to Schroeder's phase out deal, in order to protect the country from future threats of nuclear disasters.[13]

Guiding Questions

1. What are the social, economic, and political implications of phasing out Germany's nuclear program? Who will be impacted by this decision and to what extent?
2. Using the theory of planned behavior model, analyze Germany's Green Party's conflicting views on nuclear energy and clean energy?
3. What are the social, political, and economic differences that made Germany consider the tragic events in Japan as a reason to phase out their nuclear program while France did not?
4. Germany is a leader in the renewable energy sector and plays a huge role in the European Union 2020 energy strategy. What are some alternatives that could replace and improve the energy lost by the phase out deal, without increasing the carbon footprint and cost?
5. Evaluate the environmental policies and regulations (that is, the cost and benefits, liability, human health protection, and environmental security) regarding nuclear energy in Germany. Compare the results to Japan's and the United States' environmental policies and regulations regarding nuclear energy.

tise on the common good, V. Bradley Lewis brings up arguments first voiced by Aristotle and St. Thomas Aquinas. The common good has been described as a specific good that is shared and beneficial for all (or most) members of a given community; a good that is both end and means; and any action or behavior that promotes the good of the state and, by doing so, promotes the good of all its citizens.[14] The concept of Common goods is grounded in morality and ethics. Central to morality, ethics, and the concept of the common good is the notion that human beings are both *rational* and *political* beings. Unlike the rest of the animal kingdom, humans can trace their actions as resulting from a series of thoughts and reasons; it is this capacity to reason that allows us to behave individually and collectively in a manner supporting the common good. However, man acts not only for himself, but for the larger group; he recognizes that the success of the group benefits the individual. Man utilizes not only individual work for the benefit of the many, but also aspires to function in a manner that ensures that a society's work is for the benefit of the many. While it is important for people to understand the aspects that differentiate social environments, it is equally, if not more, important that we appreciate these unifying common authentic values and principles.

3.3.1. Values, Beliefs, and Behavior

Social systems are the result of groups of people acting with and on each other; individuals are nested within and make up the social structure. In order to understand social structures we must first consider the human being as an individual. Individual behavior can be seen as the result of an individual's values, belief system, and attitudes. Values refer to individually held, socially constructed ideals of desire and preference based on what the individual holds to be true or false. Belief systems are the total collection of what one holds to be true about the world, both physical and social, and about oneself.[15] One model of the interaction of these attributes is the Theory of Planned Behavior (Figure 3.3) proposed by Ajzen.[16] In this model the individual's behavior is influenced by their attitude, which is their positive or negative feeling about the behavior. Attitude is a result of the individual's belief about the consequences of the behavior and how desirable those consequences are judged to be. Behavior is also influenced by the individual's belief about other people's evaluation of the behavior and whether it should be performed, called the subjective norm. Finally, behavior is influenced by the individual's perception of the difficulty inherent in performing the behavior, measured in terms of effort, resources, and ability and weighed against the perceived value of the behavior to the individual. This is called perceived behavioral control. In this way an individual's desires, attitudes, and beliefs about the world around them and values together influence their behavior.

As demonstrated by this theory, since the individual is nested within the social structure, individual action is heavily influenced by social norms. Evaluations of both the desirability and the consequences of an action are weighed based on perceptions of social judgment. Therefore, while the society is made up of individuals, individuals are the product of society and are heavily influenced by relationships, history, and tradition. So a society's values are the product of the values of the individuals within it, and an individual's values are the product of the society in which they are embedded.

3.3.2. The Common Good and Individual Rights

This complex relationship between the individual and society finds expression in the notion of the common good. Though values differ across individuals, communities, societies and cultures, there are certain values shared across these groups. These common values, such as food, shelter, development, and happiness, can be thought of as a conception of the common good. However, individuals and indeed societies differ in the extent and the manner in which they value these goods. It is therefore difficult to define the concept of the common good and to agree on what those goods may be. The

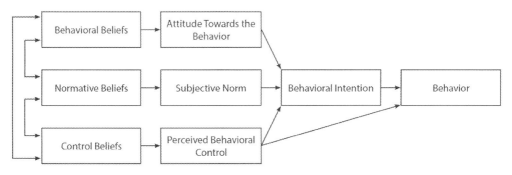

Figure 3.3. Theory of Planned Behavior.[17]

utilitarian definition of the common good is the greatest good for the greatest number. This is the simplest definition of the common good and therefore the most easily translated into action. However, this definition has two major failings. First, it falters in its conception of *common* because the greatest number is often only a section of the population (usually demographically or historically powerful groups). The utilitarian approach values the benefit or happiness of this sector at the expense of the remainder of the population. In practice this often results in some members of a society being used as a means to ensure the well-being of the rest. Second, this approach fails in its definition of good. The measures of good or happiness in this visualization are often narrowly defined and reduced to a particular set of measurable goods such as material wealth. These goods are often limited to a sector of the population and even within this sector may disregard other desirable goods such as education. In this definition the individual rights of certain members of the population are limited for the benefit of other members of the society. An example from the energy sector is the appropriation of land for building dams. Dam construction is generally viewed as being for the common good by providing power to communities, providing employment, and reducing the carbon footprint by providing renewable energy. However, those people displaced—in some cases forcibly—are disproportionately negatively affected by the loss of their land.

Since common goods are grounded in ethics and morality, another way to conceptualize them is through the prism of religion. In this context the common good is defined as a responsibility to care for others and support the general welfare, sacrificing some personal freedoms for that which is seen to benefit society. These principles of self-sacrifice and generosity can be a powerful way to support and protect struggling sectors of the populace. However, this approach to the common good also has certain pitfalls; for example often a narrow section of the population, usually religious leaders or the theologically influenced elite, is responsible for deciding what benefits society and what can be reasonably sacrificed for these benefits. In this case these espoused benefits are strongly influenced by the particular religious doctrine, making religious tradition the arbiter of individual freedoms and restricting other freedoms, especially for nonmembers of that particular religion. This is especially problematic in societies with diverse populations belonging to varying religious traditions. This approach can also be problematic in societies like the United States which value the separation of powers. In these societies, arguments rooted in religious doctrine may be derided as not belonging in the public sphere and ignored regardless of their merit.

The common good can also be defined as that state which supports the individual in the fulfillment of their potential.[18] In this conception, each individual pursues personal fulfillment to the greatest extent possible. However because individual fulfillment is limited by time, energy, and resources, individuals also value benefits arising from others' pursuit of fulfillment. Because of this, individuals come together to create a society that uses fairness as the basis to support individual fulfillment. In this conception both the participation in the society to support others and the benefits of the society to the individual are considered common goods. However, this conception is naively optimistic, assuming mutual supportiveness and individual's activities will benefit the group.[19]

In all of these discussions of the common good, there are three primary difficulties. One is the conception of the common. The number and diversity of individuals within a society make it impossible to cater to everyone, so a choice has to be made as to what portion of the society will be considered as the common. Another is the conception of good. Societies are made up of individuals who may not have the same values, belief systems, or needs. Therefore, deciding whose needs and values will be considered is a constant challenge. A third is the tension between the common good and individual rights. Considerations and responsibility for the good of others would seem to require sacrificing some individual freedoms. However, Tocqueville argues that the common good and individual rights are not in conflict. In a society where one's rights are recognized as paramount, individuals realize that the protection of their rights depends on the extent of the protection of others' rights, and define these as the common good. Therefore active and engaged participation in support of good citizenship is the best way to support individual rights.[20]

Inherent in all three of these difficulties is another important question namely, who gets to decide. A decision has to be made by the individuals within a society as to what will be valued, who will benefit, how they will benefit, and who will bear the cost. By necessity, this decision is made by a section rather than the whole population. The determination of who will have the power to make these decisions and how they will act is the essence of a political system, one of the foundational elements of the social environment.

3.3.3. The Tragedy of the Commons

One example of the tension between individual rights and social responsibilities is the notion of the tragedy of the commons. The commons are shared resources in a community, such as a public park or a lake which can be used by all. If this shared resource is destroyed by the cumulative action of individuals acting exclusively for their own benefit, the situation can be referred to as a tragedy of the commons.[21] Take for example a local lake used by fishermen. Since the resource belongs to no one, each fisherman has the right to remove as many fish as they desire from the lake. Each fisherman therefore removes as many fish as possible in order to maximize their individual profit, resulting in overfishing of the lake and the consequent destruction of the resource. In this case the common good, continued access to the fishery, was sacrificed for individual rights, pursuit of a livelihood. The converse of this would be the seizing of the lake by a local government which could, for example, mandate that it could only be used by certain fishermen or for occasional sport fishing, thus preserving it for the common good. However, this would deprive the local fishermen of the right to make a living through fishing. Other examples of the tragedy of the commons include greenhouse gas emissions and pollution from car exhaust.

Though the pursuit of the common good can seem impossible—given both the difficulties in defining both common and good and the tensions between individual and communal rights—the pursuit is not impossible. This is evidenced by other examples of the use of common resources. There are many examples all over the world where local communities have evolved systems that maximize both common access to resources and individual rights to pursue beneficial economic and entertainment objectives.

3.4. Globalization and the Common Good

Historically, interpersonal and inter-group interactions through migration, exploration, trade, and war has stimulated human development and fostered the adoption and propagation of technologies, values, and beliefs. These interactions generally result in changes in social systems for both groups, with values and beliefs altered or adopted in response to changing contexts, information, and desires. Globalization has not only significantly increased knowledge of and interactions among societies, but has also created interconnected production and consumption systems linking the fortunes of various societies. More than ever before, events in one part of the world can be felt in geographically separate places. This interconnectedness has required the development of new social, political, and economic systems spanning the globe, as well as a new understanding of the common good and individual rights. Achieving this understanding requires the identification of and agreement on common values across societies. Globalization has had the dual effect of both distributing or magnifying the negatives in the social environment (such as the manipulation of differing values to maximize benefits for certain sectors of the population) and empowering actors to positively influence social systems by leveraging global connections. For example, globalization and its contributions to the development of transnational social movements have affected hydroelectric power generation. Between 1900 and 1950 the number of dams globally increased from approximately six hundred to nearly five thousand, by 2000 the number of dams approached forty-five thousand. However, the rate of dam building has declined significantly from about one thousand dams built per year in the mid 1960s to about twenty-five per year in the 1990s. This decrease can be partly attributed to the growth of transnational non-governmental organizations and the development of communication systems that allow local people to mobilize worldwide opposition to large dam projects which would displace hundreds or millions of people.

Decisions about energy production and use must be made in the context of the political, social, and economic systems in which they exist. In order to make these decisions it is important to understand the interplay between these systems, how they affect and are affected by the agents within them and by individual and societal values, and the necessity of balancing individual and communal needs. *Case Study 2* illustrates how these differing forces may manifest in the energy sector.

Case Study 2

Eco-Energy Systems owns an ethanol plant in the heart of the United States midwest. The plant has a production capability for producing twenty million gallons of ethanol per year and employs thirty production workers. By comparison, most of the ethanol plants in the United States can produce sixty million gallons or more each year. In 2012, ethanol production dropped for the first time in sixteen years. This was primarily due to the reduced demand for gasoline since 2008 (most gasoline in the United States is blended with ethanol). The current production ceiling requires approximately thirteen billion gallons of ethanol to be blended with gasoline at ten percent (E10), as determined by a government mandate. Also, future projections by the Department of Energy show an increase in demand to fifteen billion gallons by 2015, which is the target date established by the RFS (Renewable Fuel Standards). However, the cost of biomaterials has increased substantially, undercutting profits by thirty percent. You have been assigned to lead a team of managers, engineers and scientists from Eco-Energy Systems to conduct a review of the facility and make recommendations to the company president and CEO on the future of the operation. You are expected to be sensitive to environmental and societal concerns in developing your final report.

There are some environmental and economic issues to be considered in any decisions on the future of the ethanol plant.

1. Ethanol production requires using large volumes of water. When in full production, the plant pumps approximately six million gallons of water a day from a wetland, which helps some local residents in low lying areas by keeping the water table low and basements dry. However, environmental groups are threatening legal action because they view tampering with the water levels as detrimental to local wildlife and the natural ecosystem. If these lawsuits are successful, the company would have to pay millions of dollars for remediation, leading to the closure of the plant.
2. If the plant is closed, approximately ninety jobs will be lost from the community with a population of 1200. The economic impact of closing the production facility is quite significant to a small community.
3. Any decision to dismantle the operation will require an environmental impact study which will be expensive and time consuming. There could also be issues associated with hazardous materials.
4. Based on projections of increased demand for ethanol, the President of Eco-Energy is willing to consider expanding the facility to increase production capacity to one hundred fifty million gallons per year. The expansion could also include focusing production on second-generation biofuels using cellulosic feedstocks such as corn stover or other energy crops such as miscanthus, switchgrass, and tree plantations. However, these projections of demand are uncertain.

Some guiding questions for your team to consider are:
1. How do notions of the common good play out in this case study?
2. How do you reconcile the competing interests of being a profitable business serving the interests of shareholders with being socially responsible to the community and future generations?
3. What are some of the potential financial consequences of closing the facility? What recommendations would you offer to the president?
4. What are two alternatives you would recommend to upper management? What are the consequences? Which alternative would you recommend?
5. In chapter 6, the author suggests using a "reverse auction" as one possible policy mechanism to make the production of biofuels more economically viable. Would this be a good alternative?

Notes

1. "Engineering," *Encyclopaedia Britannica*, last modified 2013, http://www.britannica.com/EBchecked/topic/187549/engineering.
2. Anthony Giddens, *The Constitution of Society: Outline of the Theory of Structure* (Berkeley: Oxford University Press, 1984).
3. Walter Carlsnaes, "The Agency-Structure Problem in Foreign Policy Analysis," *International Studies Quarterly* 36, no. 3 (1992): 245–70, http://dx.doi.org/10.2307/2600772.
4. Ibid.
5. Tina Flegel, "Public Protests Against Nuclear Power in Germany," *Turkish Policy Quarterly* 9, no. 2 (2010): 105–15.
6. Nancy Isenson, "Nuclear Power in Germany: A Chronology," *DW*, September 10, 2009, http://www.dw.de/nuclear-power-in-germany-a-chronology/a-2306337.
7. Paul Mauldin, "Germany's Nuclear Decisions: Maybe Not the Optimal Timing?" *Smart Energy Portal*, September 27, 2011, http://smartenergyportal.net/article/germany%E2%80%99s-nuclear-decisions-%E2%80%93-maybe-not-optimal-timing.
8. Axel Vorwerk, "The 2002 Amendment to the German Atomic Energy Act Concerning the Phase-Out of Nuclear Power," *Nuclear Law Bulletin* 69 (2002): 7–14.
9. BBC Monitoring, "Germany to Phase Out Nuclear Energy," *BBC News Online*, November 10, 1998, http://news.bbc.co.uk/2/hi/world/monitoring/211911.stm.
10. John Moore, "How Much Precaution is Too Much? Evaluating Germany's Nuclear Phase-Out Decision in Light of the Events at Fukushima" *The Public Sphere: Graduate Journal of Public Policy*, no. 1 (2012): 42–53; "Slowing the Phase-out: Merkel Wants to Extend Nuclear Power Plant Lifespan," *Spiegel Online*, August 30, 2010, http://www.spiegel.de/international/germany/slowing-the-phase-out-merkel-wants-to-extend-nuclear-power-plant-lifespans-a-714580.html.
11. Moore, "How Much Precaution"; World Nuclear Association. "Nuclear Power in Germany." *World Nuclear Association,* last updated November 2013, http://www.world-nuclear.org/info/Country-Profiles/Countries-G-N/Germany/#.UYxw1Mo892q.
12. Judy Dempsey and Jack Ewing, "Germany, in Reversal, Will Close Nuclear Plants by 2022," *The New York Times*, May 30, 2011, http://www.nytimes.com/2011/05/31/world/europe/31germany.html?_r=1&.
13. Ibid.; Moore, "How Much Precaution"; Nuclear Energy Institute, "Germany and Japan Rethink Nuclear Policies," *Nuclear Energy Institute,* 2012, http://www.nei.org/News-Media/News/News-Archives/germany-and-japan-rethink-nuclear-policies.

14. V. Bradley Lewis, "The Common Good in Classical Political Philosophy," *Current Issues in Catholic Higher Education* 25, no. 1 (2006): 25–41.
15. Milton Rokeach, *Beliefs, Attitudes, and Values*, San Fransisco: Jossey-Bass, 1968.
16. Icek Ajzen, "The Theory of Planned Behavior," *Organizational Behavior and Human Decision Processes* 50, no. 2 (1991): 179–211, http://dx.doi.org/10.1016/0749-5978(91)90020-T.
17. Ibid.
18. John Rawls, *A Theory of Justice* (rev. ed.) (Cambridge, MA: Harvard University Press, 1999).
19. Mary M. Keys, *Aquinas, Aristotle, and the Promise of the Common Good* (Cambridge, England: Cambridge University Press, 2006).
20. Alexis de Tocqueville, *Democracy in America*, trans. and ed. by Harvey C. Mansfield and Delba Winthrop (Chicago, IL: University of Chicago Press, 2000).
21. Garrett Hardin, "The Tragedy of the Commons," *Science* 162, no. 3859 (1968): 1243–48, http://dx.doi.org/10.2307/1724745.

Bibliography

Ajzen, Icek. "The Theory of Planned Behavior." *Organizational Behavior and Human Decision Processes* 50, no. 2 (1991): 179–211. http://dx.doi.org/10.1016/0749-5978(91)90020-T.

BBC Monitoring. "Germany to Phase Out Nuclear Energy." *BBC News Online*, November 10, 1998. http://news.bbc.co.uk/2/hi/world/monitoring/211911.stm.

Carlsnaes, Walter. "The Agency-Structure Problem in Foreign Policy Analysis." *International Studies Quarterly* 36, no. 3 (1992): 245–270. http://dx.doi.org/10.2307/2600772.

Dempsey, Judy, and Jack Ewing. "Germany, in Reversal, Will Close Nuclear Plants by 2022." *The New York Times*, May 30, 2011. http://www.nytimes.com/2011/05/31/world/europe/31germany.html?_r=1&.

Encyclopaedia Britannica. "Engineering." *Encyclopaedia Britannica*. Last modified 2013. http://www.britannica.com/EBchecked/topic/187549/engineering.

Flegel, Tina. "Public Protests Against Nuclear Power in Germany." *Turkish Policy Quarterly* 9, no. 2 (2010): 105–15.

Giddens, Anthony. *The Constitution of Society: Outline of the Theory of Structure*. Berkeley: Oxford University Press, 1984.

Hardin, Garrett. "The Tragedy of the Commons." *Science* 162, no. 3859 (1968): 1243–48. http://dx.doi.org/10.2307/1724745.

Isenson, Nancy. "Nuclear Power in Germany: A Chronology." *DW*, September 10, 2009. http://www.dw.de/nuclear-power-in-germany-a-chronology/a-2306337.

Keys, Mary M. *Aquinas, Aristotle, and the Promise of the Common Good*. Cambridge, England: Cambridge University Press, 2006.

Lewis, V. Bradley. "The Common Good in Classical Political Philosophy." *Current Issues in Catholic Higher Education* 25, no. 1 (2006): 25–41.

Lowry, William Robert. *Dam Politics: Restoring America's Rivers*. Washington, D.C.: Georgetown University Press, 2003.

Mauldin, Paul. "Germany's Nuclear Decisions: Maybe Not the Optimal Timing?" *Smart Energy Portal*, September 27, 2011. http://smartenergyportal.net/article/germany%E2%80%99s-nuclear-decisions-%E2%80%93-maybe-not-optimal-timing.

Moore, John. "How Much Precaution is Too Much? Evaluating Germany's Nuclear Phase-Out Decision in Light of the Events at Fukushima." *The Public Sphere: Graduate Journal of Public Policy,* no. 1 (2012): 42–53.

Nuclear Energy Institute. "Germany and Japan Rethink Nuclear Policies." *Nuclear Energy Institute,* 2012. http://www.nei.org/News-Media/News/News-Archives/germany-and-japan-rethink-nuclear-policies.

Rawls, John. *A Theory of Justice* (rev. ed.). Cambridge, MA: Harvard University Press, 1999.

Rokeach, Milton. *Beliefs, Attitudes, and Values*. San Fransisco: Jossey-Bass, 1968.

Spiegel Online. "Slowing the Phase-out: Merkel Wants to Extend Nuclear Power Plant Lifespan." *Spiegel Online*, August 30, 2010. http://www.spiegel.de/international/germany/slowing-the-phase-out-merkel-wants-to-extend-nuclear-power-plant-lifespans-a-714580.html.

de Tocqueville, Alexis. *Democracy in America*, translated and edited by Harvey C. Mansfield and Delba Winthrop. Chicago, IL: University of Chicago Press, 2000.

Vorwerk, Axel. "The 2002 Amendment to the German Atomic Energy Act Concerning the Phase-Out of Nuclear Power." *Nuclear Law Bulletin* 69 (2002): 7–14.

World Nuclear Association. "Nuclear Power in Germany." *World Nuclear Association,* last updated November 2013. http://www.world-nuclear.org/info/Country-Profiles/Countries-G-N/Germany/#.UYxw1Mo892q.

PART 2

ENERGY CONVERSION TECHNOLOGY

Chapter 4
Harnessing Nature:
Wind, Hydro, Wave, Tidal, and Geothermal Energy

EUGENE D. COYLE, BISWAJIT BASU,
JONATHAN BLACKLEDGE, AND WILLIAM GRIMSON

Abstract

The chapter commences with a short appraisal of current shares of generating capacity from a range of renewable energies, before proceeding to a review of selected renewable energy sources. We discuss wind, hydroelectric, wave, tidal, and geothermal energy sources, examining the history of each technology, current developments, and environmental impact. The chapter concludes with a look to the future grid, considering the impact of large scale renewable technologies on grid development.

4.1. Introduction

Given both the proven market position of fossil fuels in world energy supply and the difficulties associated with continued or increasing demand and use of coal, petroleum, and natural gas, we need to consider the current status and future potential for a range of renewable technologies: onshore and offshore wind, hydroelectric energy, wave and tidal energy, and geothermal energy. Chapters five and six will go further, exploring developments in solar energy underpinned by nanotechnology and biofuels, respectively.

Hydrogeneration remains the world's largest carbon-neutral renewable electricity resource, with global installed capacity of approximately 3.4 GWh (gigawatt hours per year). Wind installation is on a rapid growth curve, with current installation in excess of 230 TWh (terawatt hours per year). Installed wind capacity in the United States exceeded 43 GW (gigawatt) by the third quarter of 2011, making it second only

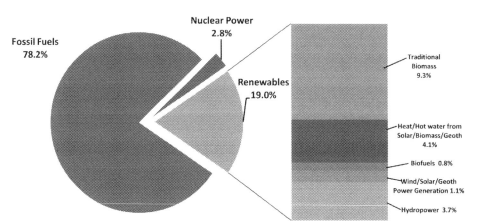

Figure 4.1. Estimated Renewable Energy Share of Global Energy Consumption, 2011.[1]

to China. The majority of wind installations to date globally have been onshore and that technology, although still developing, has reached a relatively mature status. Offshore wind farms offer greater power (watts/square meter), but such installations are more technically challenging and costly to install and maintain. Globally, biomass accounts for in excess of 200 TWh, geothermal energy for 65 TWh, solar photovoltaic energy for 12 TWh, solar thermal energy for 1 TWh, and tidal energy for 0.5 TWh. Estimated renewable energy by type is shown as a percentage of global energy consumption in Figure 4.1.

4.2. Wind Energy

Growth trends for installed wind energy capacity are on the rise around the world. Total global cumulative capacity in December 2012 stood at 282.6 GW; China tops the table at 75 GW, with USA second at 60 GW, Germany third at 31 GW; followed by Spain (23 MW), India (18.5 MW), UK (8.5 MW), Italy (8 MW), France (7.5 MW), Canada (6 MW), Portugal (4.5 MW), and the rest of the world (40 MW).

4.2.1. Historical Overview

Wind turbines, traditionally known as windmills, have been around for at least three thousand years. The main use of early machines was either for pumping water or grinding grain. Wind has been the key source of power in sailing for even longer. In the early part of the last century, wind turbines were used for electricity generation, primarily to charge batteries and facilitate the supply of power to remote locations. However, the attractiveness of these systems declined with the advent of an expanding electricity grid. An exceptional case worth noting was a fifty-three meter rotor-diameter steel wind turbine constructed in the USA in 1941.[2] This 1.25 MW Smith Putnam machine had a full-span pitch control—much like modern machines—and had flapping blades for load control. This operated as the largest wind turbine ever constructed for a period of about four years until it suffered catastrophic failure in 1945.

A record of the historical development of wind turbines recalls a 100 kW 30 m diameter Balaclava wind turbine and an Andrea Enfield 100 kW 24m diameter pneumatic design, built in the former USSR in 1931 and in the UK in the early 1950s, respectively.[3] The design of the latter turbine was based on the use of hollow blades with openings at the tip. The air was drawn up from the opening and through the tower to subsequently drive another turbine which was connected to the generator to produce electrical power. Other developments include the 200 kW 24 m diameter Gedser machine built in Denmark in 1956; testing of a 1.1 MW 35 m diameter turbine in 1963 by Electricité de France; advances made by Golding at the Electrical Research Association in the UK; and the construction of a number of lightweight turbines in Germany in the 1950s and 1960s, by Hotter.[4]

Despite this general enthusiasm and the achievement of several technical milestones, activity in the wind energy industry did not gather momentum until the dramatic surge in oil prices in 1973. This made wind energy competitive with traditional fossil fuels. It also created the necessary push for the further advancement of wind energy science and technology, helping to reduce the cost of energy (COE). This in turn stimulated a number of substantial Government-funded programmes of research, development, and demonstration both in the USA and in Europe (specifically, in the UK, Germany, Sweden and Denmark). Denmark became a pioneering nation in the development of wind turbines, a far sighted stance which placed them ahead of neighboring European countries in the ensuing decades. A series of prototype turbines were constructed in the USA. Power generation capacity of the turbines grew from 100 kW for Mod-O built in 1975 (38 m diameter) to the 2.5 MW Mod-5B built in 1987 (97.5 m diameter). A 4 MW vertical axis Darrieus wind turbine was also constructed in Canada. A similar concept 34 m diameter turbine was tested in the Sandia Vertical Axis Test Facility in the USA. Other initiatives with regard to vertical axis design involved the use of straight blades giving rise to an 'H' type rotor proposed by Peter Musgrove, resulting in a 500 kW prototype. In 1981, a 3MW horizontal axis wind turbine was developed and tested in the USA with hydraulic transmission and gear system. The entire structure was orientated into the wind instead of using a yaw controller. The optimal choices of the number of blades to be considered remained an open issue at that time and turbines with varying number of blades (one, two, or three) were constructed.

4.2.2. Wind Characteristics

The amount of energy that can be extracted from wind was eloquently determined by German physicist Albert Betz. Published in 1919, *Betz's Law* calculates the maximum power that can be extracted from the wind in open flow. The law reveals that energy created is proportional to the cube of wind speed:

$$P = 0.5 c \rho A V^3$$

where P is the power (*watts*), c is the coefficient of performance, ρ is the density of air (kg/m^3), A is the area of the turbine (m^3), and V is wind speed (m/s).

According to Betz's Law no turbine can capture more than 59.3% of the kinetic energy in wind. Modern wind turbines are designed to achieve as much as eighty percent of the Betz limit.[5]

Since the power generated from wind is highly sensitive to wind speed, an assessment of the characteristics of the wind resource is vitally important. Information on the characteristics of wind forms the basis of investigations on wind energy exploitation, including decisions on site selection, design optimization, and the best choice of turbine for a particular setting. Wind characteristics are also critical to the understanding of the effect of wind energy penetration to the electricity grid network. This is very important in the emerging context of smart grids and the ease of connectivity to local distribution or transmission networks.

Wind speed is highly variable in nature, both *spatially* (geographically) and *temporally*. Scale dependency of this variability over a wide range (both spatial and temporal) adds to the complexity. Variability or fluctuations in available energy from wind is amplified due to the consequence of the cubic relationship to speed. On a global scale, spatial variability refers to varying climatic conditions in different regions of the world, with some locations more windy than others; such fluctuations are caused by the difference in solar insolation at different latitude. Within any particular climatic region, wind speed variation will also occur due to variability in physical geography. Wind speeds are affected by the proportion of landmass and sea, the size of the land mass, the presence of mountains or plains, and the type of vegetation in the area, which governs the amount of solar reflection and radiation and, in turn, the surface temperature and humidity. On a local scale, wind speed fluctuations are affected by local topography: for example, wind speed is higher on the top of hills than on the leeward side or in a sheltered valley. Finally, at any given location wind speed is dictated by the presence of buildings, trees, or open, unobstructed land.

At any particular site, the wind speed also varies temporally. On a long term temporal scale, there may be an underlying trend in the variation of the wind speed over years or decades, but such variation is not well understood and is rather difficult to predict. On time scales shorter than one year, fluctuations can often be attributed to seasonal variations. These are better understood, however large variability can occur over short time scales and accurate prediction may not be possible beyond a few days forward. There are considerable local specific variations possible during the day (diurnal) and these can be predicted more accurately. The understanding and forecasting of these fluctuations plays a major role in the context of energy management for the grid, as they give an indication of how much can be produced from the renewable wind and hence enable decisions regarding production from other types of power plant feeding the grid. Even further refined temporal scales (on the order of minutes or seconds) indicate several aerodynamic phenomena which can contribute to fluctuations in wind speed. One such phenomenon, *turbulence*, can have a significant impact on the design, performance, and fatigue life of individual turbines and their components, and can also impact on the quality of the power produced and its effect on customers.

The total velocity of wind flow passing through the turbine rotor may be decomposed into a height dependent mean component and a stochastic (turbulent) fluctuating component. Attention to understanding turbulence in wind flow is an important field of research. One approach to tackling stochastic problems in wind turbulence is through application of power spectral density functions (PSDF).[6] A rotating blade is subject to an atypical fluctuating wind velocity spectrum, known as a rotationally sampled spectrum. Rotationally sampled spectra are used to quantify the kinetic energy as a function of frequency for rotor blades. Due to the rotation of the blades, the spectral energy distribution is altered, with variance shifting from the lower frequencies to peaks located at integer multiples of the rotational frequency.

4.2.3. Modern Wind Turbines and the Power Grid

Wind turbines can be mainly classified into two types based on the alignment of the rotor shaft: the vertical axis wind turbine (VAWT) and the horizontal axis wind turbine (HAWT), which is most widely observed today. Typical wind turbine components include turbine blades, nacelle housing, a low speed shaft, gearbox, generator, brake assembly, support tower, a cable drop to a converter and switchgear, and a local transformer for connection to the power grid. Either alternating or direct current generators, with related accompanying power electronic circuitry, may be found in turbines in operation today.

Direct drive grid connected generators are considered candidates for future wind generation, as they would eliminate the need for gear boxes and power conversion. Generator designs for wind turbines include the permanent magnet synchronous generator (PMSG), field excited synchronous generator (FESG), and the induction generator (IG). IGs may be of fixed speed (FSIG) or double fed (DFIG) design. The DFIG has a wound rotor, enabling the transfer of power from both the rotor and stator windings. The stator winding is connected directly to the three-phase electrical grid and the three-phase rotor winding is fed from the grid through a frequency converter.

With the significant penetration of wind energy into the grid, wind turbines are now integrated into the mainstream source of generating power. Commercially installed wind turbine machines of megawatt capacity have been successfully operating for a number of years. On-shore and offshore wind farms in operation in many countries around the world typically comprise ten to one hundred turbines, with some large on-shore farms containing several hundred units. On-shore wind farms are suitable for open landmass areas which have average wind speeds greater than six meters per second at a height of ten meters. An example of a wind speed power curve is shown in Figure 4.2, with an indicative cut-in speed of 3.5 m/s and a rated speed of 12.5 m/s.

Aside from the increased incremental power output, there are several beneficial effects to grouping turbines in a farm, including savings in construction costs, grid connection costs, and lower operation and maintenance (O&M) costs due to integrated management and maintenance.

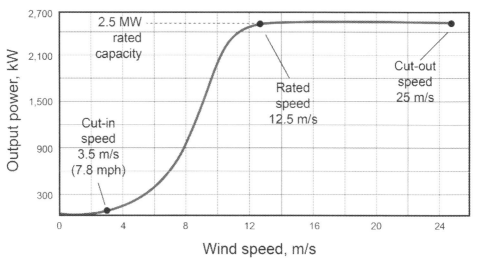

Figure 4.2. Typical Wind Speed Power Curve Characteristic: Output Power Versus Wind Speed, Indicating Cut-in Speed, Rated Speed, and Cut-out Speed.

4.2.4. Future Offshore Wind Energy Potential

Figure 4.3 provides an indication of the growth in the size of wind turbines from 1985 to the present day. Referring again to Betz's Law ($P = 0.5c\rho AV^2$), we see that the power generated by a wind turbine is linearly proportional to the swept area, and we know that area is proportional to the square of the radius (r^2 or $d^2/4$). Thus, holding other conditions constant, a doubling of the radius will increase power output fourfold. A significant increase in turbine rotor diameters is possible by locating farms offshore. In this way rotor diameters as large as 160 meters can be installed, since restriction due to aesthetic pollution or noise is not an issue. Stronger offshore wind speeds combined with less turbulence away from the shore result in increased efficiency and greater power production, especially since the size of onshore wind generators are nearing their upper limits due to social concerns.

Offshore wind turbines have greatly reduced visual impacts and lower noise constraints, allowing higher rotor speeds. However, offshore hardware and installation costs tend to be expensive, largely on account of water depth and distance from shore. Nevertheless, offshore farms can produce up to fifty percent more electricity than onshore equivalent farms. In 2006, offshore wind energy accounted for 1.8% of total installed capacity, but produced 3.3% of total wind electricity. A significant database has been established through worldwide wind atlas records data, with some eight thousand locations registering wind speeds at a standard height of eighty meters above sea level. Indications are that moving offshore is considered a likely growth area for wind technology, with potential for development in near-shore deep water zones in countries and locations including the USA, the western coast of South America, Spain, Norway, China, Japan, India, and the eastern coast of Australia. Land-based wind farm capacity is limited to fifty megawatts while offshore farms of more than one hundred megawatt capacities are possible. Offshore construction is nevertheless challenging, not least in terms of installing deep water foundations, achieving connectivity to grid and carrying out maintenance following installation of plant.

The cost of fixed mounted offshore wind turbines increases with water depth, rendering them uncompetitive in certain locations. In deep water areas, floating wind turbines may be the most cost-effective and reasonable solution.[7] Experience and expertise acquired through the offshore oil platform industry means that there is a wealth of relevant technology available for adaptation to floating offshore wind turbine platforms (FOWT). As a large platform is preferable in order to minimize ocean motion response, and the weight of a wind turbine is small compared to that of the floating platform, it is possible to install larger wind turbines with capacities of five to ten megawatts. This will enable a reduction in power generation cost.[8]

One major challenge is to design and optimize an appropriate support structure, which can contribute almost forty percent of the total cost. Ongoing research and development is essential in all areas, including assembly, installation, and decommissioning; electrical infrastructure including power transmission and HVDC; power electronic converters; monitoring of power quality; enhanced design of turbines; and operations and maintenance protocols for offshore systems, in order to reduce COE and minimize environmental impact.

FOWT systems may be divided into two groups, single turbine and multi-turbine systems.[9] Several initial concepts considered floaters supporting multiple turbines, in an attempt to reduce floater motion due to smaller thrust height to floater span ratio and to improve economy by employing a single mooring system.[10] However, such systems are subjected to high current and wave loads and their turbines suffer wake effects. The floaters

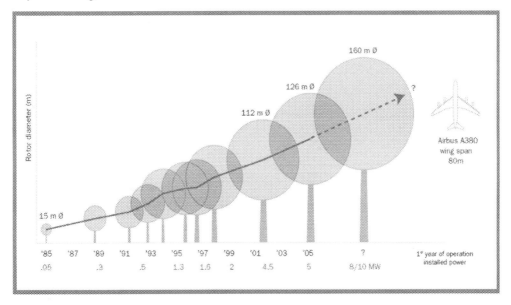

Figure 4.3. Wind Turbine Developments: Blade Span and Power Rating.

supporting a single wind turbine were considered to be more suitable for offshore wind energy.[11] Single turbine system platforms include spar (S), tension-leg platform (TLP) and barge (B) concepts as indicated in Figure 4.4.[12] The spar concept uses a long-draft spar moored by catenary or taut lines and achieves stability using ballast to lower the center of mass below the center of buoyancy. The TLP concept achieves stability through the equilibrium of tensioned taut mooring-lines and the excess buoyancy of the platform. The barge is generally moored by catenary lines and stabilized by its water-plane area. Hybrid models using features of those concepts also have potential for further development.

4.2.5 Environmental Concerns and Social Acceptance

Wind farm installation continues to grow annually in most regions, including Europe, the United States, China, and India. The EU added ten thousand megawatts of new capacity in 2010, approximately ten percent of which was offshore. Germany alone increased power capacity by two thousand megawatts in 2011, setting the mark as the leading EU member state in installation of wind energy. The United Kingdom is currently the largest market for offshore wind, with installed capacity of two thousand megawatts, fifty percent of total EU offshore installations.

Although wind turbines emit no carbon emissions and are generally positively received by local communities, they are seldom greeted with universal support. Principal objections to farm installations tend to be visual or noise related. Wind turbine noise can result owing to the passage of wind across the blade surface. Although the blades appear to move slowly, the blade tips can move at speeds in excess of one hundred miles per hour, which in turn can generate a pulsing noise. In addition, complaints sometimes reference the shadows cast by the rotation of the blades; which may be more or less pronounced depending on location and prevailing weather conditions. Objections mostly concern farms with large arrays of turbines. On the beneficial side landowners may generate income by contractually agreeing land lease for utility installation.

The International Energy Agency Task 28 working group incorporated contributions by representatives from USA, Canada, Denmark, Germany, Ireland, Japan, Norway, Switzerland, and the Netherlands to create "an interdisciplinary and cross-cultural exchange platform with goal to support the successful development of wind energy in the participating

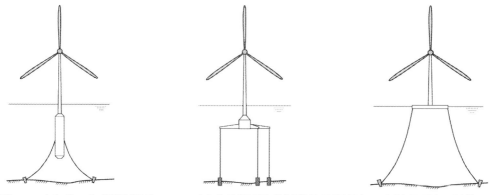

Figure 4.4. Single FOWT Systems: (a) S-FOWT, (b) TLP-FOWT, and (c) B-FOWT.[13]

countries."[14] In the context of supporting development of wind projects and understanding the opposition that can arise, social acceptance is defined as "societal consensus on the planning, construction, and operation of wind projects."[15] The group's recommendations are aimed at planners, policymakers, and practitioners of wind power development.

- Sites with potentially high conflicts, such as those close to dwellings or in protected areas, should be avoided.
- It is important to anticipate and minimize potential adverse project impacts by choosing an appropriate turbine model or by adapting to the behavior of wildlife such as migrating birds.
- Wind farm installations should maximize benefits for local communities by means of an equitable distribution of the positive and adverse impacts of a project.

Research in wind energy supported through the EU Framework program includes projects for the design of very large wind turbines (producing eight to ten megawatts) for both onshore and offshore applications.[16]

Development of wind energy has grown exponentially over recent years and many countries have surpassed expectations of the percentage wind capacity that may be safely deployed on network grids. Attention to power quality, intermittency, and strength of distribution and transmission grids have been tackled, enabling connectivity of renewable at levels well in excess of what was previously thought possible. The case for wind energy is now well established; however it is a maturing rather than a mature science and will have an important role to play in reducing carbon emissions to 2020 and beyond.

4.3. Hydroelectric Energy

4.3.1. Historical Perspective on Hydropower and Hydroelectricity

Water as a source of energy, or *hydropower*, has a long history stretching back to, and probably beyond, recorded history. In ancient times water wheels were used in milling to produce flour from wheat grain. Another use was in agriculture in the form of irrigation where some form of waterwheel was deployed to pump and feed water into distribution channels. In Mesopotamia, the use of irrigation and drainage for agriculture dates back to 5000 BC. Cities were built along the Tigris and Euphrates rivers, forming the northeastern portion of the Fertile Crescent, which also included the Jordan River Valley and the Valley of the Nile.[17] Mesopotamian innovations also included construction of water dams and the use of aqueducts.

Undoubtedly water in the form of rivers was the chief source of power during the early stages of the industrial revolution and this largely accounts for the development of various types of mills and factories in places that otherwise would be rustic, charming and idyllic. Developed in Greece, the earliest evidence of a water-driven wheel is considered to be the Perachora wheel, dating to the third century BC.[18] The Romans and Greeks were the first to operate overshot, undershot and breastshot waterwheel mills.[19] The third century AD Hierapolis water-powered stone sawmill is the earliest known machine to incorporate a crank and connecting rod mechanism.[20]

Early contributors to both the study of hydraulics and the practical realization of water wheels included Bernard Forest de Bélidor (1693–1761), who published *Architecture Hydraulique* which described vertical- and horizontal-axis hydraulic machines; Jean Victor Poncelet (1788–1867), who improved the design of turbines and waterwheels; Daniel Bernoulli (1700–1782) who wrote the theory for the conversion of water power into other forms of energy in his book *Hydrodynamica*; and mathematician and physicist Leonard Euler (1707–1783) who wrote his theory of hydraulic machines in 1750.

As the 1800s progressed a number of technological innovations such as the development of turbines led to increased efficiencies. And in general a more scientific approach was in evidence by which the energy extracted from a headrace could be maximized.

With the development of the electric generator in the latter part of the 19th century, through hydraulic coupling, hydroelectric power could be produced.[21] The first operational use of water (hydro) in the generation of electricity in the United States is attributed to Wisconsin in 1882. At the beginning of the 20th century, many small hydroelectric power plants were being constructed by commercial companies in mountains near metropolitan areas. India's first hydropower plant with a capacity of 130 kW was commissioned in 1897 at Darjeeling in West Bengal.[22] By 1920 forty percent of the power produced in the United States was hydroelectric. With the development of extensive electricity networks distribution problems in respect of waterwheel power were solved in that the location of the power generation did not have to coincide with, say, the site of the end use factory.

For this and other reasons worldwide investment in hydroelectric schemes increased dramatically. For small and less technically advanced countries that were energy poor, hydroelectric was an attractive option. For example in Ireland the waters of the Shannon (the major river in the country) were harnessed at the Ardnacrusha hydroelectric power plant and became operational under the control of the Electricity Supply Board (ESB) in the late 1920s. The Ardnacrusha hydroelectric power plant was the largest in the world until it was overtaken by the Hoover Dam, built in 1936 with power rating 1,345 MW. The economic and social impact of Ardnacrusha to Ireland was immense. The Hoover Dam was eclipsed by the 6,809 MW Grand Coulee Dam in 1942. The Itaipu Dam opened in 1984 in South America as the largest, producing 14,000 MW but was surpassed in 2008 by the Three Gorges Dam in China at 22,500 MW. Hydroelectricity eventually supplied countries including the Newfoundland-Labrador schemes (Churchill Falls) in Canada; Norway, the Democratic Republic of the Congo, Paraguay and Brazil, catering for more than eighty-five percent of their electricity needs. The United States currently has more than two thousand hydroelectric power plants which supply forty-nine percent of its renewable electricity.[23] Although some countries owing to natural topography are considered to have reached their viable potential for hydro installations, hydro remains a very important and indeed natural source of electricity generation. Examples abound throughout the world.

As the most widely used form of renewable energy, hydroelectricity accounted for sixteen percent of global generation (3,400 terawatt-hours) in 2010, and is expected to increase about three percent per year over the ensuing twenty-five years. Hydropower is produced in 150 countries, with the Asia-Pacific region generating thirty-two percent of global hydropower in 2010. It is noteworthy that China is the largest hydroelectricity producer, with 721 terawatt-hours of production in 2010, representing around seventeen percent of domestic electricity use. There are now three hydroelectricity plants larger than ten gigawatts: the Three Gorges Dam in China, Itaipu Dam across the Brazil/Paraguay border, and Guri Dam in Venezuela.

4.3.2. Hydroelectric Power Generation

The parameters governing the amount of electrical energy that may be generated using a hydropower source are the height from which the water falls and the quantity of water flowing (flow rate).

The power equation may be expressed as:

$$P = \eta p h q g$$

where P refers to the electric power output in KW, η equals a coefficient of efficiency (typically around 0.8), q is the flow rate in cubic meters per second, p equals the density or specific weight of water (1000 kg/m^3), h equals the head in meters, and g refers to the gravitational constant, 9.81 m/s^2.

In a water turbine, blades are attached to the shaft and when flowing water passes against the blades of the turbine, the shaft rotates. After transferring energy to the turbine, water is discharged via a drainage channel called the tailrace of the hydropower plant. The shaft is coupled with an electrical generator to produce energy.

An essential component of hydraulic power generation is the availability of a continuous source of water, providing a large amount of hydraulic energy. While the vertical fall (head) of water remains fairly constant once the plant has been designed, the flow rate can vary depending upon the intensity, distribution and duration of rainfall. The head of water may be available by local terrain or may be created artificially by construction of a dam. If available and adequately controlled, 'water energy' from rivers, streams, canal systems or reservoirs, can provide hydropower plant with efficiency far exceeding that of a conventional thermal power plant. As no heat is involved during hydroelectric power generation, component parts, if well maintained, can last for up to 40 years.

Water turbines may be classified based on a number of functional and operational characteristics. On the basis of head and quantity of water required, turbines can be classified as high-head (with heads ranging from several hundred meters to a few thousand meters), medium-head (with heads ranging from about 60 to 250 meters), and low-head (with heads of less than 60 meters). Francis turbines are medium-head, while Kaplan and propeller turbines are low-head. Depending on the type of flow, turbines are classified as tangential flow, axial flow, radial flow, and mixed-flow. Turbine shafts may be either of horizontal (Pelton turbine) or vertical axis alignment.

4.3.3. Social, Environmental, and Economic Impact of Hydroelectricity

There are significant advantages to the use of hydroelectric power. Based on appropriate site selection, water is a dependable source of energy which is both non-polluting and reliable. It can also be an effective source of power for remote areas. Efficiency of up to eighty percent may be achievable in transferring stored water to electrical energy. When storage facilities are effectively used, flood water may be retained and used for agricultural production, river regulation, and wildlife protection.

There can also be adverse effects to the deployment of hydropower. Over the summer months water reservoirs may have limited natural flow, diminishing the power available to the local community. There have been occasions where large numbers of people have been evacuated from their homes to enable the construction and installation of hydro power plants. Landscape clearing may also result in soil erosion and in extreme circumstances, landslides. Moreover, in normal circumstances silt may be naturally transported downstream by the flow of the river. If it is captured by the reservoir this may result in a decrease in fertility of downstream plains. Aquatic and other animal life may be adversely affected owing to reduction in dissolved oxygen levels. If not carefully monitored fish may be trapped and killed in passing through the turbine.

4.4. Wave and Tidal Energy

4.4.1. Extracting Energy from the Sea

The use of wave and tidal power has a longer history than many would suspect. The Romans built tide mills and much later tide mills were a feature to be found on the North Atlantic coast, particularly in France, Great Britain, Canada and the United States. Électricité de France (EDF) operates the largest tidal power station in the world located on the estuary of the Rance River in Brittany. Its annual output is of the order of 0.6 TWh and produces electricity at a cost nearly thirty percent below that of nuclear power. The Severn Barrage Tidal Power system in the Bristol Channel has long been talked about as a candidate for a large barrage system particularly as the River Severn has a tidal range of fourteen meters (one of the highest in the world). But like most such schemes environmental issues have prevailed to date.

This is one of the reasons why open water (especially in noted ecologically sensitive estuary environments) exploitation of tidal and wave power is dominating current considerations. It has been estimated globally that 180 TWh of economically accessible tidal energy is available. And the corresponding figure for wave energy is of the order of 500 TWh. In the 1970s Stephen Salter was one of the early advocates and developer of systems to extract energy from waves and achieved a 90–90 efficiency in tests of prototypes (i.e., ninety percent absorption of the power available in a wave and ninety percent conversion of that power into electricity). Interest in these systems waned with falling oil prices. But with current high costs of energy and concerns about climate change attention has returned to the quest of extracting energy from waves. Waves of course are a result of wind and therefore wave power systems are intermittent sources of power. In contrast the tidal system, due to a well-established earth-moon mechanism, provides a regular flow of water (stream) largely independent of the wind and thus is an attractive option for energy generation. Tidal stream generators operating in open sea situations are not without environmental challenges but they are less serious than for estuary based systems.

4.4.2. Wave Energy

4.4.2.1. Overview

The energy derived from sea waves is one of the most spectacular forms of ocean energy. Often, it leads to severe destructive effects. The waves are produced by wind action and are therefore in turn indirectly generated from solar energy. The motion of sea surface waves is principally determined by wind speed and, in particular, the gradient of the wind velocity which induces a force. Thus understanding the temporal and spatial variations of the wind force regulated through different angles of incidence upon the sea surface is a fundamental issue. Another issue is the characteristic spectrum over which the wind force is converted into wave motion. Because it is not possible to uniquely simulate such complex interactions on an entirely deterministic basis through the application of computational fluid dynamics over large scales (that is, the wind velocity cannot be known precisely as a function of time), stochastic models are required to investigate correlations between the energy associated with a sea surface wave stream and the wind velocity time series.

Yoshio Masuda is regarded as the father of modern wave energy since his research began in Japan in the 1940s. He developed a navigation system powered by wave energy and equipped with an air turbine. This device, later called an *oscillating water column* (OWC), was commercialized in Japan in 1965, and subsequently in the United States.

In Europe there has been a research thrust in wave energy since the early 1990s stimulated by the European Commission's inclusion of wave energy as a theme in their research and development programs on renewable energies. Since the start in 1992 there have been a large number of projects funded by the European Commission. The IEA-OES 2008 Annual Report provides a survey of the ongoing activities in wave energy worldwide.[24] In the recent past, there has been growing interest in wave energy in North America (USA and Canada), involving the national and regional administrations, research institutions and several companies. This has resulted in frequent meetings and conferences on ocean energy.[25]

4.4.2.2. Nature of Waves

Waves are generated from wind. Hence, as with wind, variability is the main drawback for energy generation from waves. The variation can occur over different time scales: from wave to wave, with sea state, and from month to month. There are also seasonal fluctuations in wave height. Winds generated by the differential heating of the earth pass over open bodies of water and push surface water particles along with it (whose initial conditions are established by the incident radiation) setting up a rolling motion in the water and moving the water particles in a vertical and circular path. The energy and power densities of a wave are proportional to the square of the wave amplitude and to wave period; hence knowledge of the average wave height is therefore important when considering where to place a wave farm.

Assessment of wave energy resource at a site is crucial for the purpose of design of wave energy converters and also for planning and management, in similar fashion to that described for wind.[26] For the purpose of site classification, the level of available wave energy is usually expressed as power per unit length (along the wave crest or along the shoreline direction). A good offshore location should offer the availability of an annual average ranging between 20 and 70 kW/m. These locations are mostly in areas of moderate to high latitude. Seasonal variations are less pronounced in the Southern hemisphere and hence southern coasts of South America, Africa and Australia are particularly attractive for wave energy exploitation.[27] The northern hemisphere (that is, the northern Atlantic and Pacific oceans) has large average wave heights. Further, in terms of the proximity of these waves to coast lines, there are two principal regions that stand out: the Aleutian Islands and the west coasts of Ireland and Scotland. That latter case is due to the fact that the Atlantic Ocean is characterized by prevailing winds from west to east—the Atlantic Trade Winds—and these coasts are regions with a higher population density and easier access to the infrastructure required to exploit wave energy technology. Indeed, the world's first commercial 0.5 MW wave energy plant, developed by *WaveGen*, is located in the Isle of Islay in Scotland and, on the west coast of Ireland, wave heights can vary from two to twelve meters over a week depending on seasonal changes.

The Energy Density E (energy in Joules per unit area) of a continuous sea surface wave may be approximated by

$$E = \frac{1}{8} \rho g H^2 \approx 1.23 H^2 \text{ kJm}^{-2}$$

where ρ is the density of (sea) water, g is the acceleration due to gravity and H is the wave height. This is the energy associated with the oscillation of a wave on the sea surface. For a wave period of T seconds the associated power (in watts per unit area) is given by

$$P = \frac{E}{T} \text{ Wm}^{-2}$$

Thus for an average wave height of 1 meter and an average period of 1 second the wave energy is 1.23 kJ/m² and the wave power is 1.23 kW/m². These metrics apply to wave energy conversion devices that exploit the energy of the wave at right angles to the plane of the sea surface. However, some devices exploit the power associated with the propagation of the wave front itself. In this case, the Energy Flux F (also known as the Power Density) is given by multiplying the energy density of a wave with the group velocity to give

$$F = 0.5 \, H^2 T \text{ kWm}^{-1}$$

In all cases the metrics are proportional to the square of the wave amplitude and open water waves are generated whenever wind speeds exceed ~0.5 meters per second. Large amplitude (~2 m), long period (~7–10 s) waves have power densities commonly exceeding 40–50 kW/m.

Conventional wave spectrum models are linear, and assume that the distance over which the waves develop and the duration over which the wind blows are sufficient for the waves to achieve their maximum energy for the given wind speed. It is assumed that waves can be represented by sinusoidal forms. This relies on the following: (i) waves vary in a regular way around an average wave height; (ii) there are no energy losses due to friction or turbulence, for example; and (iii) the wave height is much smaller than the wavelength. These principal assumptions provide the basis for predicting wave amplitudes on a statistical basis and it is upon this basis that many wave energy converters are designed in which the wave amplitude is taken to conform to a Rayleigh distribution. However, this distribution is known to be inaccurate which is primarily due to a lack of understanding of how, on a statistical basis, wind energy is converted into wave energy.

From a statistical point of view, what is required is a physical model that can accurately predict the distribution of sea surface waves and sea types given knowledge on the distribution of the wind velocity. A solution to this problem could then be used to estimate the 'quality of power' from a wave farm given statistical parameters that reflect the environmental conditions in which the wave farm is operating.

A common measure of the Wave Spectral Density $S(T)$ as a function of the wave period T is

$$S(T) = AT^3 e^{-bT}$$

where

$$a = 8.10 \times \frac{10^{-3} g^2}{2\pi \times 10^4}$$

And,

$$b = 0.74 \times \left(\frac{g}{2\pi v}\right)^4$$

the wind velocity v being measured at 19.5 m above the still water level.[28] However, this model does not take into account 'split spectra' as can sometimes be observed on the west coast of Ireland; that is, spectra consisting of two distinct peaks. Neither does the model, which is based on linear wave theory, take into account freak waves, which are an essentially nonlinear effect. Furthermore the model provides an estimate of the spectrum of periods of sea waves rather than a model that can be used to simulate a time series representing the vertical oscillation of a wave energy device at a location on the sea surface. The spectrum of the wave period is a measure of the power density and energy flux, but the square of the wave height is arguably a more fundamental measure, that is, the wave energy. However, the wave spectral density does at least provide a qualitative relationship between the wave properties and the wind velocity.

Another important product of the linear wave theory relates to the distribution of wave heights which is taken to conform to a Rayleigh distribution

$$P(H) = \frac{H}{\sigma^2} \exp\left(-\frac{H^2}{2\sigma^2}\right)$$

where H is the wave crest height and σ is the most probable wave height.[29] This distribution is used to define the significant wave height (SWH) which is an average of one third of the maximum wave height. High storm conditions can give wave heights of around fifteen meters and the probability of waves with twice the SWH is 0.00001. This leads to the conclusion that freak waves with heights greater than fifteen meters are effectively impossible. However, this conclusion is wrong, as freak waves are now a well known and well documented phenomenon although the reasons for their existence (especially deep water freak waves) is still a matter for debate.[30] Events of this type can cause serious damage to any wave energy conversion device and/or a wave farm that duplicates such a device unless it has been engineered to withstand such rare but extreme conditions.

4.4.2.3. Wave Energy Resources and Developments

In 2011, the UK Carbon Trust estimated that the global marine energy sector could be worth 760 billion in United States dollars by 2050. Industry estimates put annual marine energy revenues at nearly 100 billion dollars by 2025. A detailed assessment of Ireland's wave energy resource undertaken in 2005, for example, looked at the theoretical and accessible levels of wave energy in Irish waters.[31] The study indicated that a theoretical wave energy resource of up to 525 TWh exists within the total limit of Irish waters. For comparison, in 2006, the total electricity requirement for the Republic of Ireland was only 27.8 TWh of electricity.

The wave energy sector is not as far advanced as other renewable energy sectors such as wind or solar, but the concept of harnessing energy from ocean waves is not new. The first ideas were patented as far back as 1799.[32] Between 1855 and 1973, 340 patents for wave energy devices were placed. Modern research into wave energy was greatly stimulated by the oil crisis of the early 1970s, which led to a dramatic increase in oil prices. Figure 4.5 compares oil prices with the number of wave energy patents filed from 1960 to 2005.

The increasing oil prices panicked governments into stepping up research into alternative forms of power generation. Several research programs with government and private support were started mainly in the UK, Portugal, Ireland, Norway, Sweden, and Denmark. In the 1980s, however, the price of oil returned to more affordable levels and the interest in wave energy research dwindled. Funding was withdrawn from many

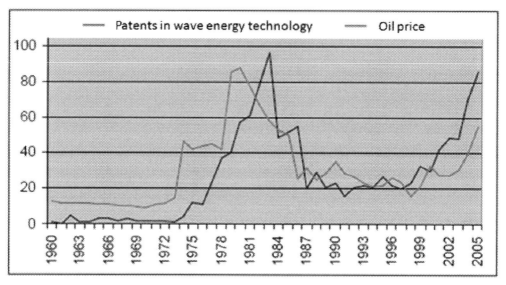

Figure 4.5. Oil Prices and Wave Energy Patents.

projects. Since the mid 1990s, the increasing levels of carbon dioxide emissions and climate change awareness has captured the attention of governments and people the world over, and, in turn, the generation of electricity from renewable sources has once again become an important area of research. In the last ten years, there has been a resurgent interest in the wave energy, particularly in Europe, as seen in the growing the number of patents illustrated in Figure 4.5 since that time. Today there are over one thousand patents relating to wave energy worldwide and an installed capacity of approximately two megawatts.

4.4.2.4. Wave Device Technology

There are a variety of devices and technologies available for the extraction of energy from waves, in contrast with the limited choices available for large wind turbines. The difference in technologies stem from different ways energy can be absorbed from waves and are also dependent on water depth and on location (shoreline, near-shore, or offshore).

There are certain differences in the way hydrodynamics affects floating wave energy converters and other similar types of bodies such as ships subjected to waves at sea. Though results and experiences from previous studies on hydrodynamics of ships carried out prior to the mid 1970s can be borrowed for application, the presence of a power takeoff mechanism (PTO) and the necessity to optimize power production are additional features which need to be considered in depth.

The obvious initial approach for dealing with theoretical developments of wave energy converters was to address the energy extraction from a regular sinusoidal wave by a floating body oscillating with a single degree of freedom with a linear PTO. The wave amplitudes were assumed to be small, enabling linearized equations of motion which facilitated the use of frequency domain analysis technique. The oscillating water columns were the first wave energy converters of their kind and were developed to full prototype stage for the purpose of energy extraction from sea waves, even prior to performance modeling and analysis studies.[33] This was primarily because the techniques from ship hydrodynamics were not applicable or easily transferrable to the study of oscillating body converters.

An oscillating water column (OWC) is made up of a chamber with an opening to the sea below the waterline. When waves approach the device, water is forced into the chamber which applies pressure to the water within the chamber. The wave action results in the captured water column within the device moving up and down like a piston which alternatively compresses and depressurizes the chamber forcing the air through an opening connected to a turbine. A low pressure Wells turbine is often used in this device a. it rotates in the same direction regardless of the air flow direction. One of the main advantages of the OWC device is its relative simplicity in design and robust construction. An example is the *W Limpet Device* by *WaveGen* which is a shore mounted OWC.[34]

An oscillating wave surge converter usually comprises a hinged deflector positioned perpendicular to the wave direction that moves back and forth exploiting the horizontal velocity of the wave. A well-known example of such a convertor is the *Oyster* device developed by Aquamarine Power, which is an oscillating device for deployment near shore on the seabed in water depth of ten to twelve meters with approximately 2 meters of the device exposed above the sea surface. The system uses multiple piston pumps to pump high-pressure seawater to the shore through subsea pipelines. The water is then used to generate electricity through a hydroelectric turbine. The cost of the pipe line is low because the device is deployed near shore. However, this decreases the efficiency of the device because a lot of the energy in the waves is depleted due to friction when the wave reaches shallow water.[35]

Point absorber devices have a small horizontal dimension comparative to the longer wavelength in which they operate. Point absorber devices utilize the rise and fall of the wave height at a single point. The devices can be floating structures that heave up and down on the sea surface or are submerged below the sea surface using the pressure differential. These devices are generally quite small and as such are not reliant on wave direction. For example, the *PowerBuoy*, developed by Ocean Power Technologies, involves a floating structure with one component relatively immobile, and a second component with movement driven by the wave motion. In essence it is a floating buoy contained within a fixed cylinder. The relative motion is used to drive an electrical generator through Faraday induction directly.[36] Many such units can easily be used in parallel to develop a wave farm.

Attenuators are long multi-segment floating structures oriented parallel to the direction of the wave front. The differing heights and force of the oncoming waves along the length of the device causes a flexing motion where the segments connect. This flexing is directly connected to hydraulic pumps or other converters. Attenuator devices have a relatively small area exposed to the face of the waves, enabling them to reduce the hydrodynamic forces of inertia, drag and slamming that have the potential to inflict significant damage to offshore devices. The *Pelamis*, designed by Ocean Power Delivery, is made up of four floating cylindrical pontoons connected via three hinged joints. The wave induced motion of these joints is resisted by hydraulic rams which pump high pressure oil through hydraulic motors via smoothing accumulators. The hydraulic motors drive electrical generators to produce electricity. Several devices can be connected together and linked to shore through a single seabed cable with a typical thirty megawatt installation. Such an installation would occupy a square kilometer of ocean and provide sufficient electricity for 20,000 homes (Figure 4.6).[37]

Figure 4.6. The *Pelamis* by Ocean Power Delivery, Ltd.

Overtopping devices have reservoirs that are used to capture sea water by impinging waves to levels above and have been tested for both onshore and floating offshore applications. The *Wave Dragon*, for example, is an offshore overtopping device. This device uses a pair of large curved deflectors that concentrate the waves toward a central raised reservoir which raises the effective wave height. Kaplan turbines are used to convert the low head of the water into mechanical energy. The turbines drive permanent magnet generators, thereby generating electricity on the same principal as conventional land based hydropower plants.[38]

4.4.2.5. Arrays, Model Testing, and Control

Generating power to feed the electricity grid from one single device will not be adequate and it is therefore preferable to have an array of devices. Hence, the hydrodynamic interaction between arrays of devices is crucially important. Studies were first made for systems of oscillating bodies.[39] Subsequent studies by Evans extended to systems of OWCs, however, as the number of devices in an array becomes large, the interaction becomes complex and approximate techniques such as multiple-scattering or the plane wave method may be applied.

For design and development of wave converters, either numerical modeling approaches or alternative physical modeling approaches, with wave tank testing, can be followed. Since the early pioneering model testing of wave energy converters in 1974 at the University of Edinburgh by Stephen Salter, significant progress has been made in experimental test studies.[40] More recently, experimental studies have progressed to the prototype development stage which has facilitated the need for larger-scale testing facilities.

As with wind and tidal energy, the extraction of energy using wave energy converters involves a number of conversion processes. It is essential that each of these processes is optimized under certain constraints in order to ensure that overall efficiency is optimized. Of particular significance in converter control is the hydrodynamic process of power absorption. Several phase-control research strategies have been proposed for simple PTO including device analysis when acted upon by irregular waves.[41]

4.4.2.6. Benefits and Disadvantages of Wave Power

The benefits of wave energy are undeniable but as with any technology at such an early stage of development there are a number of technical challenges that need to be overcome to fully realize the potential of, and most importantly, the commercial competiveness of wave power devices. Waves produce a slow (less than one hertz), random, and high-density oscillatory motion. Converting these characteristics into a useful motion to drive a generator capable of producing a quality output that will be accepted by the utility provider presents a considerable challenge. As waves vary in height and period, so does their respective energy level. In offshore locations, wave direction is highly variable and so wave devices have to be aligned accordingly. For point absorber devices such as the PowerBuoy this is less of an issue and in general becomes less of an issue the closer to the shoreline a device is installed. This is because the direction of wave travel becomes more uniformly predictable due to the refraction and reflection experienced as the water depth shallows and is in essence funneled toward the shoreline.

One of the principal disadvantages to the effective operation of wave energy converters is the environment in which they are placed. The irregular and highly unpredictable nature of the sea surface has an impact on the design of all devices. To operate efficiently, each device must be designed to operate for the most common wave levels. The device also has to be capable of withstanding the stresses induced by freak weather conditions and in the case of wave energy converters, freak waves. These conditions only occur very rarely but when they do, can deliver levels of power greater than 200 kW/m. This design requirement throws a very costly barrier in the way of developing wave energy converters as the device itself may only be rated to capture the energy from the most commonly occurring waves but has to be engineered to withstand the very high and destructive levels of power produced, albeit infrequently, by extreme weather events.

The capture of wave energy for electrical power generation is generally considered to have negligible environmental impacts. The exact nature and extent of any potential impact however remains uncertain as the technology is still in its infancy. The marine habitat could potentially be impacted depending on the nature of the device being installed, be they totally submerged structures, above sea level platforms or seabed mounted devices. Above sea floating platforms could potentially provide a resting platform for sea mammals as well as a nesting area for birds. The underwater surfaces of wave energy devices could provide for substrates or various biological systems. Changes made to the seabed for mounting devices and for the provision of submarine cables could also impose potentially negative effects on the local marine habitat. Offshore wave energy devices may be a source of conflict with commercial shipping and recreational boating. Careful consideration needs to be taken when selecting potential areas for the sitting of wave energy devices. Near shore devices have the potential to interfere with recreational activities as well as having a negative visual impact on the coastline. Any impact that may occur would also be very site specific and it is only with the development of large scale wave farms that the devices impact will become better defined.

As with any emerging technology, the goal is to eventually reduce the cost of wave energy generation. The barriers listed herein may largely be technical, but every time a barrier is overcome, wave energy generation becomes increasing economically viable and subsequently moves a step closer to widespread commercialization. Due to its variability, a 'wave climate' is difficult to measure. Wave buoys can give good estimates of the sea state but are expensive to maintain for long periods of wave climate estimation.[42] Resource assessment studies are essential when evaluating possible locations for a wave energy project, and site specific measurements and surveys are necessary before deciding on the final location for any wave farm.

In addition to the technical advantages/disadvantages, there are a number of legal incentives associated with wave energy which include the Public Utility Regulatory Policy Act (PURPA), state goals for renewable energy, renewable portfolio standards (RPS), and system benefit charges. In this context, research and development efforts are being sponsored by government agencies in Europe and Scandinavia. In the US there is little research due to lack of funding; although the US Navy, through its Office of Naval Research program, does provide some research funds. However, in general, funding levels are not currently adequate for commercially realizable projects. Although wave energy conversion technologies have significantly advanced during recent years, especially in Europe, most devices are still in the prototype-phase and there is need of technology improvements. In particular, the survivability and reliability of devices for offshore operation has still to be demonstrated. However, the combination of offshore wind, wave, and tidal energy devices which can use a common transport infrastructure provides a viable developmental route with the lead being taken by the construction of offshore wind farms.

4.4.3. Tidal Energy

4.4.3.1. Overview

Tidal energy is one of the more predictable and reliable renewable energy sources. The main objective with tidal energy is obviously to harness energy from the rise and fall of recurring tides. This may be achieved through design of tidal barrage systems. In recent years, innovative designs by companies such as Open Hydro have resulted in successful extraction of kinetic energy in tidal currents following the principles used in wind energy generation; and despite promise, this method is not yet mature.

Tidal movements are generated from the gravitational and centrifugal forces between the earth, moon and sun.[43] This results in regular rise and fall of the surface of the ocean. The causal forces are the gravitational force of the sun and moon on the earth and the centrifugal force produced by the rotation of the earth and moon about each other.[44] Because of the proximity of the moon to the earth, the gravitational force of the moon is 2.2 times larger than that of the sun. A heaving motion of water is created by the gravitational pull of the moon, which is greater on the side of the earth nearest the moon. In addition to this, the rotation of the earth-moon system, producing a centrifugal force, causes another heaving of water mass on the side of the earth furthest away from the moon. When a landmass is aligned with this earth-moon system, the water around the landmass is at high tide. Similarly, when the landmass is at ninety degrees to the earth-moon system, the water around it is at low tide. Therefore, each landmass is exposed to two high tides and two low tides during each period of rotation of the earth.[45] Since, in addition to earth spinning about its own axis, the moon also rotates around the earth; the resultant timing of the tides at a spatial point on the earth's surface varies, occurring approximately fifty minutes later each day.[46]

Tidal variations not only result in rise and fall of the ocean surface but can also lead to tidal currents. As is well known, tidal currents are experienced in coastal areas and in places where the seabed forces the water to flow through narrow channels. These currents flow in two directions; the current moving in the direction of the coast is known as the flood current and the current receding from the coast is known as the ebb current. These currents can also be used to generate electricity.

Tidal power generation facilities can be classified into two types: tidal barrages and tidal current turbines, using the potential and kinetic energy of the tides respectively.[47] Power generated from tidal barrages and tidal currents are also called *tidal range power* and *tidal stream power*, respectively.

4.4.3.2. Tidal Range Power

Tidal range power or tidal barrages utilize the potential energy of the tides. A tidal barrage is typically a dam, built across a bay or estuary that experiences a tidal range in excess of five meters.[48] Electricity generation from tidal barrages uses the same principles applying to hydroelectric generation, with the exception that tidal currents flow in both directions. A typical tidal barrage consists of turbines, sluice gates, embankments, and ship locks, and can have either a single-basin or double-basin system.

Technologies relating to generation of power using tidal barrages are both well developed and reliable. Historically, small mechanical devices were powered by tidal energy in medieval England and China. There are several sites around the world which can be considered suitable for tidal range power generation and exploitation. Among these, the 240 MWe *La Rance* system at an estuary in the Gulf of St Malo in Brittany, France, has operated reliably since 1967. This project has demonstrated the feasibility of tidal range power for large scale operation. Other notable sites include the Severn estuary in the UK, the Bay of Fundy on the eastern boundary between Canada and the United States; Mezeh Bay and Tugar Bay in Russia; and the Wash, the Mersey, the Solway Firth, Morecambe Bay, and the Humber Estuary in the UK. In addition, a number of smaller sites with potential include Garlolim Bay in Korea, the Gulf of Kachchh in India, Secure Bay in Australia, and Sao Luis in Brazil. It should also be noted, however, that environmental restrictions have limited the number of developments of tidal range power technology.

The total potential contribution of tidal water to the generation of energy is roughly 3000 GW. This quantity includes approximately 1000 GW available from shallow water depths which are accessible for large civil engineering works, with estimated generation potential of 120 GW (about twelve percent of near-shoreline and ten percent of the total world hydropower). This should be of significant interest to many countries, including the UK where the potential exists to provide twenty-five percent of total power requirements through the harnessing of tidal energy. In Canada, the Bay of Fundy (New Brunswick to Nova Scotia) is considered capable of producing 30 GW of tidal power. Unfortunately, not many tidal power plants have been constructed to this point, primarily owing of high construction cost, relatively short-term financial benefits, and the need for engagement by all stakeholders in developing long-term renewable energy strategies for tidal energy. The high cost of construction of tidal barrages is a restrictive barrier to the development of tidal range power. A high initial investment is required owing to the scale of large tidal barrage construction projects and ensuring the dam can sustain the tidal water load it will be subjected to. Nevertheless, design of tidal turbines has reached mature status and the maintenance costs accruing to existing barrages during lifetime operation, is not excessive.

4.4.3.3. Tidal Current Turbines

Like wind, tidal current technology utilizes the kinetic energy in flowing water to generate electricity; in contrast, there are differences in the operating conditions of the two. A striking difference is in the density of air and water; at typical ambient conditions, water is about 832 times denser than air. Water flow speed on the other hand is generally much slower to that of air.[49]

The two most common methods of extraction of tidal energy are based on the type of turbine used, whether horizontal or vertical.[50]
- Horizontal axis tidal current turbines: The turbine blades rotate about a horizontal axis which is parallel to the direction of the flow of water.
- Vertical axis tidal current turbines: The turbine blades rotate about a vertical axis which is perpendicular to the direction of the flow of water.[51]

Several sites worldwide offer potential for the exploitation of energy from tidal current. However, factors such as technology status, water depth, wave exposure and seabed exposure require consideration in assessing the practical energy resource availability at a particular site location. In general, tidal current sites with water flow speed greater than 2.5 m/s are considered to have significant practical energy resource and are economically viable.[52] The ideal locations for harnessing tidal energy from currents are where narrow straits occur between land masses or are adjacent to headlands. Locations with major potential include the Arctic Ocean, the English Channel, the Irish Sea, Skagerrak-Kattegat, the Hebrides, the Gulf of Mexico, the Gulf of St. Lawrence, the Bay of Fundy, the Amazon River, the Rio de la Plata, the Straits of Magellan, Gibraltar, Messina, Sicily, and the Bosporus/Istanbul Strait.[53] Indeed, a significant resource of tidal energy is situated in the Irish Sea, other regions of significance being off the west coast of Canada and the African East coast. In each case, a large increase in an accessible resource can be expected when slower current velocities can be exploited.

Tidal turbine technology is still at an early stage of development. The focus of recent developments has been on reliability with developments of both scaled-down models and full-scale prototypes.[54] The first dedicated test center, the European Marine Energy Centre (EMEC), based in Orkney, Scotland has been operational since May 2005, with a principal focus on the testing of tidal current turbines. Challenges presenting include installation issues, maintenance, electricity transmission, loading conditions and environmental impact. Installation and deployment challenges center on installation and foundation design, mooring systems design, and corrosion prevention strategies. These challenges are not unique to tidal energy installations and experience gained from other offshore marine projects will be invaluable.

Successful projects in recent years include the Open Hydro tidal energy turbine. A version of the Open Hydro device is illustrated in Figure 4.7, in which the turbine is mounted on the sea bed.

Figure 4.7. The Open Hydro Tidal Device.

Figure 4.8. Illustration of the Kite Device.

This innovative design incorporates a stator and a shaftless rotor housed for rotation within the stator, with the stator defining a channel in which the rotor is retained.

A further example of a competitive technology is the Kite Device illustrated in Figure 4.8 which uses a wing to support a turbine that is tied to the sea bed. As the tidal current flows, the wing generates lift allowing the same current to drive the turbine. The first sea going prototype of this device was prototyped in 2011 and trialed off the coast off Northern Ireland with funding by the UK Carbon Trust with the vision of producing a 10 MW array to follow in 2015.

4.4.3.4. Future Potential of Tidal Current Turbines

Extraction of marine energy using tidal current turbines has genuine potential. Several companies in the recent past have developed operational demonstration models, in both full-scale and down-scale prototype designs. Examples of scale-models include Nereus and Solon Tidal Turbines, Evopod Tidal Turbine, Gorlov Helical Turbine, TidEl Stream Generator and Stingray Tidal Energy Converter. Most devices installed are currently under test and it is expected that on continued successful testing and operation, full-scale tidal farms will soon become a commercial reality. Some prototypes have been built and tested in harsh climatic conditions.

The SeaGen and Seaflow, Open Centre Turbine, Tidal Stream Turbine and Free Flow Turbines are examples of full-scale operational tidal current turbines which are successfully generating electricity. All demonstration units operate with a horizontal axis rotor shaft. With the rapid advancement of technology, promising sites have the potential to become economically viable.

4.5. Geothermal Energy

4.5.1 Overview

Geothermal energy is the energy contained in the earth's interior in the form of heat. The inner core of the earth reaches a maximum temperature of about 4000°C. The origin of this heat is associated with the internal structure of planet Earth and the physical processes occurring therein. To extract this large quantity of heat it is necessary to have some carrier to transfer the heat to an accessible depth below the earth's surface. Generally, the heat is transferred from depth to sub-surface regions mostly through the solid submarine and land surface mainly by conduction (geothermal heat) and occasionally by active convective currents of carrier geothermal fluids such as molten magma or heated water. The heated water is essentially rainwater that has penetrated into the earth's crust from the recharge areas. The water gets heated through contact with hot rocks, and accumulates in aquifers, occasionally at high pressures and temperatures of up to above 300°C. These aquifers (reservoirs) are the essential parts of most geothermal fields.[55]

The average geothermal heat flow at the earth's surface is just 0.06 W/m^2 and hence it is not generally noticeable. This is because the temperature of rocks increases with depth, with a geothermal temperature gradient of 30°C/km. This continuous heat current is trivial when compared to other renewable supplies in the above-surface environment, that in total average about 500 W/m^2. Also, geothermal energy is unevenly distributed, seldom concentrated, and often at depths too great to be exploited industrially. However, at certain specific locations, increased temperature gradients occur, indicating significant geothermal resources. These may be commercially exploited if available within depths of approximately 5 km at fluxes of 10–20 W/m^2. Production is expected to be about 100 MW thermal km^2 in commercial supplies sustained over a period of twenty years of operation.

There are areas in earth's crust not far from the surface, that is, at a depth from the surface on the order of a few kilometers, where magma bodies present in fluid state or are undergoing cooling in the process of solidification, resulting in release of heat. Also, there may be other areas where magma is not present but due to certain geologic conditions the thermal gradient has reached an anomalously high value. In such cases, if the areas are accessible by drilling boreholes from the surface it may be commercially feasible for extracting energy.

The use of geothermal energy for electric power generation from turbines requires heat energy with higher temperature. This is occasionally the case for available geothermal energy with temperatures over 150°C. However, in general geothermal heat is of low grade and typically possesses temperatures around 50–70°C. Under such circumstances, it may better suit to directly heat buildings or it can also be used for preheating of other conventional high temperature energy facilities. For example, heat from near-surface ground or lakes is frequently used for heat pumps. Several countries have established geothermal electric power projects, including Italy, New Zealand, and the US.

Geothermal energy was first harnessed on a large scale basis in the early years of the twentieth century, with applications ranging from space heating to electricity generation. Examples include electric power generation initiated in Prince Piero Ginori Conti in 1904; geothermal steam at Larderello, Tuscany in 1913; and the first large scale municipal district heating service in Iceland in 1930. Despite the relative rarity of hydrothermal sites, it is estimated that up to six GWe in the US and seventy-two GWe worldwide could be produced with current technologies at known hydrothermal sites.[56]

Hot dry rock (HDR) is another geologic resource which, unlike hydrothermal energy, is found in abundance. These geologic structures occur beneath a large proportion of the world's landmass and exist at temperatures of 200°C. Hence, these structures will be suitable for electricity generation if energy can be extracted through use of appropriate technologies. Should more advanced extraction technologies be developed, it is estimated that energy available from HDR resources could result in electricity generation capacity of 19 GWe in the US and 138 GWe worldwide. Such technology, however, is not currently available.

Geothermal mapping has been the key to the development of geothermal energy generation worldwide. This information has been obtained through mining, oil exploration and geological surveys. Though deep drill surveys are commonly carried out reaching to a depth of six kilometers, technologies to drill boreholes to depths of fifteen kilometers or more are currently available, and will likely be exploited going forward. Energy production on the scale of hundreds of megawatts has been possible over recent decades and the use of geothermal energy for both the heating of buildings and for electric power production, is rapidly increasing.

4.5.2. Demand and Supply Management

One of the restrictions of energy produced from geothermal sources relates to the fact that heat cannot be transported easily from one point to another. In fact, distribution of heat over distances greater than about thirty kilometers is difficult. Hence, it is preferable to consume the energy generated close to the point of generation. Such concentrated usage is possible; for example in cold climates, household and business district heating schemes generate sufficient loads in regions of high population density. By way of an example, for a region with a population density greater than 350 people per square kilometer (equating to more 100 premises per square kilometer), a 100 MWth geothermal plant might serve a twenty kilometer square area with energy of approximately two kilowatts per premise.[57] Such geothermal facilities are operational in Iceland and on a smaller scale in New Zealand. Other applications are for glasshouse heating (one example is an installation in northern Europe at 60 MWth/km^2), fish farming, food drying, and factory processes.[58]

Geothermal electricity generation becomes feasible with source temperatures approaching 150°C and becomes more attractive still if the temperature from the geothermal source is in excess of 300°C. If it is feasible to generate electricity then it can be supplied to the grid, complementing energy supply on a more regional or even global scale. Heat rejected from electricity generation could be used in a combined heat and power (CHP) mode.

4.5.3. Cost of Geothermal Energy

The primary cost in a geothermal energy project is capital cost, principally for drilling of boreholes. Costs of drilling increase exponentially with depth. Since temperature increases with depth, and the value of the energy increases with temperature, there exists an economic optimum borehole depth of approximately 5 km. As a result, the scale of the energy supply output is usually greater than 100 MW (supply of electricity and heat for high temperatures; heat only for low temperatures). Reinjection of the partially cooled water from the heat exchanger can be used as a mechanism to increase the total amount of heat extracted from the geothermal source. This has the added benefit of disposing of water with increased solutes concentration (of about 25 kg/m³) and other pollutants that may be present. However, there is an extra cost associated with this process.

4.5.4. Social and Environmental Impacts

Geothermal energy is safe, reliable and competitive in terms of cost as compared to other conventional sources. Security of supply is an added value as power can be supplied continuously at full rating without any intermittency. In addition, Operating and Maintenance (O&M) costs are moderate.

The extraction of geothermal energy may have certain drawbacks with respect to its environmental impact. First, is the possibility of subsidence affecting local buildings due to removal of the hot water from the ground. Some incidents have occurred in association with the operation of the 140 MWe Wairakei power station in New Zealand.[59] This problem can be addressed by re-injecting some of the output water flow into the area. Other effects relate to the impact the geothermal plant may have on the intensity of some of the natural geysers in the vicinity of the plant, though most may not be affected significantly. These negative impacts have hindered the growth of geothermal power in Japan. Geothermal systems also emit carbon dioxide, however the level is much less than that associated with a conventional thermal power plant.

4.6. Impact of Renewable Technologies on Electricity Grid Developments

The large-scale integration of power from newly emerging sources such as wind, marine and solar on national and cross-national networks presents key technical, financial and regulatory challenges. In meeting future energy requirements it is expected there will be unprecedented increases on power demand, for instance due to electrification of transport and growth in information systems. Consumers will likely have higher expectations for both quality and quantity of service.

The European Technology Platform for Electricity Networks of the Future, also called SmartGrids ETP, is the key European forum for the crystallization of policy and technology research and development pathways for the smart grids sector. In the US, the Energy Independence and Security Act established a framework which enabled support for matching programs to states, utilities, and consumers to build smart grid capabilities, and to create a Grid Modernization Commission to assess the benefits of demand response and to recommend required protocol standards.

With the emergence of advanced power systems (*smart grids*), renewable energy, smart meters and novel storage technologies, accelerated modernization of the electricity infrastructure has commenced. It is difficult to predict what the power infrastructure of the future will look like. What is acknowledged is that it will likely be not a single network, but a network of networks, a network of smart grids. This result will present difficult network challenges with requirement to manage uncertainty on both the demand and generation side. As with existing power networks, the advanced power networks will be required to balance efficiency and reliability and maintain quality of power to customers.

Smart energy may be considered a structure where energy is part of an information vector that includes, but is not limited to, energy quantity, unit price, and exogenous costs such as emissions, and, a variety of generation and load characteristics. If energy denotes the capacity for mechanical work, smart energy may be thought of as the capacity for decision-making regarding energy. Smart energy 'agents' may have attributes that pertain not only to physical quantities of energy (e.g. kilowatt-hours) but also economic, environmental, geographic and temporal characteristics such as price and emissions.[60]

Smart energy connectivity promises to substantially undergird regional trade and development. One such political initiative, "Connecting the Americas 2022," was announced in 2012 by leaders of the Western Hemisphere and aims to increase access to reliable, clean, and affordable electricity as well as to provide new opportunities for regions with electricity surplus to exchange with regions experiencing deficit.[61]

High voltage direct current (HVDC) transmission will feature prominently in the creation of future grids. HVDC results in lower transmission losses to alternating current networks. HVDC links are long established and well proven in transmission system island grids such as between Norway and Netherlands (NorNed); a 700 MW link installed in 2008. Plans for further offshore links in the North Sea are well advanced. Under target specific need 4 of the Union of the Mediterranean chapter of cooperation, *Alternative Energies: Solar Plan*, a pathway for facilitation of a new European Supergrid was drawn.[62] The intent is to facilitate importation of large amounts of concentrating power into Europe from North Africa and the Middle East. Organizations such as the German based DESERTEC Foundation (formed in 2009) aim to create a global renewable energy plan by harnessing energy from locations where renewable energy is plentiful and transferring this energy through HVDV transmission lines, for example solar energy from North African and Middle Eastern desert locations.

There is enormous potential for countries of the Middle Eastern region to develop such solar network grids which will benefit internal economies through exportation of clean energy to Europe and beyond. Saudi Arabia is planning to install 40 GW of solar energy capacity by 2030. Masdar City in the UAE is a planned city under development which will rely entirely on solar energy and other renewable energy resources. Other countries in the region are also exploring possibilities for energy provision by solar, wind, and ocean power. In the Sultanate of Oman, the national Power and Water Procurement Company is carrying out preliminary research in solar data acquisition in preparation for a planned 200 MW concentrated solar plant which is expected to be operational by 2018. These initiatives demonstrate the seedlings for emergence of a cultural shift toward green energy projects in the region. A future scenario may result in a move from excessive reliance on oil and gas to widespread deployments in solar energy.

Notes

1. REN21, *Renewables 2013 Global Status Report* (Paris: REN21 Secretariat, 2013).
2. Palmer Cosslett Putnam, *Power from the Wind* (New York: Van Nostrand Rheinhold, 1948).
3. Edward William Golding, *The Generation of Electricity by Wind Power* (London: E. & F. N. Spon, 1976).

4. Tony Burton, David Sharpe, Nick Jenkins, and Ervin Bossanyi, *Wind Energy Handbook* (Chichester: John Wiley, 2001), http://dx.doi.org/10.1002/0470846062.
5. Albert Betz, *Introduction to the Theory of Flow Machines* (Oxford: Pergamon Press, 1966).
6. J. C. Kaimal, J. C. Wyngaard, Y. Izumi, and O. R. Cote. "Spectral Characteristics of Surface Layer Turbulence," *Quarterly Journal of the Royal Meteorological Society* 98, no. 417 (1972): 563–589, http://dx.doi.org/10.1002/qj.49709841707; and J. C. Kaimal, "Turbulence Spectra, Length Scales and Structural Parameters in the Stable Surface Layer," *Boundary-Layer Meteorology* 4, no. 1–4 (1973): 289–309, http://dx.doi.org/10.1007/BF02265239.
7. Madjid Karimirad, *Stochastic Dynamic Response Analysis of Spar-Type Wind Turbines with Catenary or Taut Mooring Systems*, doctoral dissertation, Norwegian University of Science and Technology, Norway, 2011.
8. ISSC Specialist Committee, *Ocean Wind and Wave Energy Utilization*, committee report at the 17th International Ship and Offshore Structures Congress, Seoul, Korea, August 2009.
9. Muhammad Bilal Waris and Takeshi Ishihara, "Dynamic Response Analysis of Floating Offshore Wind Turbine with Different Types of Heave Plates and Mooring Systems by Using a Fully Nonlinear Model," *Coupled System Mechanics* 1, no. 3 (2012): 1–22.
10. Nigel Barltrop, "Multiple Floating Offshore Wind Farm," *Wind Engineering* 17, no. 4 (1993): 183–188; A. Henderson and M. Patel, "Rigid-Body Motion of a Floating Offshore Wind Farm," *International Journal of Ambient Energy* 19, no. 3 (1998): 167–180, http://dx.doi.org/10.1080/01430750.1998.9675699; Pham Van Phuc and Takeshi Ishihara, A Study on the Dynamic Response of a Semi-Submersible Floating Offshore Wind Turbine System Part 2: Numerical Simulation, *Proceedings of the International Conference of Wind Engineering*, Cairns, Australia, 2007.
11. P. Bertacchi, A. Di Monaco, M. de Gerloni, and G. Ferranti, "ELOMAR: A Moored Platform For Wind Turbines," *Wind Engineering* 18, no. 4 (1994): 189–198; K. C. Tong, "Technical and Economic Aspects of a Floating Offshore Wind Farm," *Journal of Wind Engineering and Industrial Aerodynamics* 74–76 (1998): 399–410, http://dx.doi.org/10.1016/S0167-6105(98)00036-1; J. M. Jonkman, *Dynamics Modeling and Loads Analysis of an Offshore Floating Wind Turbine (Technical Report NREL/TP-500–41958)* (Golden, CO: National Renewable Energy Laboratory, 2007), http://dx.doi.org/10.2172/921803.
12. Jonkman, *Dynamics Modeling*.
13. Ibid.
14. International Energy Agency, "Social Acceptance of Wind Energy Projects," In *Expert Group Summary on Recommended Practices* (Paris: IEA, 2013), 2.
15. Ibid.
16. European Wind Energy Association, *Annual Report 2010: Powering the Energy Debate* (Brussels, Belgium: EWEA, 2011).
17. Georges Roux, *Ancient Iraq* (New York: Penguin, 1993).
18. R. A. Tomlinson, "The Perachora Waterworks: Addenda," *Annual of the British School at Athens* 71 (1976): 147–148, http://dx.doi.org/10.1017/S0068245400005864.
19. Örjan Wikander, "The Water-Mill," in *Handbook of Ancient Water Technology*, ed. Örjan Wikander, 371–400 (Leiden: Brill, 2000).
20. Kerry Rittich, Klaus Grewe, and P. Kessener, "A Relief of a Water-powered Stone Saw Mill on a Sarcophagus at Hierapolis and its Implications," *Journal of Roman Archaeology* 20 (2007): 138–164.
21. Thomas J. Blalock, "Alternating Current Electrification, 1886," *IEEE Global History Network*, last modified October 2, 2004, http://www.ieeeghn.org/wiki/index.php/Milestones:Alternating_Current_Electrification%2C_1886.
22. Gopal Nath Tiwari and Rajeev Kumar Mishra, *Advanced Renewable Energy Sources* (Cambridge: Royal Society of Chemistry Publishing, 2012).
23. Worldwatch Institute, "Use and Capacity of Global Hydropower Increases," *World Watch* (January 2012).

24. International Energy Agency Implementing Agreement on Ocean Energy Systems, *Annual Report 2008 (IEA-OES Document A08)*, ed. by A. Brito-Melo and G. Bhuyan (Lisbon, Portugal: IEA-OES, 2009).
25. Roger Bedard, Mirko Previsic, George Hagerman, Brian Polagye, Walt Musial, Justin Klure et al., "North American Ocean Energy Status: March 2007," In *Proceedings of the 7th European Wave Tidal Energy Conference, Porto, Portugal* (Southampton, UK: EWTEC, 2007); M. Previsic, A. Moreno, R. Bedard, B. Polagye, C. Collar, D. Lockard, et al., "Hydrokinetic Energy in the United States: Resources, Challenges and Opportunities," in *Proceedings of 8th European Wave Tidal Energy Conference* 2009, 76–84.
26. Wave energy resource characterization is reviewed in Stephen Barstow, Gunnar Mørk, Denis. Mollison, and João Cruz, "The Wave Energy Resource," in *Ocean Wave Energy: Current Status and Future Perspectives*, ed. by João Cruz, 93–132 (Berlin: Springer, 2008), http://dx.doi.org/10.1007/978-3-540-74895-3_4; and D. V. Evans, "Wave Power Absorption by Systems of Oscillating Surface Pressure Distributions," *Journal of Fluid Mechanics* 114 (1982): 481–499, http://dx.doi.org/10.1017/S0022112082000263.
27. Barstow et al., "Wave Energy Resource."
28. Jonathan M. Blackledge, "A Generalized Model for the Evolution of Low Frequency Freak Waves, IANEG," *International Journal of Applied Mathematics* 41 (2010): 33–35.
29. Ibid.
30. Ibid.
31. See the Wave and Tidal Energy Resource database maintained by the Sustainable Energy Authority of Ireland at http://www.seai.ie/Renewables/Ocean_Energy/Ocean_Energy_Information_Research/Irelands_Wave_and_Tidal_Energy_Resources/.
32. David Ross, *Power From the Waves* (Oxford: Oxford University Press, 1995).
33. Evans, "Wave Power Absorption."
34. For more information about Wavegen devices, visit the corporate website (http://voith.com/en/products-services/hydro-power/ocean-energies/wave-power-plants-590.html).
35. For more information on the Oyster technology, see the Aquamarine Power corporate website (http://www.aquamarinepower.com/).
36. For more information on the PowerBuoy device, see the Ocean Power Technologies corporate website (http://www.oceanpowertechnologies.com/).
37. For more information on Pelamis wave power, see the corporate website (http://www.pelamiswave.com/).
38. For more information on the Wave Dragon device, see the corporate website (http://www.wavedragon.net/).
39. Kjell Budal, "Theory of Absorption of Wave Power by a System of Interacting Bodies," *Journal of Ship Research* 21 (1977): 248–53; Johannes Falnes and Kjell Budal, "Wave Power Conversion by Point Absorbers," *Norwegian Maritime Research* 6, no. 4 (1978): 2–11; D. V. Evans, "Some Theoretical Aspects of Three Dimensional Wave Energy Absorbers," In *Proceedings of 1st Symposium on Wave Energy Utilization, Gothenberg, Sweden*, ed. Karl-Gustav Jansson, Johannes K. Lunde, and Thomas Rindby, 77–113 (Gothenberg: Chalmers University of Technology, 1980).
40. Jamie Taylor maintains information on the Edinburgh Wave Power Group at the University of Edinburgh online at http://www.mech.ed.ac.uk/research/wavepower/.
41. See Johannes Falnes, "Optimum Control of Oscillation of Wave-Energy Converters," *International Journal of Offshore and Polar Engineering* 12 (2002): 147–55, for a review.
42. Cruz, João, ed., *Ocean Wave Energy: Current Status and Future Perspectives* (Heidelberg, Germany: Springer, 2008).
43. Alan Owen, "Tidal Current Energy: Origins and Challenges," In *Future Energy: Improved, Sustainable and Clean Options for Our Planet*, ed. Trevor M. Letcher, 111–128 (Oxford: Elsevier, 2008), http://dx.doi.org/10.1016/B978-0-08-054808-1.00007-7.
44. Rajat Mazumder and Makoto Arima, "Tidal Rhythmites and Their Implications," *Earth-Science Reviews* 69, no. 1–2 (2005): 79–95, http://dx.doi.org/10.1016/j.earscirev.2004.07.004.

45. J. A. Clarke, G. Connor, A. D. Grant, and C. M. Johnstone, "Regulating the Output Characteristics of Tidal Current Power Stations to Facilitate Better Base Load Matching Over the Lunar Cycle," *Renewable Energy* 31, no. 2 (2006): 173–180, http://dx.doi.org/10.1016/j.renene.2005.08.024.
46. Godfrey Boyle, *Renewable Energy: Power for a Sustainable Future*, 2nd ed. (Oxford: Oxford University Press, 2004).
47. George Lemonis, "Wave and Tidal Energy Conversion," in *Encyclopedia of Energy*, ed. J. Cleveland Cutler, 385–396 (New York: Elsevier, 2004), http://dx.doi.org/10.1016/B0-12-176480-X/00344-2.
48. M. Sathiamoorthy and S. D. Probert, "The Integrated Severn Barrage Complex: Harnessing Tidal, Wave and Wind Power," *Applied Energy* 49, no. 1 (1994): 17–46, http://dx.doi.org/10.1016/0306-2619(94)90055-8.
49. Fergal O'Rourke, Fergal Boyle, and Anthony Reynolds, "Renewable Energy Resources and Technologies Applicable to Ireland," *Renewable & Sustainable Energy Reviews* 13, no. 8 (2009): 1975–84; and I. G. Bryden, T. Grinsted, and G. T. Melville, "Assessing the Potential of a Simple Tidal Channel to Deliver Useful Energy," *Applied Ocean Research* 26, no. 5 (2004): 198–204, http://dx.doi.org/10.1016/j.apor.2005.04.001.
50. I. G. Bryden, S. Naik, P. Fraenkel, and C. R. Bullen, "Matching Tidal Current Plants to Local Flow Conditions," *Energy* 23, no. 9 (1998): 699–709, http://dx.doi.org/10.1016/S0360-5442(98)00021-8.
51. S. Kiho, M. Shiono, and K. Suzuki, "The Power Generation from Tidal Currents by Darrieus Turbine," *Renewable Energy* 9, no. 1–4 (1996): 1242–1245, http://dx.doi.org/10.1016/0960-1481(96)88501-6.
52. Ian G. Bryden and Scott J. Couch, "ME1—Marine Energy Extraction: Tidal Resource Analysis," *Renewable Energy* 31, no. 2 (2006): 133–139, http://dx.doi.org/10.1016/j.renene.2005.08.012; I. G. Bryden, and D. M. Macfarlane, "The Utilization of Short Term Energy Storage with Tidal Current Generation Systems," *Energy* 25, no. 9 (2000): 893–907, http://dx.doi.org/10.1016/S0360-5442(00)00020-7; and W. M. J. Batten, A. S. Bahaj, A. F. Molland, and J. R. Chaplin, "Experimentally Validated Numerical Method for the Hydrodynamic Design of Horizontal Axis Tidal Turbines," *Ocean Engineering* 34, no. 7 (2007): 1013–1020, http://dx.doi.org/10.1016/j.oceaneng.2006.04.008.
53. Roger H. Charlier, "A 'Sleeper' Awakes: Tidal Current Power," *Renewable & Sustainable Energy Reviews* 7, no. 6 (2003): 515–529, http://dx.doi.org/10.1016/S1364-0321(03)00079-0.
54. Robert Gross, "Technologies and Innovation for System Change in the UK: Status, Prospects and System Requirements of Some Leading Renewable Energy Options," *Energy Policy* 32, no. 17 (2004): 1905–1919, http://dx.doi.org/10.1016/j.enpol.2004.03.017; Benoit Dal Ferro, "Wave and Tidal Energy: Its Emergence and the Challenges It Faces," *Refocus* 7, no. 3 (2006): 46–48, http://dx.doi.org/10.1016/S1471-0846(06)70574-1; Adam Westwood, "Ocean Power: Wave and Tidal Energy Review," *Refocus* 5, no. 5 (2004): 50–55, http://dx.doi.org/10.1016/S1471-0846(04)00226-4.
55. Enrico Barbier, "Geothermal Energy Technology and Current Status: An Overview," *Renewable & Sustainable Energy Reviews* 6, no. 1–2 (2002): 3–65. http://dx.doi.org/10.1016/S1364-0321(02)00002-3.
56. L. McLarty, P. Grabowski, D. Etingh, and A. Robertson-Tait, "Enhanced Geothermal Systems R&D in the United States," In: *Proceedings of the World Geothermal Congress*, Kyushu, Japan. June 10, 2000.
57. John Twidell and Tony Weir, *Renewable Energy Resources*, 2nd ed. (London: Taylor and Francis, 2006).
58. John W. Lund, Derek H. Freeston, and Tonya L. Boyd, "Direct Application of Geothermal Energy 2005: Worldwide Review," *Geothermics* 34, no. 6 (2005): 691–727, http://dx.doi.org/10.1016/j.geothermics.2005.09.003; John W. Lund, Derek H. Freeston, and Tonya L. Boyd, "Direct Utilization of Geothermal Energy 2010: Worldwide Review," in *Proceedings of the World Geothermal Congress 2010*, Bali, Indonesia, 25–29 April, 2010.
59. Twidell and Weir, *Renewable Energy Resources*.
60. Rong Gao, George Tsatsaronis, and Lefteri H. Tsoukalas, *Smart Energy: Principles and Implementation* (New York: Springer, 2012).
61. For more information see the US State Department's fact sheet regarding the "Connecting the Americas" program at http://www.state.gov/r/pa/prs/ps/2012/04/187875.htm.

62. The 'Union of the Mediterranean' (UfM), comprising 43 countries from Europe and the Mediterranean Basin (27 EU member states and 16 partner countries from North Africa, the Middle East and the Balkan States), was launched in July 2008 to promote stability and prosperity throughout the Mediterranean region. In addition to chapters in Politics and Security, Economics and Trade, Socio-Cultural, and Justice and Interior Affairs, six projects which target specific needs to the region include 1) De-pollution of the Mediterranean, 2) Maritime and Land Highways, 3) Civil Protection, 4) Alternative Energies: Mediterranean Solar Plan, 5) Higher Education and Research: Euro-Mediterranean University, and 6) The Mediterranean Business Development Initiative.

Bibliography

Barbier, Enrico. "Geothermal Energy Technology and Current Status: An Overview." *Renewable & Sustainable Energy Reviews* 6, no. 1-2 (2002): 3–65. http://dx.doi.org/10.1016/S1364-0321(02)00002-3.

Barltrop, Nigel. "Multiple Floating Offshore Wind Farm." *Wind Engineering* 17, no. 4 (1993): 183–188.

Barstow, Stephen, Gunnar Mørk, Denis Mollison, and João Cruz. "The Wave Energy Resource." In *Ocean Wave Energy: Current Status and Future Perspectives*, edited by João Cruz, 93–132. Berlin: Springer, 2008. http://dx.doi.org/10.1007/978-3-540-74895-3_4.

Batten, W. M. J., A. S. Bahaj, A. F. Molland, and J. R. Chaplin. "Experimentally Validated Numerical Method for the Hydrodynamic Design of Horizontal Axis Tidal Turbines." *Ocean Engineering* 34, no. 7 (2007): 1013–1020. http://dx.doi.org/10.1016/j.oceaneng.2006.04.008.

Bedard, Roger, Mirko Previsic, George Hagerman, Brian Polagye, Walt Musial, Justin Klure et al. 2007. "North American Ocean Energy Status: March 2007." In *Proceedings of the 7th European Wave Tidal Energy Conference, Porto, Portugal*. Southampton, UK: EWTEC, 2007.

Bertacchi, P., A. Di Monaco, M. de Gerloni, and G. Ferranti. "ELOMAR: A Moored Platform For Wind Turbines." *Wind Engineering* 18, no. 4 (1994): 189–198.

Betz, Albert. *Introduction to the Theory of Flow Machines*. Oxford: Pergamon Press, 1966.

Blackledge, Jonathan M. "A Generalized Model for the Evolution of Low Frequency Freak Waves, IANEG." *International Journal of Applied Mathematics* 41 (2010): 33–35.

Blalock, Thomas J. "Alternating Current Electrification, 1886." *IEEE Global History Network*. Last modified October 2, 2004. http://www.ieeeghn.org/wiki/index.php/Milestones:Alternating_Current_Electrification%2C_1886.

Boyle, Godfrey. *Renewable Energy: Power for a Sustainable Future*. 2nd ed. Oxford: Oxford University Press, 2004.

Bryden, Ian G., and Scott J. Couch. "ME1—Marine Energy Extraction: Tidal Resource Analysis." *Renewable Energy* 31, no. 2 (2006): 133–139. http://dx.doi.org/10.1016/j.renene.2005.08.012.

Bryden, I. G., T. Grinsted, and G. T. Melville. "Assessing the Potential of a Simple Tidal Channel to Deliver Useful Energy." *Applied Ocean Research* 26, no. 5 (2004): 198–204. http://dx.doi.org/10.1016/j.apor.2005.04.001.

Bryden, I. G., and D. M. Macfarlane. "The Utilization of Short Term Energy Storage with Tidal Current Generation Systems." *Energy* 25, no. 9 (2000): 893–907. http://dx.doi.org/10.1016/S0360-5442(00)00020-7.

Bryden, I. G., S. Naik, P. Fraenkel, and C. R. Bullen. "Matching Tidal Current Plants to Local Flow Conditions." *Energy* 23, no. 9 (1998): 699–709. http://dx.doi.org/10.1016/S0360-5442(98)00021-8.

Budal, Kjell. "Theory of Absorption of Wave Power by a System of Interacting Bodies." *Journal of Ship Research* 21 (1977): 248–53.

Burton, Tony, David Sharpe, Nick Jenkins, and Ervin Bossanyi. *Wind Energy Handbook*. Chichester: John Wiley, 2001. http://dx.doi.org/10.1002/0470846062.

Charlier, Roger H. "A 'Sleeper' Awakes: Tidal Current Power." *Renewable & Sustainable Energy Reviews* 7, no. 6 (2003): 515–529. http://dx.doi.org/10.1016/S1364-0321(03)00079-0.

Clarke, J. A., G. Connor, A. D. Grant, and C. M. Johnstone. "Regulating the Output Characteristics of Tidal Current Power Stations to Facilitate Better Base Load Matching Over the Lunar Cycle." *Renewable Energy* 31, no. 2 (2006): 173–180. http://dx.doi.org/10.1016/j.renene.2005.08.024.

Cruz, João, ed. *Ocean Wave Energy: Current Status and Future Perspectives.* Heidelberg, Germany: Springer, 2008.

European Wind Energy Association. *Annual Report 2010: Powering the Energy Debate.* Brussels, Belgium: EWEA, 2011.

Evans, D. V. "Some Theoretical Aspects of Three Dimensional Wave Energy Absorbers." In *Proceedings of 1st Symposium on Wave Energy Utilization, Gothenberg, Sweden,* edited by Karl-Gustav Jansson, Johannes K. Lunde, and Thomas Rindby, 77–113. Gothenberg, Sweden: Chalmers University of Technology, 1980.

Evans, D. V. "Wave Power Absorption by Systems of Oscillating Surface Pressure Distributions." *Journal of Fluid Mechanics* 114 (1982): 481–499. http://dx.doi.org/10.1017/S0022112082000263.

Falnes, Johannes. "Optimum Control of Oscillation of Wave-Energy Converters." *International Journal of Offshore and Polar Engineering* 12 (2002): 147–55.

Falnes, Johannes, and Kjell Budal. "Wave Power Conversion by Point Absorbers." *Norwegian Maritime Research* 6, no. 4 (1978): 2–11.

Ferro, Benoit Dal. "Wave and Tidal Energy: Its Emergence and the Challenges It Faces." *Refocus* 7, no. 3 (2006): 46–48. http://dx.doi.org/10.1016/S1471-0846(06)70574-1.

Gao, Rong, George Tsatsaronis, and Lefteri H. Tsoukalas. *Smart Energy: Principles and Implementation.* New York: Springer, 2012.

Golding, Edward William. *The Generation of Electricity by Wind Power.* London: E. & F. N. Spon, 1976.

Gross, Robert. "Technologies and Innovation for System Change in the UK: Status, Prospects and System Requirements of Some Leading Renewable Energy Options." *Energy Policy* 32, no. 17 (2004): 1905–1919. http://dx.doi.org/10.1016/j.enpol.2004.03.017.

Henderson, A., and M. Patel. "Rigid-Body Motion of a Floating Offshore Wind Farm." *International Journal of Ambient Energy* 19, no. 3 (1998): 127–134. http://dx.doi.org/10.1080/01430750.1998.9675699.

International Energy Agency. Renewable Energy Essentials: Wind. Paris: IEA, 2008.

International Energy Agency. "Social Acceptance of Wind Energy Projects." In *Expert Group Summary on Recommended Practices.* Paris: IEA, 2013. http://www.ieawind.org/index_page_postings/RP/RP%2014%20Social_Acceptance_FINAL.pdf.

International Energy Agency Implementing Agreement on Ocean Energy Systems. *Annual Report 2008 (IEA-OES Document A08).* Edited by A. Brito-Melo and G. Bhuyan. Lisbon, Portugal: IEA-OES, 2009.

ISSC Specialist Committee. *Ocean Wind and Wave Energy Utilization.* Committee report at the 17th International Ship and Offshore Structures Congress, Seoul, Korea, August 2009.

Jonkman, J. M. *Dynamics Modeling and Loads Analysis of an Offshore Floating Wind Turbine (Technical Report NREL/TP-500–41958).* Golden, CO: National Renewable Energy Laboratory, 2007. http://dx.doi.org/10.2172/921803.

Kaimal, J. C. "Turbulence Spectra, Length Scales and Structural Parameters in the Stable Surface Layer." *Boundary-Layer Meteorology* 4, no. 1–4 (1973): 289–309. http://dx.doi.org/10.1007/BF02265239.

Kaimal, J. C., J. C. Wyngaard, Y. Izumi, and O. R. Coté. "Spectral Characteristics of Surface Layer Turbulence." *Quarterly Journal of the Royal Meteorological Society* 98, no. 417 (1972): 563–589. http://dx.doi.org/10.1002/qj.49709841707.

Karimirad, Madjid. *Stochastic Dynamic Response Analysis of Spar-Type Wind Turbines with Catenary or Taut Mooring Systems.* Doctoral dissertation, Norwegian University of Science and Technology, Norway, 2011.

Kiho, S., M. Shiono, and K. Suzuki. "The Power Generation from Tidal Currents by Darrieus Turbine." *Renewable Energy* 9, no. 1–4 (1996): 1242–1245. http://dx.doi.org/10.1016/0960-1481(96)88501-6.

Lemonis, George. "Wave and Tidal Energy Conversion." In *Encyclopedia of Energy,* edited by J. Cleveland Cutler, 385–396. New York: Elsevier, 2004. http://dx.doi.org/10.1016/B0-12-176480-X/00344-2.

Lund, John W., Derek H. Freeston, and Tonya L. Boyd. "Direct Application of Geothermal Energy 2005: Worldwide Review." *Geothermics* 34, no. 6 (2005): 691–727. http://dx.doi.org/10.1016/j.geothermics.2005.09.003.

Lund, John W., Derek H. Freeston, and Tonya L. Boyd. "Direct Utilization of Geothermal Energy 2010: Worldwide Review." In *Proceedings of the World Geothermal Congress 2010,* Bali, Indonesia, 25–29 April, 2010.

Mazumder, Rajat, and Makoto Arima. "Tidal Rhythmites and Their Implications." *Earth-Science Reviews* 69, no. 1–2 (2005): 79–95. http://dx.doi.org/10.1016/j.earscirev.2004.07.004.

McLarty, L., P. Grabowski, D. Etingh, and A. Robertson-Tait. "Enhanced Geothermal Systems R&D in the United States." In *Proceedings of the World Geothermal Congress*, Kyushu, Japan, June 10, 2000.

O'Rourke, Fergal, Fergal Boyle, and Anthony Reynolds. "Renewable Energy Resources and Technologies Applicable to Ireland." *Renewable & Sustainable Energy Reviews* 13, no. 8 (2009): 1975–84.

Owen, Alan. "Tidal Current Energy: Origins and Challenges." In *Future Energy: Improved, Sustainable and Clean Options for Our Planet*, edited by Trevor M. Letcher, 111–128. Oxford: Elsevier, 2008. http://dx.doi.org/10.1016/B978-0-08-054808-1.00007-7.

Phuc, Pham Van, and Takeshi Ishihara. A Study on the Dynamic Response of a Semi-Submersible Floating Offshore Wind Turbine System Part 2: Numerical Simulation. Proceedings of the International Conference of Wind Engineering, Cairns, Australia, 2007.

Previsic, M., A. Moreno, R. Bedard, B. Polagye, C. Collar, D. Lockard, et al. Hydrokinetic Energy in the United States: Resources, Challenges and Opportunities. In *Proceedings of 8th European Wave Tidal Energy Conference* 2009, 76–84.

Putnam, Palmer Cosslett. *Power from the Wind*. New York: Van Nostrand Rheinhold, 1948.

REN21.*Renewables 2013 Global Status Report*. Paris: REN21 Secretariat, 2013.

Rittich, Kerry, Klaus Grewe, and P. Kessener. "A Relief of a Water-powered Stone Saw Mill on a Sarcophagus at Hierapolis and its Implications." *Journal of Roman Archaeology* 20 (2007): 138–164.

Ross, David. *Power From the Waves*. Oxford: Oxford University Press, 1995.

Roux, Georges. *Ancient Iraq*. New York: Penguin, 1993.

Sathiamoorthy, M., and S. D. Probert. "The Integrated Severn Barrage Complex: Harnessing Tidal, Wave and Wind Power." *Applied Energy* 49, no. 1 (1994): 17–46. http://dx.doi.org/10.1016/0306-2619(94)90055-8.

Tomlinson, R. A. "The Perachora Waterworks: Addenda." *Annual of the British School at Athens* 71 (1976): 147–148. http://dx.doi.org/10.1017/S0068245400005864.

Tiwari, Gopal Nath, and Rajeev Kumar Mishra. *Advanced Renewable Energy Sources*. Cambridge: Royal Society of Chemistry Publishing, 2012.

Tong, K. C. "Technical and Economic Aspects of a Floating Offshore Wind Farm." *Journal of Wind Engineering and Industrial Aerodynamics* 74–76 (1998): 399–410. http://dx.doi.org/10.1016/S0167-6105(98)00036-1.

Twidell, John, and Tony Weir. *Renewable Energy Resources*. 2nd ed. London: Taylor and Francis, 2006.

Waris, Muhammad Bilal, and Takeshi Ishihara. "Dynamic Response Analysis of Floating Offshore Wind Turbine with Different Types of Heave Plates and Mooring Systems by Using a Fully Nonlinear Model." *Coupled System Mechanics* 1, no. 3 (2012): 1–22.

Westwood, Adam. "Ocean Power: Wave and Tidal Energy Review." *Refocus* 5, no. 5 (2004): 50–55. http://dx.doi.org/10.1016/S1471-0846(04)00226-4.

Wikander, Örjan. "The Water-Mill." In *Handbook of Ancient Water Technology*, edited by Örjan Wikander, 371–400. Leiden: Brill, 2000.

Worldwatch Institute. "Use and Capacity of Global Hydropower Increases." *World Watch* (January 2012).

Chapter 5
Solar Power and the Enabling Role of Nanotechnology

ALI SHAKOURI, BRIAN NORTON, AND HELEN MCNALLY

Abstract

Solar power has become the world's fastest growing renewable energy source in terms of new installations in recent years. This chapter provides a brief overview of enabling solar power technologies and policies in the context of recent world markets, and factors that are likely to influence future trends. We then explore the role of nanotechnology and its significant implications for energy conversion technologies in the future. Differences between information and energy devices are highlighted followed by a more in-depth focus on solar photovoltaic and thermoelectric devices. A review is carried out of advances in nano-engineered thermoelectric materials, module designs, and topping cycle applications that can offer significant cost reductions. The chapter concludes with a brief discussion of future opportunities for nanotechnology in energy conversion, storage, and conservation such as biofuels, artificial photosynthesis, and electrochemical batteries.

5.1. Introduction

Electricity derived directly from solar and thermal sources is experiencing dramatic growth in clean energy markets, while emerging research in nanotechnology is helping to underpin broader development efforts. Solar power, including photovoltaic and solar thermal technologies, has become the world's fastest growing renewable energy source in terms of new installations in recent years. Thermoelectric conversion is enjoying growth as well, particularly in niche applications such as off-grid or mobile electricity generated from waste heat or small scale cooling via applied electrical input when run in reverse.

Nanotechnology enables the manipulation of matter and the fabrication of devices with atomic dimensions. New material properties emerge from large surface to volume ratios and the quantum mechanical nature of electrons. On a nanometer length scale, physics, chemistry, and biology converge. Precise control of a material's physical and structural properties has huge implications for energy conversion devices. Nanoscale morphologies, surface areas, and quantum scale behaviors have led to a new generation of photovoltaic devices. Improved understanding of these phenomena is being applied to generate electrical potential differences and thereby optimize conversion performance. Photovoltaic device durability aspects are discussed and how device life may be extended with nanoscale self-assembly. Currently eighty-eight percent of the world's primary energy consumed is wasted in the form of heat. To reduce the consumption of fossil fuels, it will be necessary to improve future conversion of heat into electricity.

5.2. Solar Power Overview

Before delving into the atomic scale, let's go the other direction and consider the sun for a moment. It not only represents the largest entity and namesake of our solar system, but also the direct or indirect source behind almost all forms of energy we use on earth today. It is, one way or another, lighting the words on this page right now. Electricity generated from solar photovoltaic (PV) and concentrated solar power (CSP) therefore has the potential to allay resource constraint and emissions concerns associated with the combustion of fossil fuels, and introduce more environmentally sustainable Watts to the global energy supply of tomorrow. At the end of 2012, Solar PV reached a significant milestone, exceeding one hundred gigawatts of installed capacity worldwide (Figure 5.1). Moreover, a staggering three-fourths of all new capacity has been added since 2009.[1] PV devices convert sunlight directly into electricity, whereas CSP collects the thermal energy from focused sunlight via a working fluid that subsequently provides heat or electricity.

In recent years, aggressive policy incentives in Europe have translated into a leadership role for the region in PV installations, now home to seventy percent of the global

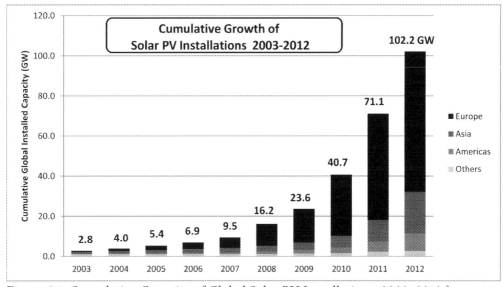

Figure 5.1. Cumulative Capacity of Global Solar PV Installations, 2003–2012.[2]

share. Asia is next at twenty percent, led by China and Japan with eight and seven percent of global shares respectively; followed by the Americas, where the United States has an eight percent global share. This has helped earn Solar PV the distinction of the fastest growing renewable energy resource in recent years as measured by new capacity additions. In 2013, solar electricity accounted for less than half a percent of the global share of electricity, whereas one recent prediction indicates US solar generation capacity could increase by one thousand percent between 2011 and 2040.[3] A similar analysis estimates solar PV could account for eleven percent of global electricity generation by 2050.[4] While recent trends have helped directly avoid CO_2 emissions and pleased many advocates of solar energy, there are three disconcerting realities confronting future prospects. The three concerns align loosely with categorical themes of our text: policy, technology, and economics.

Global policy, through a combination of clean energy capital grants and Feed-in-Tariffs, has been effective in jump-starting the nascent solar industry, but political and financial stimulus supports have begun to wane.[5] Some forecasts suggest the next five years could be relatively flat, while optimistic scenarios predict stable growth could re-emerge toward the end of the decade. A near term shift from the EU to markets elsewhere, likely in Asia and the Americas is expected. National targets, clean energy standards and emissions policies could obviously heavily influence longer term predictions.

Meanwhile, technology has made great advances in the efficiency of both individual cells and solar collection systems. Record setting cells are now exceeding 44.7% in the lab, while commensurate strides have been made to increase mass production level efficiencies which, in real-world solar panel environments, can lag world record levels by half or more. Efforts aimed at optimizing of performance specifications in view of cost for evolving technologies are helping undergird promising growth potential. Despite this, grid and market integration challenges may pose technological barriers to greater adoption. Solar and wind are intermittent renewable resources, and electrical infrastructure has not generally been designed for their seamless integration. A related issue is that the capacity factor (or effective capacity) of a solar PV or CSP plant is much lower than its nameplate rating, due to varying levels of solar incidence, geographic latitude, clouds, and night time. For example, ten gigawatts of installed PV capacity may provide only one or two gigawatts of equivalent baseload power to the marketplace;[6] whereas ten gigawatts of rated nuclear power can deliver about nine gigawatts of baseload electricity.[7] Thus, caution is advised when comparing various sources of electrical generation on a nameplate capacity basis. That said, break-through technological advances in energy storage, efficiency, and systems optimization are well within the realm of possibility to improve the overall value proposition of solar power.

The biggest economic challenge solar PV faces is to achieve *grid parity*, or competitiveness with retail electricity prices on a per kilowatt-hour basis. Current solar cell technology and installation costs result in levelized costs for PV generated electricity that are generally more costly, but can vary greatly from nearly one to more than three times market rates. This is heavily dependent on prevailing market rates, competing sources of electrical generation, incentives and subsidies, and state and local policies. Feed-in-Tariffs and subsidies have helped reduce the gap and spawn growth, but absent emissions charges or carbon-related taxes on conventional coal or natural gas generated electricity, economics will largely need to improve autonomously to be compelling. The Sun-Shot program of the US DOE's Solar Energy Technologies Office is aimed at leveraging technologies that can bring the cost of solar to grid parity by the end of the decade.[8]

Other significant solar contributions to clean energy are made by more mature but ever-evolving solar thermal technologies, such as the rooftop solar water heaters used extensively in Asia. Much of the success of these devices is attributable to the application of low-cost technology in amenable urban centers and climates, and their ease of integration with existing building infrastructure. In 2012, the equivalent energy saved and emissions avoided worldwide from this family of technologies was actually greater than that attributable to solar PV and CSP combined.[9] While long term predictions suggest PV trends will surpass passive solar thermal heating systems in deployment, it remains a salient reminder to consider local context and approach technology and policy holistically.

It is not the intent of this chapter to provide a comprehensive review of solar technologies, nor to cover all of the economic and policy considerations influencing their greater adoption. Instead, the overview is meant to convey a broader situational awareness and greater appreciation for the need to accelerate technological advances and policy dialogue. It is especially important to understand the role of nanotechnology in helping address some of these critical needs.

5.3. Nanotechnology

5.3.1. What Is Nanotechnology?

By definition, nanotechnology is simply the skill, art, or knowledge of devices or systems that manipulate phenomena at an approximate 10^{-9} m scale. In practice nanotechnology can have enormous implications; as nature operates on a nanoscale level,[10] nanotechnology provides unique opportunities for energy conversion,[11] medicine,[12] clean water,[13] food safety,[14] and environmental cleanup.[15] It has been argued that the one to one hundred nanometer length scale is the most efficient size to fabricate new materials and new devices because it is comparable to the size of atoms and molecules, the building blocks of chemistry and all materials around us. This is also the same length scale for biological building blocks such as DNA, RNA, and the various proteins inside a cell and at similar energy scales to manipulate quantum mechanics. As early as 1959, Richard Feynman described nanotechnology as a field which "might tell us much of great interest about the strange phenomena that occur in complex situations."[16] However the nanotechnology revolution did not begin until the mid 1990s,[17] and it was not until 2000 that the United States provided funding for the National Nanotechnology Initiative.[18] While nanoscale electronic and optoelectronic devices are available commercially, nanotechnology remains a relatively new area of research with much current work more aptly described as precursor nanoscience.

Nano is a multidisciplinary science and technology involving biologists, chemists and physicists, each of whom have been working at the nanoscale level for hundreds of years. At this scale, these disciplines converge with engineers and material scientists eager to exploit newly discovered properties. New material properties emerge due to the quantum mechanical nature of electrons when we approach atomic scales. High surface to volume ratios of micro- and nanoscale devices result in new electromagnetic, mechanical, chemical, thermal, and optical material properties. Recent advances in atomic and nanoscale modifications of materials and devices have enabled unprecedented control over these material properties, and are the basis for many new devices and systems used in our computing and communication infrastructure. These advances are made possible with the emergence of reliable top-down lithography,[19] bottom-up self assembly fabrication techniques,[20] nanoscale characterization of materials using techniques such as scanning probe microscopy[21] and electron microscopy.[22]

Nanoscience will continue to drive profound and broad-ranging changes to the selection of the most suitable materials, underlying functional mechanisms, and processes of manufacture for devices used for solar energy conversion applications. New materials are being conceived and well-known materials are being arranged, configured, combined, and/or coated in order to yield useful properties arising from quantum physics phenomena at a nanoscale level. A general classification of nanomaterials in applications as varied as supercapacitors and electrochromic glazings is provided in Figure 5.2.

For very small nanoparticles, all of the energy levels may be separated by finite gaps; electronically they behave as artificial molecules. However for "larger" particle sizes, the energy levels form the bands seen in a normal solid. In metals and semiconductors, the transition from finite gaps to bands occurs usually between two and six nanometers with materials such as gold and silver showing this transition at the lower end of the range, from two to three nanometers. Semiconductors such as cadmium sulphide are at

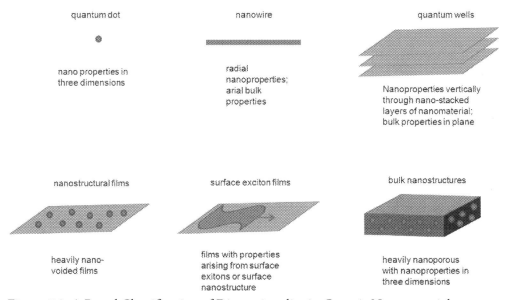

Figure 5.2. A Broad Classification of Dimensionality in Generic Nanomaterials.

the upper end of the band-gap range having energy-level schema that alter seamlessly with particle dimensions up to six nanometers. This variation leads to different wavelengths being absorbed, so as the sizes of cadmium sulphide particles change, so does their color. These properties have been employed in a range of sensor and biomarker applications. Such artificial three-dimensional structures whose properties give rise to tunable optoelectronic properties are referred to as quantum dots. They have been used to realize wavelength-selective luminescent solar energy concentrators.[23] Nanoparticles formed from good bulk electrical conductors are also of particular interest; their outermost-occupied energy levels convey charge conduction while being shielded by intervening electron change distributions from atomic nuclei. Electrons at these outermost levels thus move much more readily in response to an electrical potential. Such nanoparticles are normally less than two nanometers in diameter.

Equally as important as the nanoscale dimensions of particles are the volumes of the voids between them. Large voids expose an internal surface area that is several magnitudes larger than the external bulk surface area. Heavily voided assemblies thus obviously demonstrate enhanced performance with respect to many surface-area dependent properties. Highly nanovoided geometries can exhibit high structural strength but have little weight. These properties have evolved in nature in, for example, the nanostructures of the wings of butterflies and moths[24] and have found applications in supercapacitors. Longer path lengths and greater collision losses are associated with electrons traveling under an electric field through a heavily nanovoided medium. This can be employed to create nanoporous metals with metallic properties but high electrical resistance. The geometries and topologies possible at the nanoscale include combining metal or semiconductor nanoparticles with organic molecules leading to different charge transport and photoelectric phenomena.

5.3.2. Microelectronic and Photonic Revolutions

The microelectronic revolution started after World War II, when we first learned to make *semiconductors*. Semiconductors' electrical transport properties could be changed by six to nine orders of magnitude by doping or via the application of an external electrical field. This led to the invention of the transistor, a solid-state amplifier/switch, with great performance and scalability. With the invention of monolithic integrated circuits, thousands and, ultimately, millions of transistors could be fabricated on the same chip and mass produced. Nanotechnology research had a rebirth in the 1990s when microelectronic fabrication techniques were used to further modify other properties of matter (such as light absorption and mechanical strength) by precise control of at least one dimension of the material below ten to one hundred nanometeres. The first demonstrations, more than forty years ago, were quantum wells, achieved by confining electrons in a two dimensional sheet which resulted in electron energy quantization and enhanced electron transport and optoelectronic properties. Now quantum well lasers are used in every CD and DVD player. In the last twenty to thirty years, Quantum wires, dots, and ever more complex nanostructures have emerged. Nanotechnology has enabled exponential growth in microelectronics and optoelectronics in the second half of the twentieth century. Smaller information processing devices lead to higher integration (more transistors per chip), lower power consumption per computing operation, and faster operations due to smaller distances for charge transport. This is the basis for Moore's law, which states that the number of transistors per chip is doubled every eighteen to twenty-four months.[25]

In addition to the progress in information processing devices, we have had exponential growth in information transfer, especially in fiber optic applications. These have also benefited from miniaturization, which has enabled faster optoelectronic components (semiconductor lasers, detectors, wavelength multiplexers, and so on) as well as the development of low loss optical fibers. A corollary to Moore's Law is that the data rate for long-haul communications has doubled almost every five years between 1850 and 1950 and, since fiber optics were introduced in the late 1970s, continued to double every one or two years. Recent experiments have demonstrated the capacity to transmit one thousand terabits per second over a single twelve-core optical fiber fifty-eight kilometers long.[26]

Precise control of a material's electrical, thermal, optical, magnetic, and structural properties has huge implications for energy conversion devices such as photovoltaic devices which convert light to electricity and thermoelectric devices which convert heat to electricity. Nanotechnology is also vital to energy-related applications such as solid-state cooling applications, batteries, supercapacitors, power electronics devices, solar fuels, and many more. However, we should also be aware of major differences between information devices and energy devices. The former can be as small as possible since a bit does not have a minimum size. A property of a single electron or a single photon can define a bit. On the other hand, energy applications have a required cumulative length scale that matches human consumption (that is, the energy needed to keep the temperature of a house constant or energy needed to transport one kilogram of material by one hundred kilometeres). This distinction is important and is one of the reasons why some of the early industrial innovations of the twentieth century no longer exhibit exponential growth. For example, the speed of airplanes and the energy efficiency for transport increased almost exponentially from their advent until the middle part of last century, but in the last fifty years the speed of commercial airplanes has stayed constant and fuel efficiency has improved at more gradual rates.

A key enabling factor for the microelectronics industry has been the emergence of CAD (computer aided design) tools which enable the design of extremely complex systems with billions of building blocks. The computing power has now increased to a point that we can do first principle calculation of many material properties. This opens up the opportunity to design new materials and predict their properties. The new Materials Genome Initiative for Global Competitiveness in the US could pave the way for major applications in the energy field. For example, one could design a photovoltaic device with appropriate optical absorption and electronic transport properties from first principles and thus minimize the trial and error often used to make existing devices.

The remainder of this chapter will introduce the opportunities and challenges of nanotechnology as it can be applied to energy. A detailed overview of solar photovoltaics and thermoelectrics for direct conversion of heat into electricity will be presented as well as a discussion about opportunities for nanotechnology in energy storage and biofuels.

5.4. Solar Photovoltaics

Nanostructures are employed in many different aspects of solar energy, harnessing devices by tailoring properties through nano-morphology to produce novel meta-materials. This can lead to new paradigms for device functionality while still employing currently-used materials. For example in all-silicon tandem solar cells, nanostructures can be introduced that alter band-gaps exhibited by silicon/insulator combinations to be closer to those required to achieve optimal solar energy conversion efficiency. Such silicon nanostructures have been made by annealing sputtered silicon-rich silicon dioxide layers. The connection of fluorescent dye molecules with appropriate nanoparticles has led to new organic photovoltaic devices[27] and organic light emitting diodes.[28]

The high initial cost of wafer-based silicon solar cells has led to cost-reduction efforts to reduce required quantities of material or substitute less expensive but still suitable material combinations. For example, thin-film silicon solar cells can achieve very high open-circuit voltages.[29] The few-micron-thick layer of silicon in such cells can be realized by either etching or polymer transfer techniques.

The principal advantages of organic photovoltaic devices include wide, though presently low-efficiency, solar spectral response and low-cost manufacture via high-throughput solution processing. In contrast their disadvantages include low solar energy conversion efficiency and unproven long-term durability.

Combining a suitable lower band-gap polymer organic material and an electron-accepting inorganic material ensures broad adsorption of the solar spectrum while maximizing the short circuit current obtained. Forming an organic-inorganic hybrid solar cell in solution would lead to low-cost but reasonably efficient photovoltaic devices. Silicon nanocrystal quantum dots are an ideal electron accepting material for inclusion in such hybrid solar cells as the engineered band gap arising from electron confinement[30] is almost optimal at ~1.5 eV. Improving the energy-conversion efficiencies exhibited by organic solar cells requires a high electron charge carrier yield rate from exciton dissociation in the organic active layer. At the interfaces of materials with different electron affinities, excitons are dissociated by electric fields intensified by traps introduced from the engineered presence of "impurities." Blending conjugated polymers with molecules that have a high electron affinity, like C_{60}, leads to efficient rapid exciton dissociation in low-cost bulk-heterojunction solar cells. Bulk-heterojunction solar cells however, have a low solar energy conversion efficiency[31] of about seven percent, but achieving a power conversion efficiency of ten percent is seen as tenable.[32] Currently the most common polymer-fullerene system is based on a blend called P3HT:PCBM,[33] for which power conversion efficiencies are about 4.5%. The nanomorphology of the P3HT:PCBM blend is affected by the composition ratio of donor and acceptor materials, solvent materials and their concentration, molecular weight, and spin coating parameters.[34] Due to the small physical dimensions of the photoactive layer, absorbing a large range of wavelengths of the solar spectrum is crucial to enhancing the efficiency of organic photovoltaic devices. Since a high electromagnetic field strength exists in the vicinity of excited surface plasmons, the inclusion of plasmonic metal nanostructures enhances the solar energy absorption of organic materials. The resonance wavelength depends strongly on the metal nanostructure's dimensions, geometry, and dielectric properties. Contact electrodes represent a high interface barrier and source of contact resistance in organic solar cells. An organic cell's film is formed, usually on an indium-tin-oxide anode, either by spin coating or thermal evaporation deposition in a high vacuum. Since the optimisation of the metal-electrode Schottky barrier with PCBM and P3HT is likely to improve device performance, whichever process is employed needs to provide an anode surface that consistently presents a minimal interface barrier to the organic layer. Thermal annealing, solvent annealing, or vapor annealing processes optimize the nanomorphology of the active layer by sharply increasing hole mobility in the P3HT phase.[35]

The power efficiency of bulk heterojunction devices has seen continuous increases due to the synthesis of innovative low band-gap semiconducting polymer materials and morphology enhancement of the photoactive layer. Organic photovoltaic devices now achieve power conversion efficiencies up to 9.3%. This efficiency however, is still low compared with inorganic counterparts.

The efficiency of monolithic tandem stacks of solar cells has reached 41.6%. An efficiency of 42.7% has been achieved by partitioning the solar spectrum and doing energy conversion with five separate cells. Silicon and germanium nanostructures have potential application in photovoltaic devices including all-silicon tandem solar cells and hot carrier solar cells.

Intermediate-band solar cells have been proposed for achieving energy conversion efficiencies in monolithic devices comparable to those for tandem solar cell stacks. There are multiple designs possible with the performance of the low band gap combinations that improve as concentration increases. Intermediate-band solar cells perform differently under various concentrated spectra due to absorption disparities between bands. In spectral up-conversion devices infrared solar energy photons are absorbed and re-emitted photons are energised at the band gap that can be absorbed by a photovoltaic cell.

Concentrating photovoltaics are currently undergoing considerable commercial development.[36] With high concentration of incident solar energy certain types of solar cells exhibit high efficiency resulting in a much smaller cell area per unit output. High solar concentration is achieved usually by optical systems comprised of mirrors or lenses or, sometimes, both. At concentration ratios above two, to be effective, all optical systems must track the Sun's azimuthal motion. The latter requires moving parts that incur maintenance and replacement costs that could adversely affect economic viability or curtail useful system life. Several alternative approaches have therefore been envisaged for high concentration without the need for solar tracking.

Quantum dots can be used for luminescent concentration of diffuse solar energy,[37] by choosing quantum dots that absorb in the near infrared to provide a good match to the solar spectrum while avoiding significant reabsorption losses.[38] Low quantum yields and optical losses have limited the performance of this type of device.

Concentrating most of the solar spectrum from all incident angles without tracking may be achievable using a lens fabricated from metamaterials.[39] The latter are polymer-based nano/micro structured materials from whose materials and shapes arise; negative or extremely large electromagnetic permeability, permittivity, and refractive indices.

The use of multiple differently-oriented layers of materials that can switch from being transparent to being specularly reflective to form a single solid immobile solar tracking system.[40] The challenge here is to maintain each switched slate without significant use of energy.

In many climates, solar energy devices experience wide variations of ambient temperature on both a daily and annual basis. They are also frequently subject to rain, dust, snow, high humidity, strong winds, and atmospheric pollutants depending on location. Devices fail due to manufacturing defects, random accidents, and environmental exposure. The prevalence of each of these over a cell's life is shown in Figure 5.3.

For laminated photovoltaic cells, the principle exposure failure mechanism is delamination. Damage is usually caused by moisture penetrating through small cracks in an edge seal or, much more slowly, by water diffusing through front surface materials.[41]

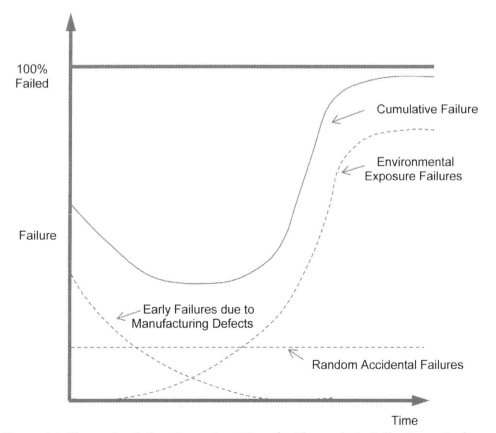

Figure 5.3. Illustrative Failure Rates Over Time for Photovoltaic Cells. Not to Scale.

Ethyl Vinyl Acetate (EVA) is the most common encapsulating material. However, thermal expansion and contraction of the EVA with diurnal variations in PV temperature causes the encapsulation integrity to fail. Without appropriate treatment, EVA can also discolor with a brown opaqueness due to exposure to the UV component of insolation. One approach that may prolong device life is to reduce the impact of some of the factors driving delamination. Delamination behaviors are complex even under controlled laboratory conditions.[42] For example, with innovative multi-layer films developed to obviate the problems inherent in the use of EVA, it has been found that delamination between the film layers of the same or very similar materials was more likely than delamination between dissimilar materials. Thermal management techniques are employed to maintain PV closer to ambient temperature to avoid temperature-dependent power output reductions. Phase change materials[43] have been used for this purpose. As a consequence the amplitude of temperature cycling endured by a PV cell is reduced with less concomitant thermal expansion and contraction. Ultimately the choices regarding system durability are determined by economic factors. Cost premiums may be associated with the quality of manufacture, appropriate selection of optimum materials, good design, exemplary operation, and scheduled maintenance. Periodic system replacement may be a preferred economic decision over selecting a more expensive system with a considerably longer life. The converse, however, may also be valid. Some PV system operations are guaranteed some revenue via a feed-in-tariff that pays back the amortised initial investment and operating costs over a specified time period. For such installations, after that period, any revenue is profit. In such circumstances long-term durability becomes very important! There may thus be scenarios where design-for-durability becomes a strong driver for innovation in PV technology. This may be seen as, an as yet un-investigated, opportunity for nanotechnology applications. For example, nanoparticles could be used as markers to identify points of local cell failure requiring repair. More fundamentally, an interesting and relevant question for future research is; is there the potential to employ nanoscale self-assembly to devise self-repairing PV cells?

5.5. Thermoelectrics

5.5.1. Overview and Motivation

Currently, photovoltaics and wind energy provide about one percent of total world energy. Biomass accounts for about eleven percent, used both in liquid fuels and, primarily, in the form of traditional biomass for cooking and heating in rural regions.[44] Most projections concur that it will take some time before a significant share of our energy is derived from renewables. More than ninety percent of primary energy is first converted to heat. Currently only twelve percent of the primary energy consumed in the world is transformed to end-use applications.[45] This is known as *exergetic efficiency*, a metric that takes into account the ability of chemical or nuclear energy sources to do useful work. In other words, one could argue we are wasting eighty-eight percent of the energy in the form of heat. The basic technologies behind our primary energy conversion devices (such as internal combustion engines and power plants) have been around for more than a century. The efficiencies are improving, but more can be done. As we try to increase energy efficiency and reduce the overall consumption of fossil fuels, we should explore opportunities to improve conversion of heat into electricity. That said, it is worth noting that the waste heat mentioned above varies greatly in terms of its potential value. It is notable that there are currently no commercially available large-scale waste heat recovery systems.

Thermoelectric (TE) power describes a solid-state means of directly converting energy in the form of heat to electricity and vice versa. If a temperature difference is maintained between the two ends of a conductor, higher thermal energy carriers will diffuse to the cold side, creating a potential difference that can be used to power an external load. This is a simplification of the Seebeck effect, which is the operative principle behind thermocouples. In addition to a large Seebeck coefficient, a high performance thermoelectric generator should minimize parasitic losses (low heat conduction between hot and cold junctions and low Joule heating as charge carriers move in the material). The energy conversion efficiency of thermoelectric devices is directly related to the thermoelectric figure of merit for the subject material, $ZT = S^2 \sigma T/\kappa$, where S is the Seebeck coefficient, σ is the electrical conductivity, T is the absolute temperature, and κ the thermal conductivity.[46] Abraham Ioffe introduced the concept of the thermoelectric figure-of-merit in the 1940s and developed some of the first thermoelectric generators to power radios during World War II. Since the working fluid is electrons or holes, thermoelectric devices have many unique advantages over conventional energy conversion systems including: no moving parts or vibration, high reliability and durability, compactness, and easy control.[47]

In the 1950s, silicon germanium and lead telluride alloys with $ZT \approx 0.5$–1 at 500–1000°C were developed. This was used to make radioisotrope thermoelectric generators (RTG) with energy conversion efficiency of four to six percent for deep space satellites and remote power generation for unmanned systems.[48] For satellites that travel far from the sun (beyond Saturn), solar radiation is too low for photovoltaic cells to be effective. Deep space missions have been using RTGs to power the imaging and communication systems. In such applications, efficiency, cost and radiation decay were obviously subordinate concerns to reliability and longevity. The main commercial success of thermoelectric technology has been in Bi_2Te_3-based Peltier modules to cool electronic and optoelectronic devices. For such small and localized applications, mechanical compressors do not scale well whereas these modules, operating under the *Peltier effect*, with modest coefficient-of-performance but fast responses, have been quite effective. Thermoelectric coolers are widely used for temperature stabilization of semiconductor lasers and for sub-ambient cooling of infrared detectors or CCD cameras.[49] Recently, thermoelectric car seat climate control systems have been commercialized[50] and applications for picnic beverage cooling is expanding. In the last ten to fifteen years, additional applications such as vehicle exhaust waste heat recovery[51] have been actively studied. In addition to niche applications, there is a huge potential for distributed power generation in poor countries. There are many communities who cannot afford the cost of power plants and an electric grid. A small amount of household electricity produced by thermoelectric modules in cooking stoves[52] or solar thermal systems[53] could improve the quality of life significantly. Consider the availability of reading lights after dusk in and otherwise "off-grid" community, as one of many examples.

If ZT is greater than three to five, the thermoelectric device can be competitive with traditional mechanical energy conversion systems. There is no fundamental limit on how large the thermoelectric figure-of-merit can be. When ZT goes to infinity, the efficiency of the thermoelectric system converges to the Carnot limit (at zero output power) or Curzon-Ahlborn limit (efficiency at maximum output power).[54] The improvement in the efficiency of thermoelectric energy conversion has been slow because all the material properties comprising ZT are coupled, and it is extremely difficult to enhance one property without affecting another. Recent advances in nano-engineered material properties have enabled breakthroughs for the enhancement of thermoelectric materials beyond ZT ≈ 1. This has been possible since we have a better microscopic understanding of heat and current transport in materials and we have the means to make artificial material composites with atomic building blocks. In the following we summarize some of the recent advances in this field.

5.5.2. Nanostructured Thermoelectric Material

A major breakthrough was the theoretical prediction of a large ZT enhancement in quantum wells and wires by Hicks and Dresselhaus in 1993.[55] This was based on the quantum confinement of electrons (or holes) enhancing the Seebeck coefficient and the electrical conductivity.[56] While this has not been demonstrated to improve the efficiency of thermoelectric generators or the cooling of Peltier modules, Hicks and Dresselhaus' paper was an important publication as it introduced researchers to the potential of nanostructuring. There are several reasons why it is hard to improve the thermoelectric power factor of quantum wells or wire materials.[57] We live in a three-dimensional world and any quantum confined structure should be imbedded in barriers. These barriers are electrically inactive but they add to thermal heat loss between the hot and the cold junctions. Sharp features in density-of-states of low dimensional nanostructures also disappear quickly as soon as there is size non-uniformity in the material.[58] Researchers have revisited the large thermoelectric power factor in extremely small quantum wells or wires and they point out that the actual improvement in the thermoelectric power factor "per conduction channel" is only twelve to forty percent.[59] This work also derives a minimum packing density for low dimensional thermoelectric material (even with ideal infinite barrier confinement) to have any improvement compared to the bulk.

It has also been proposed to use hot electron filtering to enhance the Seebeck coefficient.[60] The goal is to have potential barriers that "selectively" scatter cold electrons so that the contribution of electrons with energies larger than the Fermi level to electrical conductivity is significantly larger than the ones with energy below the Fermi level. This creates a metal-like electrical conductor with a large Seebeck coefficient. A similar concept can be applied to holes. The difficulty is in synthesizing metal/semiconductor nanocomposites with appropriate sizes on the order of an electron mean-free-path and with adequate barrier height (on the order of the thermal energy, $k_B T$, where k_B is the Boltzmann coefficient and T the absolute temperature) for electrons moving in a three-dimensional material.[61] Based on this concept, microrefrigerators on a chip have been demonstrated for localized cooling.[62] The estimated enhancement in the thermoelectric power factor (~20%) was offset by the increase in the lattice thermal conductivity.[63]

The rock salt-structured nitrides, including semiconducting ScN and metallic TiN and ZrN, have been investigated as a potential system in which energy filtering could be achieved at high temperatures for power generation applications. As a class, these materials also offer exceptionally high thermal and chemical stability, with melting points typically above 2500°C and a high degree of oxidation resistance at elevated temperatures. Preliminary study focused on ScN/(Zr,W)N semiconductor/metal superlattices grown by reactive sputter deposition.[64] The room temperature thermal conductivity is as low as 2W/m-K.[65] A high Seebeck coefficient of 840 µV/K has been measured combining the transient I-V measurement and thermal imaging.[66] This system has the potential to reach ZT values higher than two at temperatures above 1000 K.

Most of the recorded ZT values greater than 1.5 in the last ten years have been achieved by reducing the lattice thermal conductivity by increasing interface and boundary scattering of phonons in the nanostructured materials. For example, select semimetal/semiconductor nanocomposites showed enhanced ZT~1.3 at 800 K due, mainly, to the increased mid and long-wavelength phonon scattering by a few nm-sized nanoparticles.[67] Another recent demonstration showed a large power factor enhancement in Tl-doped PbTe at high temperatures up to 770 K, and attributed this enhancement to the distorted density of states by the resonant level inside the valence band of PbTe.[68] Other studies pointed out the important of convergence of bands in e.g. $PbTe_{1-x}Se_x$ and similar high ZT's have been achieved.[69] At elevated temperatures, near 600 K or higher, several PbTe-based nanostructured materials showed ZTs as high as 1.5 or even 2.2.[70] The material system, $AgPb_mSbTe_{2+m}$, also known as LAST-m, becomes spontaneously nanostructured when cooled from the melt, which helps reduce the thermal conductivity.[71] The strain field created by nano-inclusions in these material systems is believed to effectively scatter phonons to reduce the thermal conductivity. The $(PbSnTe)_x(PbS)_{1-x}$ system is phase-separated into PbTe-rich and PbS-rich regions, which become coherent nanostructures suppressing the lattice thermal conductivity. A ZT~1.5 at 640 K was reported for this material system of n-type with x ~ 0.08.[72] The n-type La-doped Ag_2Te-PbTe system has nano-scale (50–200 nm) Ag_2Te precipitates formed in PbTe, and showed ZT~1.5 at 775 K.[73] Skutterudites and clathrates have complex cage-like crystal structures with voids in which *rattler* atoms are inserted to effectively scatter acoustic phonons.[74] $CoSb_3$-based filled skutterudites such as p-type $LaFe_3CoSb_{12}$, and n-type $CeFeCoSb_{12}$ showed ZT > 1 at 800 K and higher.[75] Recently, Czochralski-grown clathrate $Ba_8Ga_{16}Ge_{30}$ showed ZT~1.6 at 1100 K.[76] Beyond 1000 K, SiGe has long been known to be a good thermoelectric material, and its ZT is typically 0.5 ~0.6 at 1100 K. After nanostructuring by hot pressing and ball milling, the ZT of p-type B-doped SiGe was improved to 0.7 at 1000 K.[77] ZT~1.3 at 1200 K was also reported for nanostructured n-type SiGe.[78] Two-phase SiGe-SiP nanocomposites had ZT > 1 at 1200 K, and this ZT enhancement was attributed to the modulation doping effect that enhanced mobility and thus electrical conductivity.[79]

Questions have been raised about some of the recently reported experimental results and theoretical concepts. This is expected for a rapidly growing and dynamic field, especially since accurate thermoelectric characterization techniques for small size samples are being developed concurrently. A major shortcoming for new researchers entering the thermoelectric field is the fact that there are very few papers which clearly describe major unsolved controversies.

5.5.3. Material Cost and Efficiency

Recently, a paper entitled *An Inconvenient Truth about Thermoelectrics* made the claim that "Despite recent advances, thermoelectric energy conversion will never be as efficient as steam engines. That means thermoelectrics will remain limited to applications served poorly or not at all by existing technology."[80] An analysis of the potential of thermoelectrics that focuses only on efficiency values is not complete. In fact, an analysis that considers optimizing the entire TE power generation system cost could lead to the exact opposite conclusion. While it is true that thermoelectrics are not likely to replace conventional Rankine cycle steam engines or Stirling engines in the near future, they could play a big role by enabling direct conversion of heat into electricity, especially for topping cycle or waste heat recovery applications. This is because the energy cost may be more important than the efficiency of energy conversion alone. For example, many groups are working on polymer solar cells even though their efficiency is much smaller than the multi-junction crystalline cells. It becomes a value judgment and optimization exercise.

A key factor to study with respect to the cost/efficiency trade off in thermoelectric power generation, is the optimization of the TE module together with the heat source and the heat sink[81] estimated the cost of material in a TE power generation system ($/W) as a function of heat source power density for different material properties, (ZT), as well as module design, (fractional area coverage of TE elements). It was shown that the module design plays a key factor in determining the cost of waste heat recovery thermoelectric systems. It is possible to bring down the material cost from $1–2/W to $0.01–0.02/W without improving ZT, if a thin film module with 100s of microns thick elements and low fractional area coverage (5–10%) with small parasitics (such as low contact resistances) can be developed.[82]

The thermal disparity between the fuel combustion temperature ~2250 K (adiabatic) and the high pressure steam temperature up to 800–900 K or high temperature combustion gases, results in a large thermodynamic losses in coal or gas turbine power plants. While some technologies exist for recuperating a portion of this lost heat, a solid-state thermoelectric placed on top of a conventional cycle may prove an effective means of directly producing additional electrical power. By selecting the right materials for the TE generator for high temperature operation, the overall energy efficiency will increase. Recent study shows that, for example, the combined TE/steam turbine topping generator system provides a lower energy cost for any period of operational life.[83] This study shows that the efficiency of the combined system increases by 10% at an interface temperature of about 800 K between the two stages for a TE material with ZT ~ 1.

5.6. Nanotechnology in Other Energy Applications

5.6.1. Biofuels

Nanotechnology can underpin biofuel development,[84] for example by using the unique nanoscale properties of biological molecules and surface area to volume ratios to create energy from agriculture products and waste. Photosynthesis is nature's example, using the sun's energy to convert water into hydrogen ions, electrons, and oxygen. Artificial photosynthesis is an alternative requiring an antenna (to absorb sunlight) or oxidizing center (to produce electrons) and a reducing center (to produce fuels). Nanotechnology is used in the antenna section to increase the surface area to volume ratio for increased energy absorption and reduce the travel distance of photo-generated minority carriers.[85] Graphitic petals[86] have recently been produced, providing yet another unique nanostructure from which to possibly specifically tune the capture of the energy spectrum. In the reducing and oxidizing centers, iron oxide nanoparticles combined with water,[87] nanotube arrays and various other nanostructured materials have been used to reduce water and CO_2 to fuels such as methane and other hydrocarbons.[88]

5.6.2. Energy Storage Devices

Energy storage[89] and transport[90] are critical research areas required to enable mobile and long term integration of many new energy technologies. Batteries are currently the most common form of electrical energy storage and nanotechnology has the potential to influence every aspect of current designs (cathode, anode and electrolyte.) Energy density, currently a barrier to wider adoption of batteries, and stability may benefit the most from the unique properties of nanotechnology. New electrode materials[91] are being developed with nanostructures including; nanocarbons, alloys, and oxides to provide greater surface area to volume ratios for increased reaction and faster energy transport. Vanadium pentoxide (V_2O_5) is being studied for its unique properties and stable form. While controlled and large scale synthesis of nanostructured materials continues to be a challenge, Parida et al.[92] recently described V_2O_5 nanoflowers with ferromagnetic and optical properties at room temperature. Specifically for lithium-ion technology, transitional metal oxides are reviewed providing high theoretical capacities for conversion reaction.[93] Within the battery electrode, nanoparticles are being used to enhance the ionic conductivity and storage capacity. Nanomaterials are even being considered for the overall structural integrity of batteries due to their enhanced mechanical characteristics.

Today's advanced lithium-ion battery technologies have made dramatic inroads in commercial applications such as portable electronic and rechargeable devices. However, due to cost, weight and other factors, they are not yet considered practical for electric vehicles or intermittent energy sources, i.e., solar, wind, tidal. Alternative forms of energy storage are thus in great demand; including biochemical, chemical, or electrochemical capacitors.[94] Nanotechnology is sure to factor in to the development of the energy storage systems of the future.

Notes

1. European Photovoltaic Industry Association (EPIA). *Global Market Outlook for Photovoltaics, 2013–2017* (Brussels, Belgium: European Photovoltaic Industry Association, 2013), 13.
2. Ibid., 30.
3. Energy Information Administration, *Annual Energy Outlook 2013* (Washington, D.C.: US Energy Information Administration, 2013), http://www.eia.gov/forecasts/aeo/.
4. International Energy Agency, *Technology Roadmap: Solar Photovoltaic Energy* (Paris: IEA Publications, 2010).
5. EPIA, *Global Market Outlook*, 33.
6. For example, using the notional solar capacity factor estimates assuming characteristics of Germany or the US desert southwest, respectively.
7. Using average capacity factors for US nuclear plants as detailed in Energy Information Administration, *Electric Power Annual 2009 [DOE/EIA-0348(2009)]* (Washington, D. C.: US Department of Energy, 2011).
8. Solar Energy Technologies Office, "SunShot Initiative," *Department of Energy*, last modified August 8, 2013, http://www1.eere.energy.gov/solar/sunshot/about.html.
9. REN21, *Renewables 2013 Global Status Report* (Paris: REN21 Secretariat, 2013), 41–48.
10. Janine M. Benyus, *Biomimicry: Innovation Inspired by Nature* (New York: Harper Collins, 1997).
11. Jason Baxter et al., "Nanoscale Design to Enable the Revolution in Renewable Energy," *Energy and Environmental Science* 2, no. 6 (2009): 559–588, http://dx.doi.org/10.1039/b821698c; P. Oelhafen and A. Schüler, "Nanostructured Materials for Solar Energy Conversion," *Solar Energy* 79, no. 2 (2005): 110–121, http://dx.doi.org/10.1016/j.solener.2004.11.004.
12. Mary-Margaret Seale-Goldsmith, and James F. Leary, "Nanobiosystems," *Wiley Interdisciplinary Reviews: Nanomedicine and Nanobiotechnology* 1, no. 5 (2009): 553–567, http://dx.doi.org/10.1002/wnan.49.
13. Nora Savage et al., eds., *Nanotechnology Applications for Clean Water* (Norwich, NY: William Andrew, 2009).
14. Timothy V. Duncan, "Applications of Nanotechnology in Food Packaging and Food Safety: Barrier Materials, Antimicrobials and Sensors," *Journal of Colloid and Interface Science* 363, no. 1 (2011): 1–24, http://dx.doi.org/10.1016/j.jcis.2011.07.017.
15. Steven T. Wereley and Carl D. Meinhart, "Recent Advances in Micro Particle Image Velocimetry," *Annual Review of Fluid Mechanics* 42, no. 1 (2010): 557–576, http://dx.doi.org/10.1146/annurev-fluid-121108-145427.
16. Richard P. Feynman, "There's Plenty of Room at the Bottom," *Engineering and Science* 23, no. 5 (1960): 22–36.
17. Stephan Herrera and Lawrence Aragon, "Small Worlds: Nanotechnology Wins over Mainstream Venture Capitalists," *Red Herring* 107 (2001): 51.
18. http://www.nano.gov/.
19. Shunri Oda and David Ferry, eds., *Silicon Nanoelectronics* (Boca Raton, FL: CRC Press, 2005).
20. A. Paul Alivisatos, Kai P. Johnsson, Xiaogang Peng, Troy E. Wilson, Colin J. Loweth, Marcel P. Bruchez, Jr., and Peter G. Schultz, "Organization of 'Nanocrystal Molecules' using DNA," *Nature* 382, no. 6592 (Aug 15, 1996): 609–611, http://dx.doi.org/10.1038/382609a0.
21. G. Binnig, C. F. Quate, and C. H. Gerber, "Atomic Force Microscope," *Physical Review Letters* 56, no. 9 (1986): 930–933, http://dx.doi.org/10.1103/PhysRevLett.56.930.
22. Weilie Zhou and Zhong Lin Wang, eds., *Scanning Microscopy for Nanotechnology: Techniques and Applications* (New York: Springer, 2006), http://dx.doi.org/10.1007/978-0-387-39620-0.

23. M. Kennedy et al., "Improving the Optical Efficiency and Concentration of a Single-plate Quantum Dot Solar Concentrator Using Near Infra-red Emitting Quantum Dots," *Solar Energy* 83, no. 7 (2009): 978–981, http://dx.doi.org/10.1016/j.solener.2008.12.010.
24. Helen Ghiradella, "Light and Color on the Wing: Structural Colors in Butterflies and Moths," *Applied Optics* 30, no. 24 (1991): 3492–3500, http://dx.doi.org/10.1364/AO.30.003492.
25. Gordon E. Moore, "Cramming More Components onto Integrated Circuits," *Electronics* 38, no. 8 (1965): 114–117.
26. D. J. Richardson, J. M. Fini, and L. E. Nelson, "Space-division Multiplexing in Optical Fibres," *Nature Photonics* 7, no. 5 (2013): 354–362, http://dx.doi.org/10.1038/nphoton.2013.94.
27. Cristoph J. Brabec and James R. Durrant, "Solution-Processed Organic Solar Cells," *Material Research Society Bulletin* 33, no. 07 (2008): 670–675, http://dx.doi.org/10.1557/mrs2008.138.
28. Sebastian Reineke et al., "White Organic Light-emitting Diodes with Fluorescent Tube Efficiency," *Nature* 459, no. 7244 (2009): 234–238, http://dx.doi.org/10.1038/nature08003.
29. B. Chhabra, C. B. Honsberg, and R. L. Opila, "High Open Circuit Voltages on 50 Micron Silicon Substrates by Amorphous-silicon (a-Si) and Quinhydrone-methanol (QHY-ME) Passivation," In *34th IEEE Photovoltaic Specialists Conference*, Philadelphia, PA, June 7–12 2009, 2187–2190, http://dx.doi.org/10.1109/PVSC.2009.5411398.
30. Chin-Yi Liu, Zacharay C. Holman, and Uwe R. Kortshagen, "Hybrid Solar Cells from P3HT and Silicon Nanocrystals," *Nano Letters* 9, no. 1 (2009): 449–452, http://dx.doi.org/10.1021/nl8034338.
31. Samuel C. Price et al., "Fluorine Substituted Conjugated Polymer of Medium Band Gap Yields 7% Efficiency in Polymer-Fullerene Solar Cells," *Journal of the American Chemical Society* 133, no. 12 (2011): 4625–4631, http://dx.doi.org/10.1021/ja112595.
32. M. C. Scharber et al., "Design Rules for Donors in Bulk Hetrojunction Solar Cells: Towards 10% Energy Conversion Efficiency," *Advanced Materials* 18, no. 6 (2006): 789–794, http://dx.doi.org/10.1002/adma.200501717.
33. Poly (3 hexylthiophene 2,5 diyl):[6,6] phenyl C61 butyric acid methyl ester.
34. Moo-Hyun Kwon, "Stability of Bulk Hetrojunction Organic Solar Cells with Different Blend Ratios of P3HT:PCBM." *Transactions on Electrical and Electronic Materials* 13, no. 2 (2012): 98–101. http://dx.doi.org/10.4313/TEEM.2012.13.2.98.
35. John A. Carr et al., "Controlling Morphology in Plastic Solar Cells," *Nanomaterials and Energy* 1, no. 1 (2012): 18–26, http://dx.doi.org/10.1680/nme.11.00010.
36. Daniel Chemisana, "Building Integrated Concentrating Photovoltaics: A Review," *Renewable & Sustainable Energy Reviews* 15, no. 1 (2011): 603–611, http://dx.doi.org/10.1016/j.rser.2010.07.017.
37. S. J. Gallagher et al., "Quantum Dot Solar Concentrator: Device Optimisation Using Spectroscopic Techniques," *Solar Energy* 81, no. 4 (2007): 540–547, http://dx.doi.org/10.1016/j.solener.2006.07.006; and S. J. Gallagher, B. C. Rowan, and P. C. Eames, "Quantum Solar Concentrator: Electrical Conversion Efficiencies and Comparative Concentrating Factors of Fabricated Devices," *Solar Energy* 81, no. 6 (2007): 813–821, http://dx.doi.org/10.1016/j.solener.2006.09.011.
38. Kennedy et al., "Improving Optical Efficiency."
39. Igor I. Smolyaninov, Yu-Ju Hung, and Christopher C. Davis, "Magnifying Superlens in the Visible Frequency Range," *Science* 315, no. 5819 (2007): 1699–1701, http://dx.doi.org/10.1126/science.1138746.
40. Brian Norton, Manus Kennedy, and Sarah McCormack, "Solar Energy Concentration Without Moving Parts in Sustainable Building Envelopes," in *Renewables in a Changing Climate: From Nano to Urban Scale*, 73–78 (Lausanne, Switzerland: Ecole Polytechnique Fédérale de Lausanne, 2009).
41. Michael D. Kempe, "Modeling Rates of Moisture Ingress into Photovoltaic Modules," *Solar Energy Materials and Solar Cells* 90, no. 16 (2006): 2720–2738, http://dx.doi.org/10.1016/j.solmat.2006.04.002.
42. G. Oreski and G. M. Wallner, "Delamination Behaviour of Multi-layer Films for PV Encapsulation," *Solar Energy Materials and Solar* 89, no. 2–3 (2005): 139–151, http://dx.doi.org/10.1016/j.solmat.2005.02.009.

43. M. J. Huang, P. C. Eames, and B. Norton, "Thermal Regulation of Building-integrated Photovoltaics Using Phase Change Materials," *International Journal of Heat and Mass Transfer* 47, no. 12–13 (2004): 2715–2733, http://dx.doi.org/10.1016/j.ijheatmasstransfer.2003.11.015.
44. Jonathan M. Cullen and Julian M. Allwood, "The Efficient Use of Energy: Tracing the Global Flow of Energy from Fuel to Service," *Energy Policy* 38, no. 1 (2010): 75–81, http://dx.doi.org/10.1016/j.enpol.2009.08.054.
45. Ibid.
46. George S. Nolas, Jeffrey Sharp, and H. Julian Goldsmid, *Thermoelectrics: Basic Principles and New Materials Developments* (New York: Springer-Verlag, 2001); D. M. Rowe, ed., *CRC Handbook of Thermoelectrics* (Boca Raton, FL: CRC Press, 1995); and D. M. Rowe, ed., *Thermoelectrics Handbook: Macro to Nano* (Boca Raton, FL: CRC Press, 2005).
47. Lon E. Bell, "Cooling, Heating, Generating Power, and Recovering Waste Heat with Thermoelectric Systems," *Science* 321, no. 5895 (2008): 1457–1461, http://dx.doi.org/10.1126/science.1158899.
48. J. C. Bass and D. T. Allen, "Milliwatt Radioisotope Power Supply for Space Applications," in *Proceedings of 18th International Conference on Thermoelectrics*, 521–524 (Piscataway, NJ: IEEE, 1999), http://dx.doi.org/10.1109/ICT.1999.843443.
49. Rowe, *CRC Handbook of Thermoelectrics*.
50. Lon E. Bell, Thermoelectric Heat Exchanger, US Patent 6,119,463, filed May 12, 1998, and issued September 19, 2000.
51. Bell, "Waste Heat with Thermoelectric Systems."
52. Paul Van der Sluis, "Improvement in Cooking Stoves," International Patent WO/2006/103613, filed March 27, 2006, and granted October 5, 2006.
53. R. Amatya and R. J. Ram, "Solar Thermoelectric Generator for Micropower Applications," *Journal of Electronic Materials* 39, no. 9 (2010): 1735–1740, http://dx.doi.org/10.1007/s11664-010-1190-8.
54. Kazuaki Yazawa and Ali Shakouri, "Cost-efficiency Trade-off and the Design of Thermoelectric Power Generators," *Environmental Science & Technology* 45, no. 17 (2011): 7548–7553, http://dx.doi.org/10.1021/es2005418.
55. L. D. Hicks and M. S. Dresselhaus, "Effect of Quantum-well Structures on the Thermoelectric Figure of Merit," *Physical Review B: Condensed Matter and Materials Physics* 47, no. 19 (1993): 12727–12731, http://dx.doi.org/10.1103/PhysRevB.47.12727.
56. Ali Shakouri and Mona Zeberjadi, "Nanoengineered Materials for Thermoelectric Energy Conversion," in *Thermal Nanosystems and Nanomaterials*, ed. Sebastian Volz, 225–299 (New York: Springer, 2009). http://dx.doi.org/10.1007/978-3-642-04258-4_9.
57. D. A. Broido and T. L. Reinecke, "Effect of Superlattice Structure on the Thermoelectric Figure of Merit," *Physical Review B: Condensed Matter and Materials Physics* 51, no. 19 (1995): 13797–13800, http://dx.doi.org/10.1103/PhysRevB.51.13797; J. O. Sofo and G. D. Mahan, "Thermoelectric Figure of Merit of Superlattices," *Applied Physics Letters* 65, no. 21 (1994): 2690–2692, http://dx.doi.org/10.1063/1.112607; and G. Chen and A. Shakouri, "Heat Transfer in Nanostructures for Solid-state Energy Conversion," *Journal of Heat Transfer-Transactions of the ASME* 124, no. 2 (2002): 242–252, http://dx.doi.org/10.1115/1.1448331.
58. Even though this makes sense intuitively, there are no detailed calculations of the effect of size non-uniformity on low dimensional thermoelectric properties.
59. Raseong Kim, Supriyo Datta, and Mark S. Lundstrom, "Influence of Dimensionality on Thermoelectric Device Performance," *Journal of Applied Physics* 105, no. 3 (2009): 034506, http://dx.doi.org/10.1063/1.3074347.
60. Ali Shakouri and John E. Bowers, "Heterostructure Integrated Thermionic Coolers," *Applied Physics Letters* 71, no. 9 (1997): 1234–1236, http://dx.doi.org/10.1063/1.119861; G. D. Mahan and L. M. Woods, "Multilayer Thermionic Refrigeration," *Physical Review Letters* 80, no. 18 (1998): 4016–4019, http://dx.doi.org/10.1103/PhysRevLett.80.4016; and Daryoosh Vashaee and Ali Shakouri, "Improved

Thermoelectric Power Factor in Metal-based Superlattices," *Physical Review Letters* 92, no. 10 (2004): 106103, http://dx.doi.org/10.1103/PhysRevLett.92.106103.

61. Ali Shakouri, "Recent Developments in Semiconductor Thermoelectric Physics and Materials," *Annual Review of Materials Research* 41, no. 1 (2011): 399–431, http://dx.doi.org/10.1146/annurev-matsci-062910-100445.

62. Xiaofeng Fan et al., "SiGeC/Si Superlattice Microcoolers," *Applied Physics Letters* 78, no. 11 (2001): 1580–1582, http://dx.doi.org/10.1063/1.1356455.

63. Ali Shakouri, "Nanoscale Thermal Transport and Microrefrigerators on a Chip," *Proceedings of the IEEE* 94, no. 8 (2006): 1613–1638, http://dx.doi.org/10.1109/JPROC.2006.879787.

64. Vijay Rawat and Timothy D. Sands, "Growth of TiN/GaN Metal/Semiconductor Multilayers by Reactive Pulsed Laser Deposition," *Journal of Applied Physics* 100, no. 6 (2006): 064901, http://dx.doi.org/10.1063/1.2337784.

65. Vijay Rawat et al., "Thermal Conductivity of (Zr,W)N/ScN Metal/Semiconductor Multilayers and Superlattices," *Journal of Applied Physics* 105, no. 2 (2009): 024909, http://dx.doi.org/10.1063/1.3065092.

66. Mona Zebarjadi et al., "Thermoelectric Transport in a ZrN/ScN Superlattice," *Journal of Electronic Materials* 38, no. 7 (2009): 960–963, http://dx.doi.org/10.1007/s11664-008-0639-5.

67. J. M. O. Zide, J.-H. Bahk, R. Singh, M. Zebarjadi, G. Zeng, H. Lu, J. P. Feser et al., "High Efficiency Semimetal/semiconductor Nanocomposites Thermoelectric Materials," *Journal of Applied Physics* 108, no. 12 (2010): 123702, http://dx.doi.org/10.1063/1.3514145.

68. Joseph P. Heremans et al., "Enhancement of Thermoelectric Efficiency in PbTe by Distortion of the Electronic Density of States," *Science* 321, no. 5888 (2008): 554–557, http://dx.doi.org/10.1126/science.1159725.

69. Yanzhong Pei et al., "Convergence of Electronic Bands for High Performance Bulk Thermoelectrics," *Nature* 473, no. 7345 (2011): 66–69, http://dx.doi.org/10.1038/nature09996.

70. Kanishka Biswas et al., "High-Performance Bulk Thermoelectrics with All-Scale Hierarchical Architectures," *Nature* 489, no. 7416 (2012): 414–418, http://dx.doi.org/10.1038/nature11439.

71. Kuei Fang Hsu et al., "Cubic AgPb(m)SbTe(2+m): Bulk Thermoelectric Materials with High Figure of Merit," *Science* 303, no. 5659 (2004): 818–821, http://dx.doi.org/10.1126/science.1092963.

72. John Androulakis et al., "Spinodal Decomposition and Nucleation and Growth as a Means to Bulk Nanostructured Thermoelectrics: Enhanced Performance in $Pb_{1-x}Sn_xTe–PbS$," *Journal of the American Chemical Society* 129, no. 31 (2007): 9780–9788, http://dx.doi.org/10.1021/ja071875h.

73. Yanzhong Pei et al., "High Thermoelectric Performance in PbTe Due to Large Nanoscale Ag2Te Precipitates and La Doping," *Advanced Functional Materials* 21, no. 2 (2011): 241–249, http://dx.doi.org/10.1002/adfm.201000878.

74. George S. Nolas, Joe Poon, and Mercouri Kanatzidis, "Recent Developments in Bulk Thermoelectric Materials," *Materials Research Society Bulletin* 31, no. 03 (2006): 199–205, http://dx.doi.org/10.1557/mrs2006.45.

75. B. C. Sales, D. Mandrus, and R. K. Williams, "Filled Skutterudite Antimonides: A New Class of Thermoelectric Materials," *Science* 272, no. 5266 (1996): 1325–1328, http://dx.doi.org/10.1126/science.272.5266.1325.

76. A. Saramat et al., "Large Thermoelectric Figure of Merit at High Temperature in Czochralski-grown Clathrate $Ba_8Ga_{16}Ge_{30}$," *Journal of Applied Physics* 99, no. 2 (2006): 023708, http://dx.doi.org/10.1063/1.2163979.

77. M. S. Dresselhaus et al., "New Directions for Low-dimensional Thermoelectric Materials," *Advanced Materials* 19, no. 8 (2007): 1043–1053, http://dx.doi.org/10.1002/adma.200600527.

78. X. W. Wang et al., "Enhanced Thermoelectric Figure of Merit in Nanostructured N-type Silicon Germanium Bulk Alloy," *Applied Physics Letters* 93, no. 19 (2008): 193121, http://dx.doi.org/10.1063/1.3027060.

79. Mona Zebarjadi et al., "Power Factor Enhancement by Modulation Doping in Bulk Nanocomposites," *Nano Letters* 11, no. 6 (2011): 2225–2230, http://dx.doi.org/10.1021/nl201206d.

80. Cronin B. Vining, "An Inconvenient Truth about Thermoelectrics," *Nature Materials* 8, no. 2 (2009): 83, http://dx.doi.org/10.1038/nmat2361.
81. Kazuaki Yazawa, and Ali Shakouri, "Efficiency of Thermoelectric Generators at Maximum Output Power," paper submitted at the International Conference on Thermoelectrics, Shanghai, China, 2010; and Kazuaki Yazawa and Ali Shakouri, "Energy Payback Optimization of Thermoelectric Power Generator Systems," In *ASME 2010 International Mechanical Engineering Congress and Exposition Volume 5: Energy Systems Analysis, Thermodynamics and Sustainability; NanoEngineering for Energy; Engineering to Address Climate Change, Parts A and B*, 569–576 (New York: American Society of Mechanical Engineers, 2010), http://dx.doi.org/10.1115/IMECE2010-37957.
82. Yazawa and Shakouri, "Cost-efficiency Trade-off"; and Kazuaki Yazawa and Ali Shakouri, "Cost-effective Waste Heat Recovery Using Thermoelectric Systems," *Journal of Materials Research* 27, no. 09 (2012): 1211–1217, http://dx.doi.org/10.1557/jmr.2012.79.
83. Kazuaki Yazawa, Yee Rui Koh, and Ali Shakouri, "Optimization of Thermoelectric Topping Combined Steam Turbine Cycles for Energy Economy," *Applied Energy* 109 (2013): 1–9, http://dx.doi.org/10.1016/j.apenergy.2013.03.050.
84. Arthur J. Ragauskas et al., "The Path Forward for Biofuels and Biomaterials," *Science* 311, no. 5760 (2006): 484–489, http://dx.doi.org/10.1126/science.1114736.
85. Joop Schoonman and Roel van de Krol, "Nanostructured Materials for Solar Hydrogen Production," *UPB Scientific Bulletin Series B: Chemistry and Materials Science* 73, no. 4 (2011): 31–44, http://scientificbulletin.upb.ro/rev_docs_arhiva/full14674.pdf.
86. Guoping Xiong, K. P. S. S. Hembram, R. G. Reifenberger, and Timothy S. Fisher, "MnO_2-coated Graphitic Petals for Supercapacitor Electrodes," *Journal of Power Sources* 227 (2013): 254–259, http://dx.doi.org/10.1016/j.jpowsour.2012.11.040.
87. Devans Gust, Thomas A. Moore, and Ana L. Moore, "Solar Fuels via Artificial Photosynthesis." *Accounts of Chemical Research* 42, no. 12 (2009): 1890–1898, http://dx.doi.org/10.1021/ar900209b.
88. Somnath C. Roy et al., "Toward Solar Fuels: Photocatalytic Conversion of Carbon Dioxide to Hydrocarbons," *ACS Nano* 4, no. 3 (2010): 1259–1278, http://dx.doi.org/10.1021/nn9015423.
89. Louis Schlapbach and Andreas Züttel, "Hydrogen-storage Materials for Mobile Applications," *Nature* 414, no. 6861 (2001): 353–358, http://dx.doi.org/10.1038/35104634.
90. W. Zhang, N. Mingo, and T. S. Fisher, "Simulation of Phonon Transport Across a Non-polar Nanowire Junction Using an Atomistic Green's Function Method," *Physical Review* 76, no. 19 (2007): 195429, http://dx.doi.org/10.1103/PhysRevB.76.195429.
91. Min-Kyu Song et al., "Nanostructured Electrodes for Lithium-ion and Lithum-air Batteries: The Latest Developments, Challenges and Perspectives," *Materials Science and Engineering* 72, no. 11 (2011): 203–252, http://dx.doi.org/10.1016/j.mser.2011.06.001.
92. Manas R. Parida et al., "Room Temperature Ferromagnetism and Optical Limiting in V_2O_5 Nanoflowers Synthesized by a Novel Method," *Journal of Physical Chemistry* 115 (2011): 112–117, http://dx.doi.org/10.1021/jp107862n.
93. Liwen Ji et al., "Recent Developments in Nanostructured Anode Materials for Rechargeable Lithium-ion Batteries," *Energy and Environmental Science* 4, no. 8 (2011): 2682–2699, http://dx.doi.org/10.1039/c0ee00699h.
94. Baxter et al., "Nanoscale Design."

Bibliography

Alivisatos, A. Paul, Kai P. Johnsson, Xiaogang Peng, Troy E. Wilson, Colin J. Loweth, Marcel P. Bruchez, Jr., and Peter G. Schultz. "Organization of 'Nanocrystal Molecules' using DNA." *Nature* 382, no. 6592 (Aug 15, 1996): 609–611. http://dx.doi.org/10.1038/382609a0.

Amatya, R., and R. J. Ram. "Solar Thermoelectric Generator for Micropower Applications." *Journal of Electronic Materials* 39, no. 9 (2010): 1735–1740. http://dx.doi.org/10.1007/s11664-010-1190-8.

Androulakis, John, Chia-Her Lin, Hun-Jin Kong, Ctirad Uher, Chun-I Wu, Timothy Hogan, Bruce A. Cook, Thierry Caillat, Konstantinos M. Paraskevopoulos, and Mercouri G. Kanatzidis. "Spinodal Decomposition and Nucleation and Growth as a Means to Bulk Nanostructured Thermoelectrics: Enhanced Performance in $Pb_{1-x}Sn_xTe$–PbS." *Journal of the American Chemical Society* 129, no. 31 (2007): 9780–9788. http://dx.doi.org/10.1021/ja071875h.

Bass, J. C., and D. T. Allen. "Milliwatt Radioisotope Power Supply for Space Applications." In *Proceedings of 18th International Conference on Thermoelectrics*, 521–524. Piscataway, NJ: IEEE, 1999. http://dx.doi.org/10.1109/ICT.1999.843443.

Baxter, Jason, Zhixi Bian, Gang Chen, David Danielson, Mildred S. Dresselhaus, Andrei G. Fedorov, Timothy S. Fisher et al. "Nanoscale Design to Enable the Revolution in Renewable Energy." *Energy and Environmental Science* 2, no. 6 (2009): 559–588. http://dx.doi.org/10.1039/b821698c.

Bell, Lon E. Thermoelectric Heat Exchanger. US Patent 6,119,463, filed May 12, 1998, and issued September 19, 2000.

Bell, Lon E. "Cooling, Heating, Generating Power, and Recovering Waste Heat with Thermoelectric Systems." *Science* 321, no. 5895 (2008): 1457–1461. http://dx.doi.org/10.1126/science.1158899.

Benyus, Janine M. *Biomimicry: Innovation Inspired by Nature*. New York: Harper Collins, 1997.

Binnig, G., C. F. Quate, and C. H. Gerber. "Atomic Force Microscope." *Physical Review Letters* 56, no. 9 (1986): 930–933. http://dx.doi.org/10.1103/PhysRevLett.56.930.

Biswas, Kanishka, Jiaqing He, Ivan D. Blum, Chun-I Wu, Timothy P. Hogan, David N. Seidman, Vinayak P. Dravid, and Mercouri G. Kanatzidis. "High-Performance Bulk Thermoelectrics with All-Scale Hierarchical Architectures." *Nature* 489, no. 7416 (2012): 414–418. http://dx.doi.org/10.1038/nature11439.

Brabec, Cristoph J., and James R. Durrant. "Solution-Processed Organic Solar Cells." *Material Research Society Bulletin* 33, no. 07 (2008): 670–675. http://dx.doi.org/10.1557/mrs2008.138.

Broido, D. A., and T. L. Reinecke. "Effect of Superlattice Structure on the Thermoelectric Figure of Merit." *Physical Review B: Condensed Matter and Materials Physics* 51, no. 19 (1995): 13797–13800. http://dx.doi.org/10.1103/PhysRevB.51.13797.

Carr, John A., Yuqing Chen, Moneim Elshobahi, Rakesh C. Mahadevapuram, and Sumit Chaudhary. "Controlling Morphology in Plastic Solar Cells." *Nanomaterials and Energy* 1, no. 1 (2012): 18–26. http://dx.doi.org/10.1680/nme.11.00010.

Chemisana, Daniel. "Building Integrated Concentrating Photovoltaics: A Review." *Renewable & Sustainable Energy Reviews* 15, no. 1 (2011): 603–611. http://dx.doi.org/10.1016/j.rser.2010.07.017.

Chen, G., and A. Shakouri. "Heat Transfer in Nanostructures for Solid-state Energy Conversion." *Journal of Heat Transfer-Transactions of the ASME* 124, no. 2 (2002): 242–252. http://dx.doi.org/10.1115/1.1448331.

Chhabra, B., C. B. Honsberg, and R. L. Opila. "High Open Circuit Voltages on 50 Micron Silicon Substrates by Amorphous-silicon (a-Si) and Quinhydrone-methanol (QHY-ME) Passivation." In *34th IEEE Photovoltaic Specialists Conference*, Philadelphia, PA, June 7–12 2009, 2187–2190. http://dx.doi.org/10.1109/PVSC.2009.5411398.

Cullen, Jonathan M., and Julian M. Allwood. "The Efficient Use of Energy: Tracing the Global Flow of Energy from Fuel to Service." *Energy Policy* 38, no. 1 (2010): 75–81. http://dx.doi.org/10.1016/j.enpol.2009.08.054.

Dresselhaus, M. S., G. Chen, M. Y. Tang, R. Yang, H. Lee, D. Wang, Z. Ren, J.-P. Fleurial, and P. Gogna. "New Directions for Low-dimensional Thermoelectric Materials." *Advanced Materials* 19, no. 8 (2007): 1043–1053. http://dx.doi.org/10.1002/adma.200600527.

Duncan, Timothy V. "Applications of Nanotechnology in Food Packaging and Food Safety: Barrier Materials, Antimicrobials, and Sensors." *Journal of Colloid and Interface Science* 363, no. 1 (2011): 1–24. http://dx.doi.org/10.1016/j.jcis.2011.07.017.

Energy Information Administration. *Annual Energy Outlook 2013*. Washington, D.C.: US Energy Information Administration, 2013. http://www.eia.gov/forecasts/aeo/.

Energy Information Administration. *Electric Power Annual 2009 [DOE/EIA-0348(2009)]*. Washington, D.C.: US Department of Energy, 2011.

European Photovoltaic Industry Association. *Global Market Outlook for Photovoltaics, 2013–2017.* Brussels, Belgium: European Photovoltaic Industry Association, 2013.

Fan, Xiaofeng, Gehong Zeng, Chris LaBounty, John E. Bowers, Edward Croke, Channing C. Ahn, Scott Huxtable, Arun Majumdar, and Ali Shakouri. "SiGeC/Si Superlattice Microcoolers." *Applied Physics Letters* 78, no. 11 (2001): 1580–1582. http://dx.doi.org/10.1063/1.1356455.

Feynman, Richard P. "There's Plenty of Room at the Bottom." *Engineering and Science* 23, no. 5 (1960): 22–36.

Gallagher, S. J., B. C. Rowan, J. Doran, and B. Norton. "Quantum Dot Solar Concentrator: Device Optimisation Using Spectroscopic Techniques." *Solar Energy* 81, no. 4 (2007): 540–547. http://dx.doi.org/10.1016/j.solener.2006.07.006.

Gallagher, S. J., B. C. Rowan, and P. C. Eames. "Quantum Solar Concentrator: Electrical Conversion Efficiencies and Comparative Concentrating Factors of Fabricated Devices." *Solar Energy* 81, no. 6 (2007): 813–821. http://dx.doi.org/10.1016/j.solener.2006.09.011.

Ghiradella, Helen. "Light and Color on the Wing: Structural Colors in Butterflies and Moths." *Applied Optics* 30, no. 24 (1991): 3492–3500. http://dx.doi.org/10.1364/AO.30.003492.http://www.ncbi.nlm.nih.gov/entrez/query.fcgi?cmd=Retrieve&db=PubMed&list_uids=20706416&dopt=Abstract.

Gust, Devans, Thomas A. Moore, and Ana L. Moore. "Solar Fuels via Artificial Photosynthesis." *Accounts of Chemical Research* 42, no. 12 (2009): 1890–1898. http://dx.doi.org/10.1021/ar900209b.

Heremans, Joseph P., Vladimir Jovovic, Eric S. Toberer, Ali Saramat, Ken Kurosaki, Anek Charoenphakdee, Shinsuke Yamanaka, and G. Jeffrey Snyder. "Enhancement of Thermoelectric Efficiency in PbTe by Distortion of the Electronic Density of States." *Science* 321, no. 5888 (2008): 554–557. http://dx.doi.org/10.1126/science.1159725.

Herrera, Stephan, and Lawrence Aragon. "Small Worlds: Nanotechnology Wins over Mainstream Venture Capitalists." *Red Herring* 107 (2001): 51.

Hicks, L. D., and M. S. Dresselhaus. "Effect of Quantum-well Structures on the Thermoelectric Figure of Merit." *Physical Review B: Condensed Matter and Materials Physics* 47, no. 19 (1993): 12727–12731. http://dx.doi.org/10.1103/PhysRevB.47.12727.

Hsu, Kuei Fang, Sim Loo, Fu Guo, Wei Chen, Jeffrey S. Dyck, Ctirad Uher, Tim Hogan, E. K. Polychroniadis, and Mercouri G Kanatzidis. "Cubic AgPb(m)SbTe(2+m): Bulk Thermoelectric Materials with High Figure of Merit." *Science* 303, no. 5659 (2004): 818–821. http://dx.doi.org/10.1126/science.1092963.

Huang, M. J., P. C. Eames, and B. Norton. "Thermal Regulation of Building-integrated Photovoltaics Using Phase Change Materials." *International Journal of Heat and Mass Transfer* 47, no. 12–13 (2004): 2715–2733. http://dx.doi.org/10.1016/j.ijheatmasstransfer.2003.11.015.

International Energy Agency. *Technology Roadmap: Solar Photovoltaic Energy.* Paris: IEA Publications, 2010.

Ji, Liwen, Zhan Lin, Mataz Alcoutlabi, and Xiangwu Zhang. "Recent Developments in Nanostructured Anode Materials for Rechargeable Lithium-ion Batteries." *Energy and Environmental Science* 4, no. 8 (2011): 2682–2699. http://dx.doi.org/10.1039/c0ee00699h.

Kempe, Michael D. "Modeling Rates of Moisture Ingress into Photovoltaic Modules." *Solar Energy Materials and Solar Cells* 90, no. 16 (2006): 2720–2738. http://dx.doi.org/10.1016/j.solmat.2006.04.002.

Kennedy, M., S. J. McCormack, J. Doran, and B. Norton. "Improving the Optical Efficiency and Concentration of a Single-plate Quantum Dot Solar Concentrator Using Near Infra-red Emitting Quantum Dots." *Solar Energy* 83, no. 7 (2009): 978–981. http://dx.doi.org/10.1016/j.solener.2008.12.010.

Kim, Raseong, Supriyo Datta, and Mark S. Lundstrom. "Influence of Dimensionality on Thermoelectric Device Performance." *Journal of Applied Physics* 105, no. 3 (2009): 034506. http://dx.doi.org/10.1063/1.3074347.

Kwon, Moo-Hyun. "Stability of Bulk Hetrojunction Organic Solar Cells with Different Blend Ratios of P3HT:PCBM." *Transactions on Electrical and Electronic Materials* 13, no. 2 (2012): 98–101. http://dx.doi.org/10.4313/TEEM.2012.13.2.98.

Liu, Chin-Yi, Zacharay C. Holman, and Uwe R. Kortshagen. "Hybrid Solar Cells from P3HT and Silicon Nanocrystals." *Nano Letters* 9, no. 1 (2009): 449–452. http://dx.doi.org/10.1021/nl8034338.

Mahan, G. D., and L. M. Woods. "Multilayer Thermionic Refrigeration." *Physical Review Letters* 80, no. 18 (1998): 4016–4019. http://dx.doi.org/10.1103/PhysRevLett.80.4016.

Moore, Gordon E. "Cramming More Components onto Integrated Circuits." *Electronics* 38, no. 8 (1965): 114–117.

Nolas, George S., Joe Poon, and Mercouri Kanatzidis. "Recent Developments in Bulk Thermoelectric Materials." *Materials Research Society Bulletin* 31, no. 03 (2006): 199–205. http://dx.doi.org/10.1557/mrs2006.45.

Nolas, George S., Jeffrey Sharp, and H. Julian Goldsmid. *Thermoelectrics: Basic Principles and New Materials Developments*. New York: Springer-Verlag, 2001.

Norton, Brian, Manus Kennedy, and Sarah McCormack. "Solar Energy Concentration Without Moving Parts in Sustainable Building Envelopes." In *Renewables in a Changing Climate: From Nano to Urban Scale*, 73–78. Lausanne, Switzerland: Ecole Polytechnique Fédérale de Lausanne, 2009.

Oda, Shunri, and David Ferry, eds. *Silicon Nanoelectronics*. Boca Raton, FL: CRC Press, 2005.

Oelhafen, P., and A. Schüler. "Nanostructured Materials for Solar Energy Conversion." *Solar Energy* 79, no. 2 (2005): 110–121. http://dx.doi.org/10.1016/j.solener.2004.11.004.

Oreski, G., and G. M. Wallner. "Delamination Behaviour of Multi-layer Films for PV Encapsulation." *Solar Energy Materials and Solar* 89, no. 2–3 (2005): 139–151. http://dx.doi.org/10.1016/j.solmat.2005.02.009.

Parida, Manas R., C. Vijayan, C. S. Rout, C. S. Suchand Sandeep, Reji Philip, and P. C. Deshmukh. "Room Temperature Ferromagnetism and Optical Limiting in V2O5 Nanoflowers Synthesized by a Novel Method." *Journal of Physical Chemistry* 115 (2011): 112–117. http://dx.doi.org/10.1021/jp107862n.

Pei, Yanzhong, Jessica Lensch-Falk, Eric S. Toberer, Douglas L. Medlin, and G. Jeffrey Snyder. "High Thermoelectric Performance in PbTe Due to Large Nanoscale Ag2Te Precipitates and La Doping." *Advanced Functional Materials* 21, no. 2 (2011): 241–249. http://dx.doi.org/10.1002/adfm.201000878.

Pei, Yanzhong, Xiaoya Shi, Aaron LaLonde, Heng Wang, Lidong Chen, and G. Jeffrey Snyder. "Convergence of Electronic Bands for High Performance Bulk Thermoelectrics." *Nature* 473, no. 7345 (2011): 66–69. http://dx.doi.org/10.1038/nature09996.

Price, Samuel C., Andrew C. Stuart, Liqiang Yang, Huaxing Zhou, and Wei You. "Fluorine Substituted Conjugated Polymer of Medium Band Gap Yields 7% Efficiency in Polymer-Fullerene Solar Cells." *Journal of the American Chemical Society* 133, no. 12 (2011): 4625–4631. http://dx.doi.org/10.1021/ja1112595.

Ragauskas, Arthur J., Charlotte K. Williams, Brian H. Davison, George Britovsek, John Cairney, Charles A. Eckert, William J. Frederick, Jr. et al. "The Path Forward for Biofuels and Biomaterials." *Science* 311, no. 5760 (2006): 484–489. http://dx.doi.org/10.1126/science.1114736.

Rawat, Vijay, Yee Kan Koh, David G. Cahill, and Timothy D. Sands. "Thermal Conductivity of (Zr,W)N/ScN Metal/Semiconductor Multilayers and Superlattices." *Journal of Applied Physics* 105, no. 2 (2009): 024909. http://dx.doi.org/10.1063/1.3065092.

Rawat, Vijay, and Timothy D. Sands. "Growth of TiN/GaN Metal/Semiconductor Multilayers by Reactive Pulsed Laser Deposition." *Journal of Applied Physics* 100, no. 6 (2006): 064901. http://dx.doi.org/10.1063/1.2337784.

REN21. Renewables 2013 Global Status Report. Paris: REN21 Secretariat, 2013.

Reineke, Sebastian, Frank Lindner, Gregor Schwartz, Nico Seidler, Karsten Walzer, Björn Lüssem, and Karl Leo. "White Organic Light-emitting Diodes with Fluorescent Tube Efficiency." *Nature* 459, no. 7244 (2009): 234–238. http://dx.doi.org/10.1038/nature08003.

Richardson, D. J., J. M. Fini, and L. E. Nelson. "Space-division Multiplexing in Optical Fibres." *Nature Photonics* 7, no. 5 (2013): 354–362. http://dx.doi.org/10.1038/nphoton.2013.94.

D. M. Rowe, ed. *CRC Handbook of Thermoelectrics*. Boca Raton, FL: CRC Press, 1995.

Rowe, D. M., ed. *Thermoelectrics Handbook: Macro to Nano*. Boca Raton, FL: CRC Press, 2005.

Roy, Somnath C., Oomman K. Varghese, Maggie Paulose, and Craig A. Grimes. "Toward Solar Fuels: Photocatalytic Conversion of Carbon Dioxide to Hydrocarbons." *ACS Nano* 4, no. 3 (2010): 1259–1278. http://dx.doi.org/10.1021/nn9015423

Sales, B. C., D. Mandrus, and R. K. Williams. "Filled Skutterudite Antimonides: A New Class of Thermoelectric Materials." *Science* 272, no. 5266 (1996): 1325–1328. http://dx.doi.org/10.1126/science.272.5266.1325.

Saramat, A., G. Svensson, A. E. C. Palmqvist, C. Stiewe, E. Mueller, D. Platzek, S. G. K. Williams, D. M. Rowe, J. D. Bryan, and G. D. Stucky. "Large Thermoelectric Figure of Merit at High Temperature in Czochralski-grown Clathrate $Ba_8Ga_{16}Ge_{30}$." *Journal of Applied Physics* 99, no. 2 (2006): 023708. http://dx.doi.org/10.1063/1.2163979.

Savage, Nora, Mamadou Diallo, Jeremiah Dunchan, Anita Street, and Richard Sustich, eds. *Nanotechnology Applications for Clean Water*. Norwich, NY: William Andrew, 2009.

Scharber, M. C., D. Mühlbacher, M. Koppe, P. Denk, C. Waldauf, A. J. Heeger, and C. J. Brabec. "Design Rules for Donors in Bulk Hetrojunction Solar Cells: Towards 10% Energy Conversion Efficiency." *Advanced Materials* 18, no. 6 (2006): 789–794. http://dx.doi.org/10.1002/adma.200501717.

Schlapbach, Louis, and Andreas Züttel. "Hydrogen-storage Materials for Mobile Applications." *Nature* 414, no. 6861 (2001): 353–358. http://dx.doi.org/10.1038/35104634.

Schoonman, Joop, and Roel van de Krol. "Nanostructured Materials for Solar Hydrogen Production." *UPB Scientific Bulletin Series B: Chemistry and Materials Science* 73, no. 4 (2011): 31–44. http://scientificbulletin.upb.ro/rev_docs_arhiva/full14674.pdf.

Seale-Goldsmith, Mary-Margaret, and James F. Leary. "Nanobiosystems." *Wiley Interdisciplinary Reviews: Nanomedicine and Nanobiotechnology* 1, no. 5 (2009): 553–567. http://dx.doi.org/10.1002/wnan.49.

Shakouri, Ali. "Nanoscale Thermal Transport and Microrefrigerators on a Chip." *Proceedings of the IEEE* 94, no. 8 (2006): 1613–1638. http://dx.doi.org/10.1109/JPROC.2006.879787.

Shakouri, Ali. "Recent Developments in Semiconductor Thermoelectric Physics and Materials." *Annual Review of Materials Research* 41, no. 1 (2011): 399–431. http://dx.doi.org/10.1146/annurev-matsci-062910-100445.

Shakouri, Ali, and John E. Bowers. "Heterostructure Integrated Thermionic Coolers." *Applied Physics Letters* 71, no. 9 (1997): 1234–1236. http://dx.doi.org/10.1063/1.119861.

Shakouri, Ali, and Mona Zeberjadi. "Nanoengineered Materials for Thermoelectric Energy Conversion." In *Thermal Nanosystems and Nanomaterials*, edited by Sebastian Volz, 225–299. New York: Springer, 2009. http://dx.doi.org/10.1007/978-3-642-04258-4_9.

Smolyaninov, Igor I., Yu-Ju Hung, and Christopher C. Davis. "Magnifying Superlens in the Visible Frequency Range." *Science* 315, no. 5819 (2007): 1699–1701. http://dx.doi.org/10.1126/science.1138746.

Sofo, J. O., and G. D. Mahan. "Thermoelectric Figure of Merit of Superlattices." *Applied Physics Letters* 65, no. 21 (1994): 2690–2692. http://dx.doi.org/10.1063/1.112607.

Solar Energy Technologies Office. "SunShot Initiative." *Department of Energy*. Last modified August 8, 2013. http://www1.eere.energy.gov/solar/sunshot/about.html.

Song, Min-Kyu, Soojin Park, Faisal M. Alamgir, Jaephil Cho, and Meilin Liu. "Nanostructured Electrodes for Lithium-ion and Lithium-air Batteries: The Latest Developments, Challenges and Perspectives." *Materials Science and Engineering* 72, no. 11 (2011): 203–252. http://dx.doi.org/10.1016/j.mser.2011.06.001.

Van der Sluis, Paul. "Improvement in Cooking Stoves." International Patent WO/2006/103613, filed March 27, 2006, and granted October 5, 2006.

Vashaee, Daryoosh, and Ali Shakouri. "Improved Thermoelectric Power Factor in Metal-based Superlattices." *Physical Review Letters* 92, no. 10 (2004): 106103. http://dx.doi.org/10.1103/PhysRevLett.92.106103.

Vining, Cronin B. "An Inconvenient Truth about Thermoelectrics." *Nature Materials* 8, no. 2 (2009): 83–85. http://dx.doi.org/10.1038/nmat2361.

Wang, X. W., H. Lee, Y. C. Lan, G. H. Zhu, G. Joshi, D. Z. Wang, J. Yang et al. "Enhanced Thermoelectric Figure of Merit in Nanostructured N-type Silicon Germanium Bulk Alloy." *Applied Physics Letters* 93, no. 19 (2008): 193121. http://dx.doi.org/10.1063/1.3027060.

Wereley, Steven T., and Carl D. Meinhart. "Recent Advances in Micro Particle Image Velocimetry." *Annual Review of Fluid Mechanics* 42, no. 1 (2010): 557–576. http://dx.doi.org/10.1146/annurev-fluid-121108-145427.

Xiong, Guoping, K. P. S. S. Hembram, R. G. Reifenberger, and Timothy S. Fisher. "MnO2-coated Graphitic Petals for Supercapacitor Electrodes." *Journal of Power Sources* 227 (2013): 254–259. http://dx.doi.org/10.1016/j.jpowsour.2012.11.040.

Yazawa, Kazuaki, Yee Rui Koh, and Ali Shakouri. "Optimization of Thermoelectric Topping Combined Steam Turbine Cycles for Energy Economy." *Applied Energy* 109 (2013): 1–9. http://dx.doi.org/10.1016/j.apenergy.2013.03.050.

Yazawa, Kazuaki, and Ali Shakouri. "Efficiency of Thermoelectric Generators at Maximum Output Power." Paper submitted at the *International Conference on Thermoelectrics*, Shanghai, China, 2010.

Yazawa, Kazuaki, and Ali Shakouri. "Energy Payback Optimization of Thermoelectric Power Generator Systems." In *ASME 2010 International Mechanical Engineering Congress and Exposition Volume 5: Energy Systems Analysis, Thermodynamics and Sustainability; NanoEngineering for Energy; Engineering to Address Climate Change, Parts A and B*, 569–576. New York: American Society of Mechanical Engineers, 2010. http://dx.doi.org/10.1115/IMECE2010-37957.

Yazawa, Kazuaki, and Ali Shakouri. "Cost-efficiency Trade-off and the Design of Thermoelectric Power Generators." *Environmental Science & Technology* 45, no. 17 (2011): 7548–7553. http://dx.doi.org/10.1021/es2005418.

Yazawa, Kazuaki, and Ali Shakouri. "Cost-effective Waste Heat Recovery Using Thermoelectric Systems." *Journal of Materials Research* 27, no. 09 (2012): 1211–1217. http://dx.doi.org/10.1557/jmr.2012.79.

Zebarjadi, Mona, Zhixi Bian, Rajeev Singh, Ali Shakouri, Robert Wortman, Vijay Rawat, and Timothy Sands. "Thermoelectric Transport in a ZrN/ScN Superlattice." *Journal of Electronic Materials* 38, no. 7 (2009): 960–963. http://dx.doi.org/10.1007/s11664-008-0639-5.

Zebarjadi, Mona, Giri Joshi, Gaohua Zhu, Bo Yu, Austin Minnich, Yucheng Lan, Xiaowei Wang, Mildred Dresselhaus, Zhifeng Ren, and Gang Chen. "Power Factor Enhancement by Modulation Doping in Bulk Nanocomposites." *Nano Letters* 11, no. 6 (2011): 2225–2230. http://dx.doi.org/10.1021/nl201206d.

Zhang, W., N. Mingo, and T. S. Fisher. "Simulation of Phonon Transport Across a Non-polar Nanowire Junction Using an Atomistic Green's Function Method." *Physical Review* 76, no. 19 (2007): 195429. http://dx.doi.org/10.1103/PhysRevB.76.195429.

Zhou, Weilie, and Zhong Lin Wang, eds. *Scanning Microscopy for Nanotechnology: Techniques and Applications*. New York: Springer, 2006. http://dx.doi.org/10.1007/978-0-387-39620-0.

Zide, J. M. O., J.-H. Bahk, R. Singh, M. Zebarjadi, G. Zeng, H. Lu, J. P. Feser et al. "High Efficiency Semimetal/semiconductor Nanocomposites Thermoelectric Materials." *Journal of Applied Physics* 108, no. 12 (2010): 123702. http://dx.doi.org/10.1063/1.3514145.

Chapter 6

Biofuel Prospects in an Uncertain World

WALLY TYNER AND RICHARD A. SIMMONS

Abstract

Biofuels have grown from almost nothing in the mid-1970s to over 10 percent of gasoline consumption in some countries. Biofuels are essentially a government created industry in that when they were introduced, they required government subsidies to compete with fossil fuels. Governments saw several benefits for biofuels: 1) reducing dependence on foreign oil and reducing supply disruptions such as those of the 1973 and 1979 oil crises; 2) providing income and employment in rural areas; 3) improving air quality through lower tailpipe emissions; and 4) reducing greenhouse gas (GHG) emissions because CO_2 is sequestered during the production of the feedstock. In some measure, biofuels have contributed to achieving all these objectives, but questions have been raised regarding the effectiveness of biofuels in doing so, as well as on unintended consequences of large-scale biofuel production. We will explore all these issues in this chapter, and our scope will remain limited to liquid vehicle fuels. First, we provide a brief history of biofuels production. Then we explore the implications of the development of first generation biofuels with emphasis on corn ethanol in the United States. Most of the chapter is devoted to second-generation biofuels; that is, biofuels produced from cellulosic feedstocks including crop residues such as corn stover, dedicated energy crops such as miscanthus and switchgrass, tree plantations such as poplar or willow, and forest residues. The early hope was that cellulosic biofuels would be developed fairly quickly, and that first generation biofuels (mainly corn and sugarcane ethanol) would plateau with much of the growth coming from second-generation biofuels. However, to date there has been almost no commercial development of cellulosic biofuels. Much has been invested in research, but most of the development has been at the pilot or demonstration plant level. We will explore five areas of uncertainty facing potential investors in cellulosic biofuels. Finally, we will summarize the major challenges and opportunities for cellulosic biofuels.

6.1. Biofuels History

Biofuels have been a part of the global energy picture since the mid-1970s. Brazil was the first major producer beginning in 1975 with the launch of its PROALCOOL program, which provided subsidies for sugarcane ethanol production.[1] This policy was mainly motivated by the 1973 oil crisis, and Brazil saw sugarcane ethanol as a means of becoming more independent from the rest of the world for liquid fuel. Production began in the United States in the early 1980s, stimulated by the National Energy Conservation Policy Act of 1978.[2] That legislation initially provided a subsidy for ethanol of $0.40 per gallon, and the subsidy continued at a level between $0.40 and $0.60 per gallon until it finally expired at the end of 2011. Similarly, European Union countries began biofuel production supported by government subsidies about the same time.

Over time, as biofuel production grew, governments in all three regions moved from reliance on subsidies as the major policy instrument to mandates. In Brazil and the EU, targets are established for blending certain percentages of biofuels in the total fuel mix. In the US, there is a volumetric mandate known as the Renewable Fuel Standard (RFS),[3] which sets steadily increasing volume requirements for different types of biofuels through 2022. By that terminal year, thirty-six billion gallons of ethanol equivalent are prescribed, of which twenty-one billion will need to come from sources other than grains or corn starch as indicated in Figure 6.1. The EPA has been forced to waive most of the cellulosic biofuel portion of the RFS every year because the biofuels do not exist, and will need to continue cellulosic biofuel waivers through 2022, as conversion facilities could not be developed at the pace of increase included in the RFS. Thus, maintaining the mandated trajectory is in serious doubt.

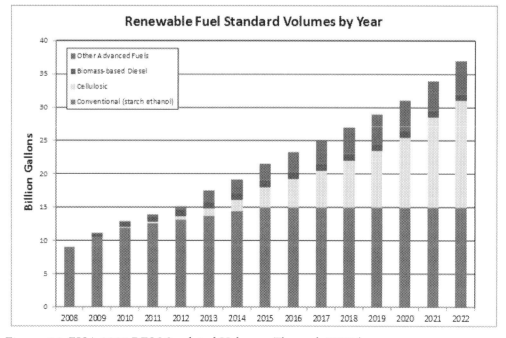

Figure 6.1. EISA 2007 RFS Mandated Volumes Through 2022.[4]

The main difference between a mandate and a subsidy is that the cost of a subsidy is directly born by the government budget, whereas the cost of a mandate is born by consumers through higher costs of the blended fuels. That is, presumably the mandated product (biofuel in this case) would not be used in the absence of the mandate; otherwise, there would be no need for the mandate. Thus, the projected higher cost of the biofuel is passed on to consumers through higher cost of the blended fuel. In a sense the cost is not visible to consumers, whereas the government budget cost of a subsidy is very visible. This visibility became more of an issue as the subsidized volumes grew over time. Today the dominant biofuels policies are mandates and not subsidies. The EU targets were set in the 2009 directive on mandated use of renewable energy, which requires ten percent of transportation fuels to be renewable by 2020.[5] Most authorities expect the liquid fuels part of that to be around 5.6 percent, with electric vehicles and other options taking the rest.[6] Estimates drawn from a variety of sources indicate that the United States, Brazil and the European Union account for nearly ninety percent of global biofuels production, which today remains almost exclusively first generation. Small but increasing quantities of biofuels also have been developed in other regions as indicated in Figure 6.2.

6.2. First Generation Biofuels

Ethanol production capacity has grown substantially in both Brazil and the United States. For the past decade, most of the automobiles sold in Brazil have been flex-fuel, meaning that they can consume any blend of ethanol and gasoline. This provides great flexibility both for consumers and government policy. For example, in a recent year, there was a poor sugarcane harvest in Brazil, and at the same time world sugar prices were abnormally high. The Brazilian Government responded by lowering the mandated ethanol average blend level first from twenty-five to twenty percent and then down to eighteen percent. The level was subsequently increased again when conditions improved for sugar supply. Consumers in Brazil can choose whatever percentage of ethanol they want (above the minimum mandated) because many gasoline stations are equipped

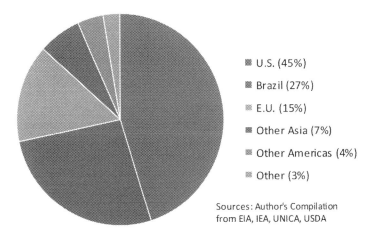

Sources: Author's Compilation from EIA, IEA, UNICA, USDA

Figure 6.2. Global Share of Biofuels Production, 2010 (Energy Equivalent Basis).

with what are called blender pumps. Ethanol contains about two-thirds as much energy as gasoline, so ethanol must be priced no more than sixty-seven percent of gasoline price to induce consumers to use ethanol blends. If the ethanol price is below that fraction, consumers may opt for high ethanol blends or even pure ethanol. Thus, because Brazil made a decision to invest in flex-fuel vehicles and flexible fuel dispensing infrastructure a decade ago, their supply chain today is quite flexible.

The US is not in that situation. In the early 1980s, the US decided on a ten percent ethanol blend as the standard, and it became known as gasohol. All vehicles sold were required to be able to use the ten percent blend, also known as E10. A second blend of up to eighty-five percent ethanol (E85) was also developed, and to consume this fuel requires flex-fuel vehicles. To date very few of the vehicles on the road in the US are flex-fuel (about 11 million out of 250 million). Also, E85 often is not cost competitive with E10 on an energy basis, which determines mileage, so most ethanol is consumed as E10. This was not a problem until recently. Today the US consumes about 133 billion gallons of gasoline type fuel per year. At a ten percent blend, this means the maximum ethanol consumption would be 13.3 billion gallons. The RFS mandates consumption of fifteen billion gallons of conventional (corn-based) ethanol per year by 2015, and the US has very nearly that much ethanol production capacity today. Thus the maximum ethanol consumption given the E10 blend level (called the *blend wall*) is less than the existing production capacity.[7] The blend wall is the most important issue faced by the US ethanol industry today. The Environmental Protection Agency (EPA) in 2012 gave final approval for increasing the blend limit to fifteen percent, but the approval was only for vehicles built since 2001. About one-third of the US vehicle fleet is older, and is not approved to use the E15 blend. Despite the fact that the newer vehicles account for a greater proportion of the miles driven, fueling stations have been reluctant to switch to E15 because they would still lose a significant fraction of their customer base. Thus, at present, the blend wall is binding, and is a major impediment to growth of the ethanol industry.

The European biofuels market is largely for biodiesel, not ethanol. Biodiesel can generally be blended up to twenty percent without adverse effects. With more than ninety percent of the energy content of petroleum-derived diesel, biodiesel impinges vehicle range substantially less than bioethanol does in comparison with gasoline. In this case, the major impediment to biodiesel development is the high cost of the raw vegetable oil, commonly from rapeseed, soybeans, or palm.

Clearly one of the objectives of government support for biofuels was increasing farm and rural incomes.[8] However, an unintended consequence has been that the biofuel demand for corn, sugar, and oilseeds has contributed to higher global prices for these commodities. Biofuels are not the only driver of higher commodity prices, but they are important.[9] Agricultural commodity prices have increased because of increased global demand for commodities due partly to dietary transition especially in developing countries. As income rises, people demand more animal protein in their diets, which increases demand for animal feed ingredients. In the decade prior to 2008, global consumption of agricultural commodities grew faster than global production in all but one year. That trend meant that global stocks to use ratios were quite low going into 2008, the first large commodity price surge. Also, there were regional production shortfalls that accentuated the shortage. Furthermore, the United States dollar fell sixty-seven percent between 2002 and 2008, which led to increased prices for commodities sold in that currency. Then, on top of these important drivers, there was an increase in use particularly of corn for biofuels.

For the biofuels driver, we need to distinguish between biofuels demand driven by government mandates or subsidies and demand driven by the higher price of crude oil and gasoline. In 2008, we estimated that the surge in crude oil price actually had been more important as a driver of increased ethanol demand for corn than had been the government subsidy.

In 2011, there was another surge in many agricultural commodity prices. Our analysis then indicated that the major drivers were a concurrent increase in Chinese demand for soybean imports and a surge in demand for corn for ethanol. Corn and soybeans are generally grown in the same areas in the US, and the combination of these two drivers led to a 189% increase in the acreage needed in the US to supply these two demands between 2005 and 2010. Thus, a big part of the surge in commodity prices in 2011 was this perfect storm of demand surges.

We began 2012 with low stocks, and then were hit by the drought of 2012. The drought caused agricultural commodity prices, especially corn, to surge again in 2012. This surge led to calls for reduction or elimination of the US RFS for corn ethanol.[10] However, US policy is no longer a key driver of the corn ethanol industry. While government subsidies, some of which have now expired, and the RFS were critical to the development of the industry, the economic reality today is that corn ethanol is cheaper per gallon than gasoline, and blenders today have an economic incentive to continue blending ethanol even in the absence of government mandates. This is true despite the fact that ethanol is more expensive on an energy basis, as most gasoline is sold as E10, and there is no competition on an energy basis. Thus, removal or reduction of the RFS in 2012 or 2013 likely would have little impact on the demand for corn for ethanol, and thus on the corn price.[11]

There is no doubt that urban consumers are adversely impacted by higher agricultural commodity prices (due to any cause), particularly in developing countries. However, there is another dimension to the higher prices that gets little attention. The higher commodity prices can help increase rural incomes in developing countries.[12] The World Bank says seventy percent of the world's poor live in rural areas and derive their primary livelihood from agriculture. To the extent that developing country farmers have access to the higher commodity prices, their incomes could be increased and poverty reduced. There is already evidence that cropland area has increased significantly in the past six years mainly in developing countries with the higher commodity prices.[13]

6.3. Second Generation Biofuels

Second generation biofuels can be produced from a variety of advanced feedstocks including cellulosic sources, and perhaps eventually algae. Cellulosic feedstocks can be agricultural residues such as corn stover, wheat straw, or forest residues, or they can be dedicated energy crops such as miscanthus, switchgrass, poplar, and willow. These feedstocks can be converted to biofuels via biochemical or thermochemical or hybrid processes. The end product of biochemical conversion is typically ethanol. Thermochemical or hybrid conversion processes can produce a wide range of biofuels including green diesel, jet fuel, bio-gasoline, and others. Because of the blend wall discussed above and for many other reasons, there is today much more interest in the conversion of cellulosic feedstocks to drop-in biofuels that are more compatible with existing infrastructure and fuel supply chains.

Though extremely complicated to measure definitively, two indicative measures have been used to assess the relative energy and emissions benefits of biofuels as compared with the petroleum fuels they replace. Overall energy ratio represents the ratio of energy contained in a given quantity of finished fuel to the energy needed to produce that same quantity of fuel. For example, it is estimated that corn ethanol has an approximate energy ratio in the range of about 1.3 or 1.6 to one,[14] whereas estimates for sugarcane ethanol range from about five to one to greater than nine to one.[15] The energy ratio for sugarcane ethanol is favorable, but varies greatly depending on harvesting method (such as manual vs. mechanized) and how efficiently agricultural residue (*bagasse*) is converted to process steam and surplus electricity. Biodiesel produced from soy or rapeseed oil is estimated to have an energy ratio above two to one but below five to one.[16] Advanced and cellulosic biofuels produced from a host of new pathways are striving to exceed the energy ratio of the most energy efficient first generation biofuels, but this is highly case sensitive and remains to be confirmed at commercial scales. It should be noted that energy ratios alone do not tell the whole story. The form of energy is critical. When solid biological feedstocks are converted into liquid fuels, the form of energy becomes much more useful for the transportation system. Also, any transformation of energy from one form to another (such as from crude oil to gasoline) involves a loss in

Table 6.1. Renewable Fuel Categories and GHG Thresholds as Defined in the Renewable Fuels Standard.[17]

Renewable Fuel Category	Minimum Lifecycle GHG Reduction of Subject Biofuel Category Relative to a Baseline of Gasoline or Diesel	Example
Conventional Biofuel	20%	Starch feedstocks including corn, sorghum, wheat
Biomass-Based Diesel	50%	Biodiesel such as from soy or rapeseed and non-ester renewable diesel
Cellulosic Biofuel	60%	Any fuel derived from cellulose, hemicellulose or lignin non-food feedstocks
Other Advanced Biofuel	50%	Any fuel derived from renewable feedstocks including sugarcane ethanol

energy. The point is that it becomes more useful energy. Electricity is more useful for lighting than coal. To a real world marketplace then, it is not so much the joules themselves as the form of those joules that matters.

Another critical measure of a biofuel's relative performance is reduction in greenhouse gas (GHG) emissions as compared to gasoline or diesel fuels. This too is a complicated metric to pinpoint, however, in the context of the Renewable Fuels Standard, the EPA has established key default criteria to help frame objective GHG targets for primary biofuel categories (Table 6.1). It is of note that the GHG emissions improvements for cellulosic and advanced biofuels must exceed those of first generation biofuels to qualify toward the US RFS volumetric requirements.

With the billions invested in biofuels research, substantial progress has been made in advancing feedstock development and conversion processes for second-generation biofuels. In general, the economic reality is that cellulosic biofuels are not competitive with fossil fuels on a market basis. Some form of government support is needed to elicit private sector investment in cellulosic biofuels. In addition, there is considerable uncertainty facing potential investors in cellulosic biofuels.[18] In the following section, we will examine five major sources of uncertainty: future oil price, feedstock cost and availability, conversion technology yield and cost, environmental impacts, and government policy.

6.3.1. Oil Price Uncertainty

Anytime an investment is being considered, one must forecast future input and output product prices to do an economic assessment of project viability. For biofuels, the key price uncertainty is future oil prices since all the biofuel competing products are derived from crude oil. Furthermore, petroleum represents a major input to agricultural crop production in the form of diesel fuel and fertilizer, and this impact is thereby manifested in biofuel feedstock costs. Each year, the US Department of Energy (DOE) forecasts future oil prices using a reference case, low oil price case, and high oil price case (Figure 6.3).[19] The DOE crude oil reference forecast for 2040 is $163/bbl. in 2011 dollars. That value is somewhat less than our estimate of the crude oil price needed to render cellulosic biofuels economic without government intervention. The low and high price forecasts for 2040 are $75 and $237 respectively. DOE does not assign probabilities to the three cases, but clearly the reference case is the one they deem most likely. A potential investor facing this huge range of possible prices and knowing that the breakeven price is likely a bit higher than the reference case would be unlikely to make the investment. Future oil price uncertainty deters investment in cellulosic biofuels.

6.3.2. Feedstocks Cost and Availability

There are potential issues with both the quantity and cost of cellulosic feedstocks. The good news, however, is that most of the assessments of feedstock availability have determined that there is plenty of cellulosic material available in the US to meet and exceed the RFS.[21] The problem is one of cost. A National Research Council report estimated feedstock costs ranging from $75 to $133 per dry ton.[22] Generally, residues, such as corn stover, wheat straw, and forest residues, had lower costs ($75–$92) than

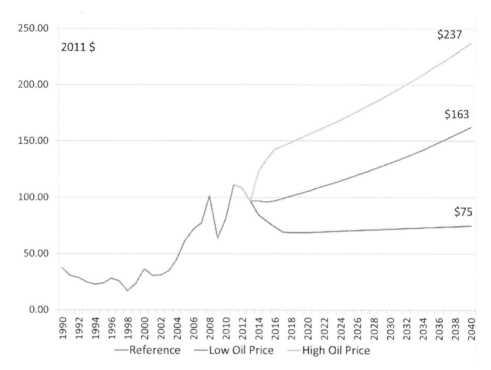

Figure 6.3. DOE (Brent) Crude Oil Price Forecasts 2013–2040.[20]

dedicated energy crops, like miscanthus, switchgrass, and short rotation woody crops ($89–$133). These costs are all much higher than had been expected a few years ago. The high cost of feedstock is one of the major factors driving the projected higher cost of biofuels. Current market conditions for certain dedicated crops are also not attractive to farmers, who can realize much greater returns from conventional agricultural crops. In the near term, increased use of cover crops and agricultural residues for feedstocks may help mitigate this challenge, as they afford farmers supplementary profit from currently allocated and revenue-generating land.

Beyond cost, there are also some logistical issues involved in developing the feedstock supply system to biorefineries. For the dedicated energy crops, long term contracts will be needed, as farmers would not be willing to make the up-front investment required to establish the crops without some assurance of a market for the product. Biorefineries also may want longer term contracts for crop residues as well to assure they have a locked-in feedstock supply before investing the hundreds of millions of dollars in a facility. To develop these long term contracts, appropriate risk sharing mechanisms will have to be designed. In addition, most cellulosic biomass has low bulk density, meaning its transportation over distances greater than one hundred miles can render biofuel production economics no longer viable.

6.3.3. Conversion Technology Yield and Cost

Many of the conversion technologies briefly described below and in the accompanying sidebar on *Second Generation Biofuel Conversion Technologies*, have been shown to work in that they produce the desired biofuel. The question again is at what cost. Since there is very little commercial production, and none without some form of government support, we do not really know what cellulosic biofuels will cost. Our best estimate is that using the range of feedstock prices above, biofuels are likely to cost at least $4.50/gallon of gasoline equivalent. Costs could be considerably higher, or they could be lower with a major technical breakthrough. This gasoline equivalent level implies crude oil breakeven price of around $150/barrel.[23]

Some of the technologies being considered have been around for a very long time. For example, the Fischer-Tropsch (FT) process of producing synthetic gas, which can then be converted into a range of hydrocarbon products, was used by Germany during World War II. At that time, German oil supply was restricted or cut off, and coal was therefore converted into liquid fuels via the FT process. There are plants today that use the process with coal or shale as a raw material, and the process has been proven with biomass. Again, the question is one of cost. Biomass is not as energy dense as coal, making the economic viability more sensitive to transportation and logistics. What is more, experience has shown that the end fuel cost is high even using coal as a feedstock, in part due to the large scale and high capital cost structure typical of a typical Fischer-Tropsch gasification plant.

Second Generation Biofuel Conversion Technologies

As noted, great interest and expectations, but slow and costly development characterize efforts to commercialize advanced biofuel technologies which rely on non-food feedstocks and sustainable approaches. Recall that leading sources of cellulosic biomass may include agricultural residue, wood waste, tree trimmings, switchgrass and miscanthus. An objective of biofuel innovators on this frontier is to deliver energy balances, energy ratios and GHG reductions approaching or exceeding those typical of sugarcane ethanol, which is considered an Advanced Biofuel by the EPA. Two leading conversion technologies are being pursued via substantial public and private investments, and are summarized briefly below:

Biochemical conversion involves a multi-step process beginning with the pre-treatment of biomass with some combination of chemical agents and/or catalysts, whose function is to begin breaking down the cellular structure of its three primary constituents: cellulose, hemi-cellulose and lignin. Conditioning and enzymatic hydrolysis typically follow, which balance the pH and help free the sugars of the cellulose and hemi-cellulose. Finally, microorganisms such as yeasts and bacteria are introduced to biologically convert these sugars into alcohols via fermentation. It should be noted here that other low temperature, non-biological processing routes, referred to simply as *chemical conversion pathways* represent viable technologies as well. In these, catalytic and mechanical systems are the principle means of producing fuel from the sugars and intermediates in biomass. In the United States, several demonstration or pre-commercial plants based upon variants of the biochemical conversion pathway are scheduled to come online in the 2013–2015 timeframe, and assist in validating process parameters at scale. Major areas of technical uncertainty and continuing research and development include: feedstock variability, cost and effectiveness of enzymes, production of enzymes at scale, fermentation processing parameters such as rates and yields, and process integration.[24] None alone is considered insurmountable, but interactions exist which can complicate and delay technical and economic optimization.

Thermochemical conversion involves heating biomass in the absence of oxygen or air, followed by extraction of hydrocarbons in either liquid (pyrolysis) or gaseous (synthesis gas) form. In pyrolysis, the resultant decomposed biomass, or *bio-oil* can be treated with chemical agents to fractionate the liquid molecules and be further refined into an array of bioproducts, sugars, and cellulosic fibers. In gasification, the resultant syn-gas composed of primarily hydrogen and carbon monoxide, is reformed, cleaned, and conditioned to form the building blocks of synthetic hydrocarbon molecules. This technology is generally less sensitive to variations in quality and type of biomass than biochemical pathways. At the same time, thermochemical processes are extremely capital-intensive, so efficiency and yield are more dependent upon scale and location. Primary technical challenges focus on making use of the full complement of molecules in the biomass, improving catalysts, and in upgrading of synthetic hydrocarbons.[25] Demonstration and pre-commercial scale production facilities are pending in both the US and Europe to deliver advanced biofuels via thermochemical pathways. The well-known Fischer-Tropsch process is an example of a thermochemical conversion. Not limited to biomass, commercial FT processes employ gasification to convert both coal and natural gas to liquids such as CTL and GTL.

In a 2013 report, the IEA identified more than one hundred advanced biofuel pilot, demonstration or commercial projects being pursued worldwide. Biofuels production from second-generation cellulosic sources in 2011 was estimated at 137,000 tons, or about 0.15% of all biofuels currently produced. Though the global output of second-generation biofuels has tripled since 2010, it remains far below expectations due to project delays, closures, and a host of other technical and non-technical problems. If all known plants either announced or under construction come to fruition, cellulosic biofuels would still comprise less than one percent of all biofuels by 2018. Of the seventy projects for which data was provided, forty-three were biochemical, twenty were thermochemical and seven were chemical.[26]

6.3.4. Environmental Issues

First generation biofuels have been characterized as having environmental problems such as increased soil erosion and chemical runoff due to the higher level of corn production necessary for ethanol production.[27] Also, when corn is diverted from current uses to ethanol production, most current needs, such as feeding animals, must still be met. That means corn or some substitute crop area must be increased somewhere in the world. This increased area can result in deforestation or converting pasture to cropland. Such land conversion releases greenhouse gases (GHG) rendering corn ethanol (and other first generation biofuels) less attractive from the perspective of reducing GHGs. Estimating these induced land use changes is very complicated, is uncertain, and is controversial.

The environmental picture is somewhat different for second generation biofuels. The environmental impacts likely differ depending on whether the feedstock is a residue or perennial crop. For residues, the main concern is the possibility of increased soil erosion and loss of soil organic matter. For example, when corn stover is removed, it is possible that soil erosion would increase as there would be less residue to hold the soil. Similarly for organic matter, to the extent that residue added soil organic carbon (still being debated), removing residue could reduce soil carbon. However, there are no land use changes associated with residue use for feedstock as the residue is a co-product with the crop being grown. For dedicated energy crops, we expect a reduction in soil erosion and chemical runoff. Since the crops remain for ten years or more, they will hold soil better than an annual crop. Also, little fertilizer is normally applied to the dedicated crops, so there would be less runoff. In addition, the perennial crops provide wildlife habitat. There is some concern with possible loss of biodiversity if most of the land surrounding a biorefinery were to be planted in one crop. There could be land use change if substantial amounts of dedicated crops were used for biofuels, but likely, it would be less than first generation crops.

It is likely that second-generation feedstocks would cause less of a food-fuel problem than first generation. However, at the end of the day, land is the limiting resource, so even dedicated crops that are not consumed by humans use land that could be used for livestock feed.

6.3.5. Government Policy

The final major uncertainty is government policy. In the US, EU, and Brazil, government policies have changed over time. In Brazil, the required minimum blending level has changed with evolving economic conditions. In the EU, the biofuel targets and associated sustainability criteria change over time. Also, in the EU the renewable energy targets are difficult to enforce, and the private sector cannot be assured they will be enforced.

In the US, the RFS for cellulosic biofuels has an out-clause that permits blenders to purchase a credit from EPA and another advanced biofuel certificate in lieu of actually blending the cellulosic biofuel. For example, in 2013 the EPA credit cost is $0.42, and an advance biofuel blending certificate from sugarcane ethanol is about $0.77. So a blender could avoid blending a gallon of cellulosic biofuel by paying about $1.19. In May 2013, the wholesale price of gasoline was about $2.86 per gallon. If cellulosic gasoline is $4.50, the blender would have to choose between buying the cellulosic gasoline and blending it at a net cost of $4.50 – 2.86 = $1.64, or paying the $1.19 to avoid blending. Clearly under these conditions, the blender would choose not to blend. In this case the RFS is not an iron-clad mandate, and, in fact, turns out not to be a mandate at all. This out-clause in the RFS applies to cellulosic biofuels for any year in which EPA waives any part of the RFS. In reality, EPA will be forced to waive some part of the cellulosic RFS every year because the RFS grows faster than cellulosic capacity could possibly grow. Thus, there is no binding cellulosic RFS in the US. With no effective government mandate for cellulosic biofuels, and cellulosic biofuels not yet competitive with fossil fuels on the purely commercial market, it will be difficult to obtain investment and financing for plants.

Given this reality, what government policy options might get the industry moving? In the US, the Department of Defense, most notably the Navy and Air Force, are keenly interested in using renewable jet fuel for a significant fraction of their fleet.[28] They are motivated by both environmental factors and energy security. One policy mechanism that could be used to help the military procure biofuels is a reverse auction. In a reverse auction, firms bid for long term contracts to supply a fixed quantity of certified biofuel to the military. For example, the bid request could specify fifty million gallons per year delivered to Air Force Base X for the next fifteen years. Bidders would then estimate their costs of supplying the biofuel and submit bids to the Air Force. The lowest priced bid wins the contract. This mechanism reduces or eliminates uncertainty in several ways. First, oil price uncertainty is eliminated because the firm has a fixed price contract regardless of what happens to the oil price. Second, presumably any firm bidding would know their technology at least well enough to place a bid. Third, bidders would presumably have provisional contracts with feedstock suppliers before entering a bid, so they would know their feedstock cost. Thus, many of the uncertainties described above would be reduced via the reverse auction. Suppose that the projected price of equivalent fossil fuels over the 15 year period was $4.00 per gallon, and the winning cellulosic fuel bid was $4.50. That amounts to an implicit subsidy for the biofuel of $0.50 per gallon, but the subsidy is determined by a market mechanism—the reverse auction process. If the government wants to get second-generation biofuel plants built, this mechanism likely would be cost effective because the level of the subsidy is competitively determined in the market.

6.4. Conclusions: Major Challenges and Opportunities

First generation biofuels are now well established in Brazil, the US, and the EU. Sugarcane ethanol is likely to continue growing in Brazil as there are ample land resources available to expand production even while ensuring the protection of sensitive areas including the Amazon. In the US, the RFS level of fifteen billion gallons of conventional corn ethanol is already in place, and the capacity is not expected to grow. This level of corn ethanol has been one, but certainly not the only, contributor to higher agricultural commodity prices. However, as corn yields increase over time, the fraction of the total corn crop going to ethanol will begin to decline, and any price pressure brought on by biofuels will begin to diminish.

For second-generation biofuels, the technology has been slow to develop. The US RFS mandated level of cellulosic biofuel for 2013 is one billion gallons, but the EPA has waived all but fourteen million gallons of that mandate because, by their estimate, only that much will be available in 2013 from all the small commercial and demonstration plants in the US. The five uncertainties described above—oil price, feedstock, technology, environment, and government policy—have impeded investment in the industry. Absent changes in government policy or significant technical breakthroughs, it is unlikely that we will see large scale development in the near future. There is plenty of biomass feedstock, the technologies are becoming increasingly viable, and environmental issues can be managed. However, economics will have to improve to entice substantial capital investment into the sector.

Notes

1. Peter Zuurbier and Jos Van de Vooren, eds., *Sugarcane Ethanol: Contributions to Climate Change and the Environment* (Wageningen, Netherlands: Wageningen Academic Publishers, 2008), http://dx.doi.org/10.3920/978-90-8686-652-6.
2. National Energy Conservation Policy Act, Pub. L. No. 95-619, 92 Stat. 3206 (1978).
3. Energy Independence and Security Act of 2007, Pub. L. No. 110-140 (2007).
4. U.S. Environmental Protection Agency, *Renewable Fuel Standard Program (RFS2) Regulatory Impact Analysis*, EPA-420-R-10-006 (Washington, D.C.: Office of Transportation and Air Quality, 2010).
5. European Commission, "Directive 2009/28/EC of the European Parliament and of the Council," *Official Journal of the European Union*, L 140/16–62.
6. Perrihan Al-Riffai, Betina Dimaranan, and David Laborde, *Global Trade and Environmental Impact Study of the EU Biofuels Mandate* (Brussels, Belgium: Directorate General for Trade of the European Commission, 2010).
7. Wallace E. Tyner, Frank Dooley, and Daniela Viteri, "Alternative Pathways for Fulfilling the RFS Mandate," *American Journal of Agricultural Economics* 93, no. 2 (2010): 465–472, http://dx.doi.org/10.1093/ajae/aaq117; Wallace E. Tyner and Daniela Viteri, "Implications of Blending Limits on the US Ethanol and Biofuels Markets," *Biofuels* 1, no. 2 (2010): 251–253, http://dx.doi.org/10.4155/bfs.09.24; and Wallace E. Tyner, "The Integration of Energy and Agricultural Markets," *Agricultural Economics* 41, no. S1 (2010): 193–201, http://dx.doi.org/10.1111/j.1574-0862.2010.00500.x.
8. Wallace E. Tyner, "The US Ethanol and Biofuels Boom: Its Origins, Current Status, and Future Prospects," *Bioscience* 58, no. 7 (2008): 646, http://dx.doi.org/10.1641/B580718.
9. Philip Abbott, Christopher Hurt, and Wallace E. Tyner, *Farm Foundation Issue Report: What's Driving Food Prices?* (Oakbrook, IL: Farm Foundation, 2008); Philip Abbott, Christopher Hurt, and Wallace E. Tyner, *Farm Foundation Issue Report: What's Driving Food Prices? March 2009 Update* (Oakbrook, IL: Farm Foundation, 2009); Philip Abbott, Christopher Hurt, and Wallace E. Tyner, *Farm Foundation Issue Report: What's Driving Food Prices in 2011?* (Oakbrook, IL: Farm Foundation, 2011); and Farzad Taheripour, Christopher Hurt, and Wallace E. Tyner, "Livestock Industry in Transition: Economic, Demographic, and Biofuel Drivers," *Animal Frontiers* 3, no. 2 (2013): 38–46, http://dx.doi.org/10.2527/af.2013-0013.
10. Colin A. Carter and Henry I. Miller, "Corn for Food, Not Fuel," *New York Times*, July 30, 2012.
11. Wallace E. Tyner, Farzad Taheripour, and Chris Hurt, *Potential Impacts of a Partial Waiver c the Ethanol Blending Rules* (Oakbrook, IL: Farm Foundation, 2012).
12. Wallace E. Tyner, *National and Global Market Implications of the 2012 U.S. Drought*, Global Agricultural Development Initiative Issue Brief Series (Chicago, IL: The Chicago Council on Global Affairs, 2013).
13. Farzad Taheripour, and Wallace E. Tyner, "Biofuels and Land Use Change: Applying Recent Evidence to Model Estimates," *Applied Sciences* 3 (2013): 14–38, http://dx.doi.org/10.3390/app3010014.
14. Hosein Shapouri, James A. Duffield, and Michael Q. Wang. *The Energy Balance of Corn Ethanol: An Update (Agricultural Economic Report No. 813)* (Washington, D.C.: U.S. Department of Agriculture,

2002); and Michael Q. Wang et al., "Well-to-Wheels Energy Use and Greenhouse Gas Emissions of Ethanol from Corn, Sugarcane and Cellulosic Biomass for US Use," *Environmental Research Letters* 7, no. 4 (2012): 045905, http://dx.doi.org/10.1088/1748-9326/7/4/045905.
15. Wang et al., "Well-to-Wheels Energy Use"; and Isaias C. Macedo, Joaquim E.A. Seabra, and Joao E.A.R. Silva, "Green House Gases Emissions in the Production and Use of Ethanol from Sugarcane in Brazil: The 2005/2006 Averages and a Prediction for 2020," *Biomass and Bioenergy* 32, no. 7 (2008): 582–595, http://dx.doi.org/10.1016/j.biombioe.2007.12.006.
16. Elwin G. Smith, H. H. Janzen, and Nathaniel K Newlands, "Energy Balances of Biodiesel Production from Soybean and Canola in Canada," *Canadian Journal of Plant Science* 87, no. 4 (2007): 793–801, http://dx.doi.org/10.4141/CJPS06067; and A. Prahdan et al., *Energy Life-Cycle Assessment of Soybean Biodiesel: Agricultural Economic Report Number 845* (Washington, D.C.: U.S. Department of Agriculture, 2009).
17. U.S. Environmental Protection Agency, *Renewable Fuel Standard Program (RFS2) Final Rule*, EPA-420-F-10–007 (Washington, D.C.: Office of Transportation and Air Quality, 2010).
18. Wallace E. Tyner, "Policy Update: Cellulosic Biofuels Market Uncertainties and Government Policy," *Biofuels* 1, no. 3 (2010): 389–391, http://dx.doi.org/10.4155/bfs.10.15; and Wallace E. Tyner, "Biofuels and Agriculture: A Past Perspective and Uncertain Future," *International Journal of Sustainable Development & World Ecology* 19, no. 5 (2012): 389–394, http://dx.doi.org/10.1080/13504509.2012.691432.
19. Energy Information Administration, *Annual Energy Outlook 2013* (Washington, D.C.: US Energy Information Administration, 2013), http://www.eia.gov/forecasts/aeo/.
20. EIA, *Annual Energy Outlook 2013*.
21. National Research Council, *Renewable Fuel Standard: Potential Economic Effects of U.S. Biofuel Policy* (Washington, D.C.: National Academy of Sciences, 2011); and U.S. Department of Energy, *U.S. Billion Ton Update: Biomass Supply for a Bioenergy and Bioproducts Industry*, R. D. Perlack and B. J. Stokes (Leads), ORNL/TM-2011/224 (Oak Ridge, TN: Oak Ridge National Laboratory, 2011).
22. National Research Council, *Renewable Fuel Standard*.
23. Price is for Brent crude oil.
24. "Biomass Research," National Renewable Energy Laboratory, last modified September 26, 2012, http://www.nrel.gov/biomass/.
25. "Research and Development: Processing and Conversion," Bioenergy Technologies Office, U.S. Department of Energy, last modified June 18, 2013, http://www1.eere.energy.gov/bioenergy/processing_conversion.html.
26. Dana Bacovsky et al., *Status of Advanced Biofuels Demonstration Facilities in 2012: A Report to IEA Bioenergy Task 39* (Paris: International Energy Agency, 2013), http://demoplants.bioenergy2020.eu/files/Demoplants_Report_Final.pdf.
27. National Research Council, *Renewable Fuel Standard*.
28. Secretary of the Air Force for Installations and Equipment, *Air Force Energy Plan 2010* (Washington, D.C.: US Air Force, 2009); Wallace E. Tyner, "Policy Update: Biofuels: The Future is in the Air," *Biofuels* 3, no. 5 (2012): 519–520, http://dx.doi.org/10.4155/bfs.12.45.

Bibliography

Abbott, Philip, Christopher Hurt, and Wallace E. Tyner. *Farm Foundation Issue Report: What's Driving Food Prices?* Oakbrook, IL: Farm Foundation, 2008.

Abbott, Philip, Christopher Hurt, and Wallace E. Tyner. *Farm Foundation Issue Report: What's Driving Food Prices? March 2009 Update*. Oakbrook, IL: Farm Foundation, 2009.

Abbott, Philip, Christopher Hurt, and Wallace E. Tyner. *Farm Foundation Issue Report: What's Driving Food Prices in 2011?* Oakbrook, IL: Farm Foundation, 2011.

Al-Riffai, Perrihan, Betina Dimaranan, and David Laborde. *Global Trade and Environmental Impact Study of the EU Biofuels Mandate*. Brussels, Belgium: Directorate General for Trade of the European Commission, 2010.

Bacovsky, Dana, Nikolaus Ludwiczek, Monica Ognissanto, and Manfred Wörgetter. *Status of Advanced Biofuels Demonstration Facilities in 2012: A Report to IEA Bioenergy Task 39.* Paris: International Energy Agency, 2013. http://demoplants.bioenergy2020.eu/files/Demoplants_Report_Final.pdf.

Carter, Colin A., and Henry I. Miller. "Corn for Food, Not Fuel." *New York Times*, July 30, 2012.

Energy Information Administration. *Annual Energy Outlook 2013.* Washington, D.C.: US Energy Information Administration, 2013. http://www.eia.gov/forecasts/aeo/.

European Commission. "Directive 2009/28/EC of the European Parliament and of the Council." *Official Journal of the European Union*, L 140/16–62.

Macedo, Isaias C., Joaquim E.A. Seabra, and Joao E.A.R. Silva. "Green House Gases Emissions in the Production and Use of Ethanol from Sugarcane in Brazil: The 2005/2006 Averages and a Prediction for 2020." *Biomass and Bioenergy* 32, no. 7 (2008): 582–595. http://dx.doi.org/10.1016/j.biombioe.2007.12.006.

National Renewable Energy Laboratory. "Biomass Research." Last modified September 26, 2012. http://www.nrel.gov/biomass/.

National Research Council. *Renewable Fuel Standard: Potential Economic Effects of U.S. Biofuel Policy.* National Academy of Sciences, 2011.

Prahdan, A., D. S. Shrestha, A. McAloon, W. Yee, M. Haas, J. A. Duffield, and H. Shapouri. *Energy Life-Cycle Assessment of Soybean Biodiesel: Agricultural Economic Report Number 845.* Washington, D.C.: U.S. Department of Agriculture, 2009.

Secretary of the Air Force Installations and Equipment. *Air Force Energy Plan 2010.* Washington, D.C.: US Air Force, 2010.

Shapouri, Hosein, James A. Duffield, and Michael Q. Wang. The Energy Balance of Corn Ethanol: An Update (Agricultural Economic Report No. 813. Washington, D.C.: U.S. Department of Agriculture, 2002.

Smith, Elwin G., H. H. Janzen, and Nathaniel K Newlands. "Energy Balances of Biodiesel Production from Soybean and Canola in Canada." *Canadian Journal of Plant Science* 87, no. 4 (2007): 793–801. http://dx.doi.org/10.4141/CJPS06067.

Taheripour, Farzad, Christopher Hurt, and Wallace E. Tyner. "Livestock Industry in Transition: Economic, Demographic, and Biofuel Drivers." *Animal Frontiers* 3, no. 2 (2013): 38–46. http://dx.doi.org/10.2527/af.2013-0013.

Taheripour, Farzad, and Wallace E. Tyner. "Biofuels and Land Use Change: Applying Recent Evidence to Model Estimates." *Applied Sciences* 3 (2013): 14–38. http://dx.doi.org/10.3390/app3010014.

Tyner, Wallace E. "Biofuels and Agriculture: A Past Perspective and Uncertain Future." *International Journal of Sustainable Development & World Ecology* 19, no. 5 (2012): 389–394. http://dx.doi.org/10.1080/13504509.2012.691432.

Tyner, Wallace E. "The Integration of Energy and Agricultural Markets." *Agricultural Economics* 41, no. S1 (2010): 193–201. http://dx.doi.org/10.1111/j.1574-0862.2010.00500.x.

Tyner, Wallace E. "Policy Update: Biofuels: The Future is in the Air." *Biofuels* 3, no. 5 (2012): 519–520. http://dx.doi.org/10.4155/bfs.12.45.

Tyner, Wallace E. "Policy Update: Cellulosic Biofuels Market Uncertainties and Government Policy." *Biofuels* 1, no. 3 (2010): 389–391. http://dx.doi.org/10.4155/bfs.10.15.

Tyner, Wallace E. "The US Ethanol and Biofuels Boom: Its Origins, Current Status, and Future Prospects." *Bioscience* 58, no. 7 (2008): 646. http://dx.doi.org/10.1641/B580718.

Tyner, Wallace E. *National and Global Market Implications of the 2012 U.S. Drought.* Global Agricultural Development Initiative Issue Brief Series. Chicago, IL: The Chicago Council on Global Affairs, 2013.

Tyner, Wallace E., Frank Dooley, and Daniela Viteri. "Alternative Pathways for Fulfilling the RFS Mandate." *American Journal of Agricultural Economics* 93, no. 2 (2010): 465–472. http://dx.doi.org/10.1093/ajae/aaq117.

Tyner, Wallace E., Farzad Taheripour, and Chris Hurt. *Potential Impacts of a Partial Waiver of the Ethanol Blending Rules.* Oakbrook, IL: Farm Foundation, 2012.

Tyner, Wallace E., and Daniela Viteri. "Implications of Blending Limits on the US Ethanol and Biofuels Markets." *Biofuels* 1, no. 2 (2010): 251–253. http://dx.doi.org/10.4155/bfs.09.24.

U.S. Department of Energy. "Research and Development: Processing and Conversion." Bioenergy Technologies Office. Last modified June 18, 2013. http://www1.eere.energy.gov/bioenergy/processing_conversion.html.

U.S. Department of Energy. *U. S. Billion Ton Update: Biomass Supply for a Bioenergy and Bioproducts Industry*, R. D. Perlack and B. J. Stokes (Leads), ORNL/TM-2011/224. Oak Ridge, TN: Oak Ridge National Laboratory, 2011.

U.S. Environmental Protection Agency. *Renewable Fuel Standard Program (RFS2) Regulatory Impact Analysis*, EPA-420-R-10-006. Washington, D.C.: Office of Transportation and Air Quality, 2010.

U.S. Environmental Protection Agency. *Renewable Fuel Standard Program (RFS2) Final Rule, EPA-420-F-10–007*. Washington, D.C.: Office of Transportation and Air Quality, 2010.

Zuurbier, Peter, and Jos Van de Vooren, eds. *Sugarcane Ethanol: Contributions to Climate Change and the Environment*. Wageningen, Netherlands: Wageningen Academic Publishers, 2008. http://dx.doi.org/10.3920/978-90-8686-652-6.

Chapter 7

A Future for Nuclear Energy?

LEFTERI H. TSOUKALAS, RONG GAO, AND EUGENE D. COYLE

Abstract

Nuclear energy is an ultra-concentrated source of energy; one tonne of natural uranium is capable of producing forty-four million kWh of electricity. By comparison, to produce the same amount of electricity would require twenty thousand tonnes of coal or 8.5 million cubic meters of natural gas. Nuclear energy is meanwhile controversial in the public arena, principally due to its related association to atomic weaponry, its operational safety records and the radioactive waste materials it produces. These concerns have severely undermined the progress of nuclear power over recent decades. It is imperative today that we reevaluate these concerns in view of the emerging global energy picture. In evaluating today's energy production options, lifecycle environmental costs must be equitably factored in. Such determinations may shift the economic feasibility from conventional sources of energy, not least coal and gas, to other sustainable and renewable energy sources, including hydro, wind and solar energy. The case for including nuclear power as a prominent aspect of the new energy paradigm has become a legitimate question deserving of analysis and exploration.

In this chapter the story of nuclear energy is explored commencing with an introduction to the essential elements in nuclear power generation followed by a brief historical recall of nuclear energy, estimates of current generation status, nuclear energy safety, nuclear accidents and their after impact, and challenges in dealing with nuclear waste management. The future role for nuclear 'fission' energy is discussed with commentary on the need for public engagement. This is followed by a brief review of research and development in nuclear 'fusion' energy and the case for its future deployment.

7.1. Introduction: Essentials of Nuclear Energy

Nuclear power is based on a fundamental principle discovered about seventy years ago. The nucleus of a fissile isotope, such as uranium$_{235}$ (^{235}U) or plutonium$_{239}$ (^{239}Pu), becomes an unstable compound after capturing an extra neutron and it will promptly split into two smaller fragments, releasing enormous amounts of heat in the process. The heat is carried away by a coolant, typically water, gas or liquid metal, which subsequently converts water into steam that drives a turbine to generate electricity.

As with other material commodities, nuclear fuels experience three phases during their life cycle: acquisition, utilization and disposal (Figure 7.1).

The diagram simplistically represents a best-case scenario for nuclear fuels in which total recycling is achieved. In the ideal total recycling closed fuel-cycle scenario, the generation and disposal of nuclear waste will no longer pose a major problem. In practice, significant economic, technical and indeed political barriers remain to be addressed and overcome.

Uranium is the major fuel used in nuclear power reactors. Even though uranium is a rather abundant resource in the earth, the fissile (useful) isotope, ^{235}U, accounts for only 0.7% of natural uranium; the majority is mostly ^{238}U, a stable isotope. To achieve the required efficiency to sustain nuclear fissions, commercial nuclear power reactors use enriched uranium, with a ^{235}U concentration of three percent and above.[1] An enrichment process is therefore required to make natural uranium useable in nuclear power reactors. Transforming natural uranium to nuclear fuel involves three typical processes: mining, milling, and fabrication.

Uranium is a weakly radioactive element found in the Earth's crust. It is approximately five hundred times more abundant than gold. Since 2008, the world's total identified uranium resources have grown by 12.5%, of which over 30% is located in Australia. Uranium ore extracted from the earth is usually in the form of a compound known as triuranium octoxide, U_3O_8. A leaching process using sulfur acid is needed to separate uranium from other waste. The final product, generally referred to as *yellow cake*, consists of at least eighty percent pure uranium, which is transported to a processing facility where nuclear fuel is fabricated.

Enrichment is necessary in nuclear fuel production owing to the fact that the uranium$_{235}$ concentration in natural uranium ore is too low to be used in commercial nu-

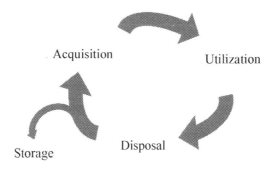

Figure 7.1. A Simplified Nuclear Fuel Cycle.

clear power plants. Prior to the enrichment process, U_3O_8 is first converted to a gas form of uranium, UF_6. The objective of enrichment is to separate ^{235}U from ^{238}U, which can be done by virtue of an obvious physical property: that ^{235}U is slightly lighter than ^{238}U. In practice, there are two commonly used techniques: gaseous diffusion and gas centrifuge. In the gaseous diffusion process, UF_6 gas is pumped through layers of special porous membranes. Since ^{235}U is lighter, it tends to diffuse faster and will thus be separated from ^{238}U at the end of the pipeline. Because the weight difference between these two isotopes is small, hundreds of filtering membranes are required. The gas centrifuge approach uses a large number of cylinders rotating at very high speeds. Since gas modules with uranium$_{238}$ are heavier, the centrifugal force will drive them to the outer part of the cylinders while uranium$_{235}$ will remain closer to the center. Again, multiple units are required to achieve a desirable level of enrichment.

The nuclear fuels used in power generating reactors are arranged in a bundle form called a *fuel assembly* comprising hundreds of *fuel rods*. To produce a fuel rod, uranium dioxide is first baked into cylinder ceramic pellets at temperatures up to 1400°C and is then inserted into zirconium metal tubes. The dimensions and arrangements of fuel rods are characteristic to specific reactor designs.

The working principle for a power generation nuclear reactor was discovered more than seventy years ago and has remained largely unchanged during that time. As in a fossil fuel plant, nuclear fuels are burned to produce heat through *nuclear fission*, rather than a chemical combustion process. The fission of the uranium$_{235}$ nucleus produces two smaller fragments, called *fission products*, of two to three free neutrons and some gamma rays. The total mass of the fission products is smaller than the mass of the original uranium nucleus; the loss in the mass appears as kinetic energy of the fission products, typically 200 MeV (or 3.2×10^{-11} J) per fission event.

If all conditions are favorable, the free neutrons released during the initial (first generation) fission event may be subsequently captured by other fissile nuclei which in turn trigger further fissions. If at least one such event happens on average, a sustainable chain reaction is achieved. Regulating the neutron population is crucial in controlling the operation of a nuclear reactor. This is achieved primarily through the use of control rods. Control rods are comprised of material that can absorb neutrons very efficiently, such as graphite.

As the fission process progresses, fissile materials are continuously consumed and fission products build up inside the fuels. Some fission products have high neutron absorption capacity and will reduce the number of free neutrons available to the chain reaction process. This eventually leads to a situation where fission cannot be sustained. At this point, the reactor has to be shut down and refueled. A burn-up factor is used to measure the quantity of fuel that has been consumed. The burn-up factor is usually expressed as the total thermal energy generated per unit mass, typically in gigawatt days per metric ton of enriched uranium (GWd/tU). The average designed burn-up for second-generation reactors is about forty GWd/tU. Later technologies in fuel design have improved this number upwards to sixty GWd/tU and higher. A higher burn-up allows longer operation cycles and greatly improves in both cost and safety.

Nuclear Fuel Disposal: Once removed from their reactors, nuclear fuels become *spent fuels*. They are sometimes incorrectly labeled as waste; but in fact, more than ninety-seven percent of the fissionable material is still contained within. Nuclear fission chain reactions cease from the moment of reactor shutdown, however the nuclear fuels remain highly radioactive. The main radiation comes from two sources: fission products and actinides.

The fission of the uranium nucleus creates two smaller nuclei, called fission products. The resultant nuclei are normally unstable and will undergo beta decay to more stable nuclei, releasing beta or gamma radiation. One possible reaction is the following:

$$^{235}U + n \rightarrow {}^{236}U \rightarrow {}^{140}Xe + {}^{94}Sr + 2n$$

Most of the fission products have very short half-lives. $Xenon_{140}$ has a half-life of fourteen seconds and $Strontium_{94}$ a half-life of seventy-five seconds. Other possible fission products have longer half-lives and are called *long-lived fission products* (LLFP). For example, $Iodine_{131}$ has a half-life of eight days, while $Cesium_{137}$ and $Strontium_{90}$ of have half-life of about thirty years. Long lived fission products pose a bigger environmental hazard and need to be closely monitored. Nevertheless, most fission products have half-lives of less than ninety years, making them relatively easy to handle. In fact, after forty years, their radioactivity will have reduced to a thousandth of their original level.

Besides the fission products referred to above, nuclear spent fuels also contain actinides such as $Plutonium_{239}$, $Plutonium_{240}$, $Americium_{241}$, $Americium_{243}$, $Curium_{245}$, and $Curium_{246}$, which are mainly produced by neutron capturing of $Uranium_{238}$. Most actinides have very long half-lives, typically thousands or millions of years, presenting a gigantic challenge for long term storage of nuclear waste.

The existence of long-lived fission products and actinides demands careful handling of nuclear spent fuels. When they are initially discharged from the reactor, the highly radioactive isotopes in the fuels are still decaying, thus generating enormous amounts of heat that requires appropriate cooling. Current practice is to submerge the spent fuels in pools of water. The water serves as both a coolant and as a layer of radiation shielding. After a minimum of one year of cooling inside the pool, the spent fuels may be removed from the water and inserted into a gas-filled steel cylinder container, called a cask. Following removal, there are two possible destinies for the spent fuels: direct disposal or recycling. In the US, direct disposal is the only legal option. The spent fuels would preferably be transferred to a permanent repository for long-term storage but such repositories are not currently available.[2] Spent fuels are therefore (temporarily) stored at plant site locations. The direct disposal approach treats spend fuels as waste and immediately creates a huge challenge for nuclear waste management since these wastes need to be stored securely for at least one thousand years. Other countries, such as the United Kingdom and France, have chosen the recycling approach to alleviate the local waste management problem. The recycling of nuclear spent fuels takes place in two steps. First, the small amount (three percent) of highly radioactive material is separated and concentrated into a special glass which is stored securely. The remainder of the spent fuels, consisting of most of the unused uranium and the newly generated plutonium, can be reused to fabricate fresh fuel rods.

7.2. History of Nuclear Engineering

The harnessing of nuclear energy was made possible by the discovery of the nuclear fission reaction in 1938. German chemists Otto Hahn and Fritz Strassmann observed that when uranium nuclei are bombarded with neutrons they may split into smaller fragments. They also estimated that the energy released during the reaction was about two hundred MeV, which was later confirmed experimentally by Otto Frisch in 1939. In addition, their research also showed that extra neutrons were created during the fission process, indicating that a self-sustainable chain reaction might be possible if the newly created neutrons could trigger further fission reactions.[3]

The mechanism of nuclear fission was further investigated by other scientists and in a very short period of time some quite significant discoveries were made. First, it was found that uranium$_{235}$ had a better chance for a fission reaction than uranium$_{238}$. Second, slow neutrons (or thermal neutrons) had higher probability of being captured by the uranium nuclei than fast neutrons. Since the neutrons released from the fission process have very high kinetic energy, they need to be slowed down using a moderator to increase the potential for a chain reaction. Third, natural uranium contained only 0.7% of uranium$_{235}$ and thus an enrichment process was necessary. These findings were the base of using nuclear power. At that time, making atomic bombs was the primary objective as World War II had just began.

However, peaceful utilization of nuclear power had never been overlooked. In fact a group of scientists in Britain suggested in 1941 that besides building atomic bombs, nuclear fission could be used in a controlled fashion to produce useful heat. Following the war, this option was pursued with intensified efforts. The first reactor used to produce electricity was brought online in December 1951 in Idaho in the United States. Other countries followed quickly. The Soviet Union put into operation the world's first nuclear power generator in 1954, a five gigawatt unit. These projects convincingly demonstrated that peaceful utilization of nuclear power was feasible. The first fully commercial nuclear power plant was designed by Westinghouse in the US and commenced operation in 1962. The Westinghouse unit was comprised of a pressurized water reactor (PWR) with a capacity of 250 MWe. Since then nuclear power generation has significantly increased, with growth pattern illustrated in Figure 7.2.

7.3. Current Status of Nuclear Energy

Recent estimates by the International Atomic Energy Agency (IAEA) reveal that annual generation of nuclear power is currently on a slight downward trend, decreasing 1.8% in 2011 to 2558 TWh. In 2009, nuclear energy accounted for approximately 13.5% of world electricity demand. IAEA reported a significant increase in projected nuclear generating capacity. In 2009, 130 power reactors with net capacity of 150 GWe (gigawatt electric) were planned. China is the largest growth country with an expected fifty gigawatt nuclear installation by 2020. India is also planning a large increase, with installation of up to twenty reactors by 2020. The contribution of nuclear energy to

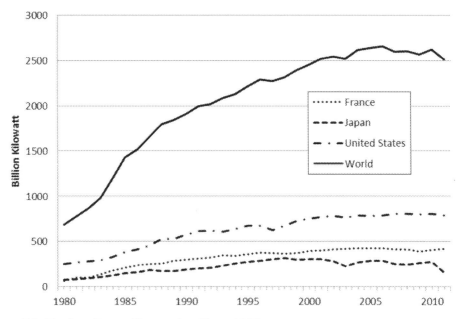

Figure 7.2. Nuclear Power Generation Since 1980.

electricity generation is significant and is likely to grow in terms of percentage contribution to increasing global energy demand. In December 2011, the US Nuclear Regulatory Commission granted approval to Toshiba Corporation's Westinghouse newest reactor design (type AP1000), clearing the path for a sale of this reactor in the US and a revival of domestic nuclear power construction. US utilities are seeking permission to build up to twelve of the new reactors. In Germany, on the other hand, nuclear power had accounted for twenty-three percent of national electricity consumption, prior to the permanent shutdown of eight plants in March 2011. German nuclear power commenced with development of research reactors in the 1950s and 1960s, resulting in the first commercial plant going online in 1969. It has been high on the political agenda in recent decades, with continuing debate regarding when the technology should be phased out. The topic received renewed attention due to the political impact of the Russia-Belarus energy dispute in 2007 and following the Fukushima nuclear accident in 2011.

According to IAEA, as of 2012, there are 435 nuclear power reactors in operation with 370,049 MWe installed capacity.

Efficiency in electricity generation is of key importance and nuclear fission is by far the most efficient source of energy. Nuclear fusion may result in even higher efficiency than nuclear fission, should fusion prove to be controllable. An energy density logarithmic scale comparison, summarized in Figure 7.3, shows that uranium fuels are many orders of magnitude higher than other fuels in terms of energy density.

Owing to its fuel energy density, a nuclear power reactor can operate without interruption for up to two years before refueling is required. High energy density also results in lower fuel cost in energy production. The US Nuclear Energy Institute estimates that for a coal-fired plant, seventy-eight percent of the cost is fuel, for natural gas eighty-

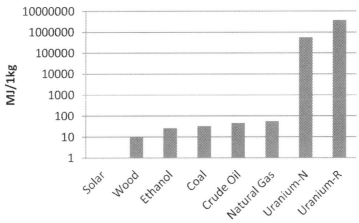

Figure 7.3. Energy Densities of Major Fuel Sources.[4] Uranium-N and Uranium-R Stand for Natural Uranium and Reactor-graded Uranium, Respectively.

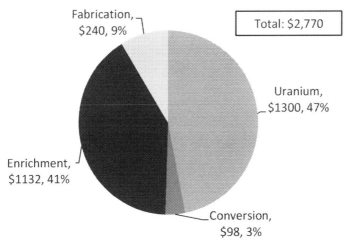

Figure 7.4. Cost Components of Getting One Kilogram of UO_2 in 2011.

nine percent and for nuclear a mere fourteen percent. A study by the World Nuclear Association suggested that the cost of nuclear fuel to generate one kilowatt of electricity was about $0.0077 USD in 2011. Their analysis indicated that enrichment of uranium accounts for about half the cost (Figure 7.4). Lower percentage fuel cost makes nuclear power largely resistant to market fluctuation.

One of the most significant advantages of nuclear power is that it is carbon emission free. The commitment to low carbon economics makes the nuclear option cost attractive since other non-carbon free sources, such as coal-fired and gas-fired plants, will necessitate development of more advanced (and hence more expensive) technologies in order to achieve a reduced carbon emission footprint. An EIA report (Figure 7.5) suggests that the nuclear power remains economically competitive among all possible alternatives.

With low fuel cost and uninterrupted availability nuclear power can be a reliable and predictable source for base-load electricity. Unlike other types of energy, electricity

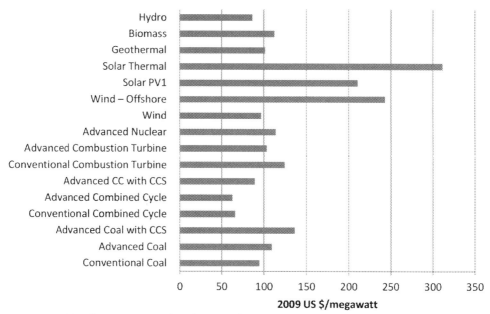

Figure 7.5. Total System Levelized Cost of New Generation Resources, 2016.[5]

is unique in that it cannot be stored efficiently. Generation and consumption must be balanced at all times. Base-load power is the minimum requirement to meet this balance. Nuclear power meets the necessary requirements that base-load power be safe, economically viable and reliable. In some countries nuclear power produces a significant portion of electricity (Figure 7.6). France, for example, generated more than seventy-seven percent of its electricity with nuclear power in 2011.

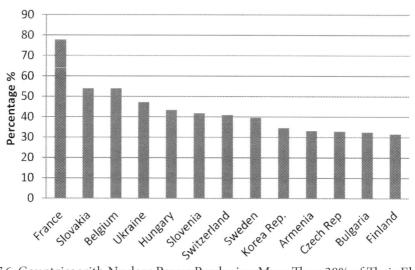

Figure 7.6. Countries with Nuclear Power Producing More Than 30% of Their Electricity in 2011.[6]

7.4. Nuclear Energy Safety

Safety is of paramount concern in the construction and operation of all nuclear power plants. Safety in careful processing and disposal of nuclear waste is also of critical importance. In this section, we discuss how to achieve these goals in practice. A review of fundamental concepts is made with omission of in-depth technical detail. We also examine how a masterly engineered system can collapse owing to a series of cascading events.

Nuclear power plants are designed, constructed and operated under very rigorous internationally agreed safety standards. Through adoption of defense-in-depth design and operation procedures, protocols are put in place with intent to ensure that serious malfunction may arise subject only to failure of multiple consecutive and independent safety measures. Table 7.1 lists five essential levels of protection in plant design and operation, from accident prevention to disaster mitigation. Required measures to achieve these objectives are also listed.

We shall not attempt to discuss these measures in detail. Instead we propose reducing them down to three basic components: the intrinsic safety feature of a chain reaction, the control of reactivity, and the residual heat removal mechanism.

7.4.1. Chain Reaction

The chain reaction is central to both nuclear power reactors and atomic energy, in order to achieve sustainable fission. This parallel has given rise to the misunderstanding that an out-of-control nuclear reactor power plant may turn itself into an atomic bomb. This scenario will never occur as the design of a nuclear power reactor is vastly different than that of a nuclear bomb. Creating a nuclear explosion is physically impossible in a nuclear power reactor.

Table 7.1. Defense-in-depth Concept of a Nuclear Power Plant.

Level	Objectives	Measures
1	Failure prevention	Adequate site selection, high quality design, construction and operation to reduce or prevent failures
2	Failure detection and control	Control and protection system with built-in surveillance features to detect and manage abnormal operations before they become significant
3	Design basis accident control	Built-in engineering safety feature and accident response procedures to prevent core damage and manage abnormal operations before they become significant
4	Severe accident control	Reactor containments to limit the impact of severe accidents that are not anticipated in the design basis
5	Severe accident mitigation	Off-site emergency plan to mitigate the radiological consequence of significance release of radioactive material

Almost all currently operating nuclear power reactors are thermal reactors, which use thermal neutrons to sustain chain reactions. Neutrons initially released from nuclear fission carry very high kinetic energy (~2 MeV), and are termed fast neutrons. As fast neutrons move at very high speed, they have less probability of being captured by other fissionable nuclei. By comparison, thermal neutrons have kinetic energy of less than 1 eV. From a practical viewpoint, slowing fast neutrons down to thermal neutron speed provides a means for lower enriched uranium to be used as fuel. The material used to slow down fast neutrons is termed a moderator. The moderator in a thermal reactor consists of light material such as water. Using water as the moderator introduces an important safety feature: it provides a negative void coefficient. A common feature in reactor design is to use water as both moderator and coolant.[7] When reactivity levels rise the moderator/coolant temperature will rise accordingly. As temperature increases, the density of moderator/coolant decreases, reducing its effectiveness as moderator and resulting in fewer number of thermal neutrons for fission. This negative feedback mechanism is an inherent stabilizing factor in reactor operation. In an extreme accident scenario, the moderator/coolant will vaporize and effectively stop the chain reaction.

7.4.2. Reactivity Control

The primary control of reactor reactivity is through control rods. Control rods are made with material that can efficiently absorb neutrons. Reactivity control is achieved through mechanically changing the position of control rods inside the reactor core. Moving deeper within the core results in more neutrons being absorbed, thus reducing reactivity; moving toward core edge has the opposite result. Over ninety-nine percent of the neutrons (prompt neutrons) are released within a very short time (half-life of 10^{-22} second) during fission, which is much too quick for mechanical systems to respond. Therefore, the remaining one percent of the delayed neutrons is crucial for reactor controllability. Reactors are carefully designed to be sub-critical for ninety-nine percent of the prompt neutrons, ensuring that these neutrons cannot themselves sustain chain reactions, but may achieve criticality with the addition of the remaining one percent of delayed neutrons. Control rods are designed to function sufficiently quickly to counter any power surge and eliminate the possibility of an unwanted criticality.

7.4.3. Residual Heat Removal

The problem of residual heat removal is unique to nuclear power generation. During normal operation, heat generated from nuclear fission is continuously carried away by the coolant, maintaining steady- state conditions. Following shutdown, scheduled or unscheduled, nuclear fission stops but heat generation does not. Fission products generated through nuclear fissions are radioactive and will keep decaying regardless. The amount of heat generated by decay of the fission product is significant, typically five percent of the power generated prior to the shutdown. If this heat is not removed efficiently, the reactor core can be heated to thousands of degrees and reach meltdown. In a worst case scenario, extreme heat can cause structural damage to the reactor containment, resulting in the release of radioactive material to the environment. It is therefore vital for a nuclear power plant to maintain residual heat removal capability at all times, even under that most severe event condition. A major portion of a reactor's residual heat removal systems are active systems, which means that they rely on electrical pumps to produce a constant flow of coolant to circulate the residual heat away from the reactor core. Systems are designed such that external power is available during an emergency event. On the other hand, in the event of a natural disaster such as in the event of an earthquake, local power from the grid infrastructure could be rendered unavailable. To militate against such an emergency it is a requirement that nuclear power plants be equipped with onsite backup generators and battery power supplies.

7.5. Nuclear Accidents and Impacts

In ensuring maximal safety, the design, construction and operation of nuclear power plants are required to conform to very rigorous standards and procedures. The US Nuclear Regulatory Commission (NRC) mandates that rector designs must meet a requirement of core damage one in ten thousand years. Most of the current commercial reactors in the US are designed for one in one hundred thousand year damage compliancy. The intent for next generation plant is to push this figure to one in ten million years. As with other manmade products these theoretical safety parameters cannot totally eliminate the risk of accidents. Human designed engineering systems, however well designed, can fail subject to unforeseen circumstantial occurrences. Safety regulations in the nuclear power industry are particularly stringent as the potential human, physical and psychological impacts of system failure are inordinately great. Over the past fifty years there have been three major nuclear accidents: Three Mile Island, Chernobyl, and Fukushima. Upon close examination of nuclear accidents it is customary to discover that a chain of events led to the critical event, including: flawed design, human error and unexpected events. Learning from previous accidents and making necessary adjustments to break the chain of events are integral components of nuclear energy safety culture.

7.5.1. Three Mile Island, 1979

The Three Mile Island nuclear facility is located near Harrisburg, Pennsylvania, in the United States. The accident occurred at four in the morning on March 28, 1979. Initially there was a minor malfunction in the secondary cooling system. The malfunction caused a rise of temperature in the primary reactor coolant and later triggered automatic reactor shutdown sequences. During the shutdown, a pilot operated relief valve opened to avoid potential over pressure, as per design. However, it failed to close properly as programmed after ten seconds, resulting in lost coolant. Unfortunately there was no instrumentation in situ to detect the position of the relief valve, thus operators were not aware of this situation. The built-in safety mechanism detected the loss of coolant and responded to it by injecting replacement water into the reactor, resulting in a rise of water level in the pressurizer, a tank designed to maintain proper pressure level in the reactor. Operators noticed the anomaly in the pressurizer but were unable to properly diagnose its root cause: the relief valve. Instead, they were under the impression that the pressurizer was over-filled. To correct this artificial problem, operators reduced the flow of replacement water according to operating manual instructions. Without sufficient coolant, steam formed and caused excessive vibration of the cooling pumps. To avoid damage resulting from the vibration, operators shut down the pump station, with effect of worsening the situation. Without coolant, the reactor core overheated and started to melt. This continued for over two hours before operators finally closed the relief valve and started to restore the cooling system. It took almost one month for the damaged reactor to fully shut down (cold shutdown) on April 27.

The Three Mile Island accident was one of the worst case scenarios in terms of reactor core damage. Later inspection revealed forty-five percent of the core had melted. Fortunately, its consequence was limited since the damaged core was well confined inside the reactor vessel. During the event, the amount of radioactive material released to the environment was negligible. In a study carried out by the State Health Agency over a period of eighteen years, no abnormal health issues were reported for the population of thirty thousand residing in the surrounding locality. The Three Mile Island accident was thoroughly studied by scientists, engineers and regulatory agency authorities. Additional rigorous safety measures were enforced following the accident.

7.5.2. Chernobyl, 1986

The Chernobyl accident was the most devastatingly destructive in the history of nuclear power. The chain of events, ironically, starts with an electrical engineering experiment designed to test a safety feature. In a nuclear power plant, uninterrupted electricity supply is required to drive coolant pumps. In case of total loss of station power, on-site backup generators are on hand to immediately come on line. At Chernobyl, however, there was a sixty to seventy-five second gap before the diesel generators were able to reach full power. To bridge this gap, engineers hypothesized that following the loss of power the slowing down steam turbine might be able to spin long enough and continue providing sufficient electricity for the coolant pumps before the backup generators took over. They needed to verify this idea experimentally and decided to take advantage of a routine shutdown to perform the test. For some reason, safety officers were not consulted concerning the test, and the required approval was not secured. This lack of communication was of critical import and proved a deadly mistake. In carrying out the test some major flaws in reactor design were unwittingly unearthed. The Chernobyl reactor was of graphite-moderated water cooled design. Unlike many other reactor designs, this one has an important characteristic: a positive void coefficient of reactivity. This means reactivity increases as coolant temperature increases, the opposite effect to a reactor with negative void coefficient. In normal circumstances a positive void coefficient (PVC) does not indicate that a reactor is unstable as there are other mechanisms inherently available to stabilize the overall reactivity. However, under certain circumstances, especially at low power level and with certain fuel configurations, PVC will dominate and may result in very unstable conditions. The test crew at Chernobyl were not aware of this potential threat. However, even with this flaw, the test could have been safely completed if everything had gone to plan, and the same test had been carried out multiple times previously without incident.

The test was scheduled to finish during the day shift of April 25. However, an unexpected outage at a local power station forced the power grid operator to order that the Chernobyl reactor remain in service for the duration of the power shortage. This unexpected development resulted in delaying the test to midnight of April 25, as the test procedure would require reducing the power output of the reactor. The delay was critical; the night shift crew were less prepared for the test. A number of changes were made, putting the reactor into a dangerous positive feedback mode. This positive feedback was successfully compensated for by the automatic control system throughout the test duration. The disaster started when an emergency shutdown procedure, called SCRAM (Safety Control Rod Axe Man), was initiated. This standard procedure inserted control rods into the core to quickly stop the reaction. There was a major flaw in Chernobyl's control rod design, with the result that for the first few seconds after the control rods were inserted, reactivity increased rather than decreased. The combination of positive void coefficient and a flawed control rod design led to a rapid power surge which destroyed the reactor within seconds. Violent fires and explosions expelled tons of highly radioactive material into the environment. Two operators were killed in the explosions and another twenty-eight emergency workers died within three months owing to acute radiation poisoning. About four thousand cases of radiation-exposure-related thyroid cancers were diagnosed in the affected population. A much larger number of people were psychologically affected. The Chernobyl disaster was unique in the sense that its flawed design was not adopted anywhere in the West. However, the demonstration of a lax safety culture and poor crisis management during the event resulted in severe damage to the reputation of nuclear power, leading to decades of stagnation in the nuclear power industry.

7.5.3. Fukushima, 2011

At 2:46 p.m. on March 11, 2011, a major earthquake of magnitude 9.0 hit the east coast of Japan. The Fukushima Daiichi nuclear power plant was located about 180 kilometers south of the epicenter. The nuclear plant had six reactor units. Units 1, 2, and 3 were operating when the earthquake hit, however all were successfully shut down. Units 4, 5, and 6 were not in service at the time of the quake. These reactors received no physical damage from the earthquake, even though its intensity exceeded the plant design limit. Since the earthquake also disabled the external electricity supply to the power plant, back up on-site generators, located in the basement of the reactor buildings, started automatically to operate the residual heat removal system. About fifty minutes after the earthquake, a major tsunami of up to fifteen meters in height arrived.

All nuclear power plants located along the coastline are required to implement tsunami countermeasures. However, the one in Daiichi was only able to handle tsunamis of up to 5.7 meters in height. This standard was established in the 1960s based on limited scientific data regarding the likelihood of super tsunamis and earthquakes. The tsunami submerged the basements of the plant and caused many critical safety components to malfunction, including sea water pumps, diesel power generators, and batteries. Without heat sinks, the core temperature rose and water vaporized. Emergency workers tried to restore cooling capability but without sufficient power supply and lacking essential equipment, their efforts were not effective. A few hours later, the cladding of the fuel elements was damaged due to high heat and the nuclear fuel was exposed, releasing radioactive substances to the atmosphere. Hot steam interacted with zirconium cladding and produced hydrogen, igniting hydrogen explosions inside the containment buildings.

The recovery was a lengthy process: it took months to achieve cool shutdown conditions. No casualties were reported relating to the nuclear accident but radioactive material released to the air and water posed great environmental and health concerns. The exact impact, however, will not be available until more data has been accumulated. The Fukushima accident has forced a reconsideration of the required safety standards for nuclear power plants. Rare events do occur in nature and safety measures and protocol must be continually updated.

7.6. Challenges in Nuclear Waste Management

Handling the spent fuel (waste) discharged from a nuclear reactor is a rather complicated issue. Highly radioactive material generated from nuclear fissions poses serious health and environmental threats if not managed properly. A typical 1 GMWe (1,000 MWe) nuclear reactor produces about twenty-seven tonnes of spent fuel annually. From a technical viewpoint this translates to a relatively small volume of waste material, however political and environmental considerations have had great impact on the problem.

There are two possible options for handling these spent fuels: either through direct disposal or via recycling. The United States, Canada and a number of other countries

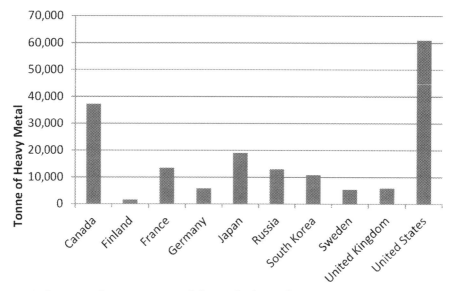

Figure 7.7. Spent Fuel Inventory as of the End of 2007.[8]

have chosen the direct disposal approach. Nuclear spent fuels are treated as waste immediately after they are removed from the reactor. Eventually the waste is buried, as is, deep underground. Other countries, including the UK and France have chosen to recycle. This is a preferable solution both from an economic and environmental concern viewpoint. About ninety-seven percent of the material could potentially be recovered from the spent fuel and reused in the production of fresh fuels. The remaining three percent of high level waste must be vitrified and permanently buried.

The inventory of nuclear spent fuels will vary depending on the generation capacity and waste management approach of a particular country. An indication of the global nuclear spent fuel inventory by country at the end of 2007 is shown in Figure 7.7. The USA and Canada are the two largest holders of spent fuel.

Multiple layers of protection are designed to safely dispose high level nuclear waste material. The waste is first solidified in an insoluble matrix such as borosilicate glass and then sealed inside stainless canisters. The canisters are further surrounded by clay to prevent ground water penetration before they are put into a deep underground repository site. Although these protective measures should be adequate in reliably isolating nuclear waste from the environment for a long period of time (typically one hundred thousand to one million years), lack of absolute confidence in the technology still prevails in the general public persona, resulting in strong resistance to siting waste repositories in community localities. Proposed projects are often delayed or abandoned. Thus far, no geological repository has been licensed for retaining of civilian nuclear wastes. Dry cask storage has often been implemented as a temporary storage solution, in the hope of a more permanent solution being agreed upon going forward. Currently, spent fuels are stored in canisters and surrounded by heavy reinforced concrete walls. Dry-cask storage is a relatively inexpensive solution but as the volume of the stockpile increases the security threat is also augmented accordingly.

The Nuclear Waste Policy Act passed by the United States Congress in 1982[9] required the Department of Energy to identify and construct an underground facility for permanent storage of high level nuclear waste. Initially Yucca Mountain was selected as one of three candidate sites and the US congress commissioned a thorough evaluation study of the site in 1987. The proposed site is located in Nevada, 120 kilometers northwest of Las Vegas. The Yucca repository is a deep underground facility, about 350 meters below surface level, with capacity to accommodate more than 63,000 tonnes of spent fuel. It was scheduled to receive spent fuels by 1998. In spite of the fact that the Yucca Mountain project was thoroughly studied dating back to 1978, debates and counter-debates on project efficacy were never successfully resolved or agreed upon. One of the central issues is whether the facility is environmentally safe and would remain so for up to 1 million years. Following many delays, the project was officially terminated in 2011. The failure of the Yucca Mountain project has not just been excessively costly but has also left the US in a very difficult situation, lacking long term storage for the country's nuclear waste into the foreseeable future.

7.7. Future Role for Nuclear Power

Awareness and understanding of both the historical and contemporary contexts of nuclear power are essential inputs to any attempt to answer questions regarding the future of nuclear power: will it rebound as an important source of energy or will nuclear energy be gradually phased out? In endeavoring to answer this question a number of variables need to be considered. These include the global future energy outlook, innovation in nuclear technological developments, and public acceptance or non-acceptance of nuclear energy. These variables are somewhat interdependent and resolution and decision making is a subjective process. In this section we consider the variables independently. We also encourage the reader to independently consider the issues and to participate in a much required public debate on this important topic.

Global Energy Outlook is largely shaped by two parameters: energy security and climate change. Energy security addresses the availability of resources. Currently fossil fuels such as oil, gas, and coal comprise about eighty percent of world total energy use, servicing transportation, electricity generation, industry, and the residential and commercial built environment. As noted in Chapter 1, abundant and easy access to these resources has supported unprecedented growth rates of human civilization for more than a century. The presumption that these resources remain abundant has changed since 2005 when global petroleum production became inelastic with respect to prices, indicating that conventional oil production had plateaued and will begin to decline. In "oil's tipping point has passed" Murray and King argue that the oil market has "tipped into a new state; production is now inelastic, unable to respond to rising demand, and leading to wild price swings."[10] The study reveals that while production of crude oil increased to meet demand between 1988 and 2005, since that time production has been roughly constant, despite an increase in price of around fifteen percent per year. The harnessing of available oil reserves is also proving more costly and difficult, contributing to the tail-off in levels of production. Furthermore, nearly seventy percent of the world's remaining conventional oil reserves are located in five Persian Gulf countries.

Recent studies also suggest that coal supplies are less abundant than previously believed. Over reliance on fossil fuels, high concentration of resources and declining supply create great concerns in respect of security and availability of energy supply. Energy Watch indicate that US coal production peaked in 2002 and that world coal production is projected to peak by 2030.[11] Supplies of natural gas are abundant and offer the best production potential of the fossil-fuels coal, oil and gas. Recent discoveries include significant finds in Israel and Mozambique. Twenty-five percent of electricity generation in the United States is supplied by natural gas power plants. Although production of conventional natural gas in North America peaked in 2001,[12] recent years have seen the rise in hydraulic fracturing of shale rock as a new means of further exploitation of natural gas, with significant potential finds both in the US and in many other countries. While fracking offers significant promise in non-conventional natural gas exploration, it also carries environmental concerns, in particular relating to stress on surface-water and groundwater supplies. Air pollution owing to volatile organic compounds and other hazardous compounds are also of environmental concern.[13] There is also controversy surrounding the quantity and abundance of supply required to meet demand going forward.[14]

Climate change will increasingly factor into the determination of the future role for nuclear energy in national grids. Accepting that greenhouse gases are a major contributing factor to global warming, in endeavoring to avoid catastrophic consequences to the ecosystem resulting from warmer climate, it is imperative that enhanced efforts be made toward limiting the rate and pace of average global temperature increase. As discussed earlier in this book, the Kyoto Protocol has set goals to limit temperature rise to not greater than two degrees Celsius above preindustrial levels. To achieve this objective an aggressive shift from the use of carbon-emitting resources to generate energy is required. Whether such a shift is possible through introduction of renewable, non-carbon emitting fuel sources alone, is highly questionable. It is therefore essential that a wider role for nuclear energy be discussed between nuclear energy scientists, policy makers and within the broader public domain. Security of energy supply is an equally challenging and deterministic input to this debate. Energy security arguments suggest that reductions in fossil-based energy resources, which can require transportation over several thousand kilometers across land and sea boundaries, could help address overdependence and inefficiency, factors that would have significant economic and social implications.

Technological innovations in reactor design will be key to the future role for nuclear power. Evolutionary new reactor designs have made nuclear power safer and more cost-effective to currently operating plants. The adoption of advanced fuel cycle technologies will greatly relieve the pressure of waste disposal and make nuclear power a virtually sustainable resource.

7.7.1. Advanced Reactors to Improve Safety and Efficiency

Since the introduction of nuclear reactors in the 1950s there have been four generations of innovation in reactor design. The first generation (I) was the early prototype reactor, developed during the 1950s and 1960s as a proof of concept design. The majority of commercial nuclear power reactors that are currently in operation belong to the second-generation (II), designed to be economical and reliable for large scale generation. Many of these reactors are nearing the end of their operational lives and will be replaced by generation III/III+ reactors, now mature following decades of research and innovation. By using state-of-the-art technologies through incorporation of over fifty years of experience, the new generation of reactors offers a major upgrade in terms of energy efficiency and operational safety. The new reactor plants are certified to operate for sixty years. Older power reactors were designed to operate for forty years.[15] The new generation reactors will possess higher thermal efficiency as they operate at higher temperatures. The new reactors also feature standardized and modular designs, a major improvement which will be beneficial in expediting licensing procedures and reducing construction costs. Advanced reactors are more robust and less complicated in structure, with fewer components facilitating operation and maintenance. Most appealingly, a feature of some new reactors is the adoption of passive safety systems and other inherent safety features, representing a major leap over the old reactors that rely on active residual heat removal systems. New designs use passive safety features such as gravity, providing natural circulation which helps to stabilize reactors and to keep them intact in the event of an emergency for an extended period of time (up to seventy-two hours), during which external resources can be arranged and put into operation. With such passive safety systems in place, a catastrophic event such as the Fukushima accident could be avoided.

These emerging generation reactors may also be used to produce hydrogen economically. This will be of particular benefit as a transportation fuel, an industry which currently consumes more than forty percent of global petroleum supplies. If hydrogen can replace petroleum in transportation, reliance on petroleum will be greatly reduced. This will in turn enable significant reductions in greenhouse gases emissions. A further attractive feature of some generation IV reactors is that they operate a closed fuel cycle: they can burn nuclear waste. Nuclear waste produced from generation II/III reactors can be fed as fuel to generation IV reactors, reducing the burden on waste management.

7.7.2. Closed Fuel Cycle to Increase Uranium Utilization and Reduce Uranium Proliferation Risk

The simple once-through fuel cycle utilizes less than one percent of available uranium; proposed advanced once-through systems will utilize less than two percent uranium. This severe underutilization of uranium gives rise to two major problems: cost efficiency and waste management. Sources of uranium, like petroleum, are limited on earth. If uranium continues to be used with such poor efficiency, the current known sources will be depleted within one hundred years. Although further exploration may result in the discovery of new resources, a more reliable and desirable solution would be to improve utilization efficiency. Underutilization also leads to higher volumes of nuclear waste which requires treatment and safe storage. Some radioactive actinides have very long half-lives and need to be safely stored for over ten thousand years. Finding a place for this purpose is extremely difficult, as we have learned from the Yucca Mountain Project.

Scientists and engineers are working on a more promising and preferable solution, that of nuclear waste recycling. Recycling could solve two problems at once, through achieving much higher utilization efficiency and producing significantly less quantities of radioactive waste to be disposed. The most mature recycling technique currently available is through using fast reactors. Fast reactors use fast neutrons rather than thermal neutrons to sustain chain reactions. Using fast neutrons presents a great technical challenge in reactor design since fast neutrons are difficult to capture by fissionable nuclei. Construction costs are also high making them an unattractive economic proposition at the current time. However, should scientists and engineers succeed in mastering the technological challenges while reducing implementation costs, fast reactors will be very attractive, possessing a number of unique features. Fast neutrons have the ability to split the problematic long-lived actinides and transmute them to isotopes with manageable half-lives of typically hundreds, rather than millions of years. A further advantageous feature of fast reactors is that they can become breeder reactors; they can be used to produce fuel while generating power. Through incorporation of these features, closed fuel cycle reactors would be achievable, with many generation IV reactors operating as fast reactors.

7.8. Social Engagement

Public acceptance is vitally important for successful deployment of new technologies in the energy and industrial processing sectors. In respect of nuclear power acceptance is of even greater import. For technologies such as coal, wind and solar, risks and benefits have been well aired and are quite well understood, with mixed reaction and public acceptance. A consensus does not exist regarding the risks associated with nuclear power, there is widespread public anxiety relating to radiation risk and to the management of nuclear waste. Building on this chapter's earlier discussion about nuclear accidents, it is understandable that concerns can have an overriding bearing on acceptance of nuclear energy, adding to the complexity in development and implementation of required solutions.

7.8.1. Radiation Risk

Fear of radiation in the public mindset is well founded, not least as a result of the dramatic effects of nuclear radiation which resulted from the bombing of Hiroshima and Nagasaki at the end of World War II. Radiation can kill cells and can incur inadvertent changes to DNA structure. The degree of damage to the human body relates on the dose the body has been subjected to. The effective dose received by a human being is measured using the Sievert (Sv) unit. Exposure to high a dose of radiation (more than three Sievert) over a short period of time can result in acute radiation sickness and can indeed be fatal. In the 1986 Chernobyl accident, 134 emergency workers were subjected to an extremely high radiation dose (between 800 and 16,000 milliSievert). Twenty-eight of these workers died from radiation sickness within months of the accident. These losses could have been significantly reduced had the radiation measurement equipment been operational. False readings from the faulty equipment had misled the emergency workers into believing the reactor was intact and the plant safe.

Potential long-term health problems resulting from low-level radiation include cancers and other heritable diseases. The International Commission on Radiation Protection suggests that the chance of developing a cancer increases by 5.5% for every Sievert (Sv) exposure and by 0.2% for heritable effects. These figures have been extrapolated from high-level dose observations and their validity to low-level dose equivalence may not have direct correlation. It is believed that less than 100 mSv over a long period of several years poses no measurable health effects. For a period of twenty years (1986–2005) following the Chernobyl accident, more than five million people lived in the contaminated area, and were subject to doses of ten to fifty mSv. Citizen health status was closely monitored by international agency organizations during this time. The Chernobyl Forum, organized by agencies including IAEA, WHO, and the UN, and with input from three countries directly affected by the Chernobyl disaster, Belarus, Ukraine, and Russia, published their most recent findings in 2005. They concluded that no measurable increase was found in incidence of radiation-induced leukemia or solid cancers, other than thyroid cancer. The thyroid gland accumulates radioactive ioline-131 from the food stream, making it vulnerable to radiation exposure. Fortunately, thyroid cancer is a treatable disease. Following the recent Fukushima accident, a preliminary report published by WHO in 2012 estimated that people residing in the contaminated zone received radiation doses of between ten to fifty mSv. The long term health effects resulting from this exposure may or may not be observable but will necessitate a long time frame to carry out observations, analyses, and publication of consequential results. Health effects resulting from low-level dose radiation exposure can be difficult to measure owing to many interrelated concurring factors. For example, about forty percent of the general population will eventually develop cancers regardless of exposure to radiation during their lifetime. Small percentages of radiation-related superimposed cancers, if any, can therefore be difficult to measure and quantify. It is also important to note that the human body has its own defense system that may repair some minor damage caused by low level radiation, thus minimizing and masking its effect. More scientific data is needed in order to establish a guideline for low level radiation risk. Such studies are imperative to building public confidence and to establishing proper evacuation zones should accidents happen.

It is also important to point out that the above figures typify worst case scenarios. On average, if one happens to reside in the near vicinity of a nuclear power plant, resulting radiation doses of typically less than 0.1 mSv per annum will add to the 2.4 mSv dose accruing as a result of natural sources, and is not considered a sizable increase.

7.8.2. Change of Nuclear Energy Culture

The nuclear power community also bears responsibility for lack of public acceptance of nuclear energy. Failure to engage with the public in providing relevant information can result in failure, no matter how technically sound and safe a new innovation may be. The Yucca Mountain Project is a typical example where technical decisions were overruled by political expediency, due to lack of sufficient public support. A thorough study of the Yucca Mountain project was carried out and the project was highly commended from a technical viewpoint. However, a lack of strong support from local residents and state government, combined with political and financial uncertainties, made project survival very difficult. In contrast, the Waste Isolation Pilot Plant (WIPP) in the state of New Mexico, a geologic repository for military radioactive waste, was successfully received by the local community. This was achieved through committed public engagement, resulting in strong support from all sectors. This example indicates that through appropriate engagement and information viewpoint exchange, public opinion can sway in favor of high priority energy installations.

Unfortunately transparency and openness has not always been the hallmark, and this invariably has damaged the reputation and credibility of nuclear power. The general public in the Soviet Union was not informed of the Chernobyl accident until two days post the initial plant disaster. Evacuation of the nearby city of Pripyat did not commence until thirty-six hours after the accident. Evacuees were informed that the clean-up would last only a few days. Poor management of the crisis and the release of false information resulted in severe damage to public confidence in nuclear power, both regionally and throughout the world.

Initiating change in nuclear culture toward a more open forum of engagement and discussion is complex owing to the necessity of striking a balance to the need for securing sensitive and confidential information. Information which may be shared and that which may not can form sensitive boundaries, with decisions which can have significant consequence. At Fukushima, initially some vital plant information was not available to the emergency work force, because the utility deemed it sensitive. The lack of information delayed the rescue efforts and was one of factors which contributed to the escalation of the tragic event.

An important question remains: what is the financial outlay for a nuclear reactor? Costing nuclear power comprises four major components: construction, operation, waste disposal, and plant decommission. Waste disposal and decommission costs are normally calculated into the generation cost, comprising only a small fraction of the total when compared to construction cost. The real cost of constructing a new nuclear plant is a hot topic of debate. Various estimates in overall cost invariably arise and have a tendency to change rapidly over time. The estimated overall per unit construction cost in the US in 2003 was approximately \$2,000/kWe. This increased to \$4,000/kWe by 2009. Based on these numbers, the overall cost of nuclear power today has increased to \$84/MWh, exceeding that of coal (\$62/MWh) and gas (\$65/MWh). At first glance, in terms of cost analysis the future looks pessimistic for nuclear power. However, a closer examination will reveal that the majority of the cost surge is the result of regulation changes, on account of public resistance and political intervention. For example, the next generation French reactor, Flamanville 3, experienced repeated delays with resulting cost increase of more than two billion euro. Much of the additional costs incurred are a result of design modifications requested to comply with stricter safety regulations which have been stipulated post the Fukushima accident. The Shoreham nuclear power plant located in Long Island, New York, is another example of how nuclear power can become expensive. Construction commenced in 1973 and was projected for completion in 1979. The initial cost was estimated at \$217 million. During construction however additional regulation requirements were established, necessitating significant design changes which resulted in ensuing project delay. The plant construction was finally completed in 1984. Unfortunately the most costly delays were just about to begin. The Three Mile Island accident in 1979 and the Chernobyl accident in 1986 altered public opinion to nuclear power, making it very difficult for the newly built plant to obtain an operating license. As a result the Shoreham plant never came into operation and the project was finally abandoned in 1989. The final cost exceeded six billion US dollars. This constituted excessive and unnecessary cost inflation, and ultimate waste of public and private funding. Such waste may be mitigated going forward by movement to modular plant design and construction.

In evaluating the future potential for nuclear power it is also necessary to consider the competing technologies. In order to replace conventional coal, oil and natural gas, the principal competitors to nuclear energy are hydro, wind and solar. Hydro is capable of providing clean, reliable base-load energy; however its availability is confined by location. Solar is currently expensive and may not be feasible for widespread utilization in the near future. Wind power is relatively cheap, estimated at between $70/KWh and $120/KWh, which is on par with nuclear power. Cost will continue to drop as technology improves and the industry matures. However, important as wind, solar and other alternative renewable energy sources are, owing to intermittency and scale it is unlikely they will succeed as base-load providers. In maintaining balance between generation and consumption, power grids rely on certainty of supply. A small amount of uncertainty can be offset by other reliable resources, such as through use of coal and natural gas. Should the share of wind power increase to significant levels, some major innovations must be introduced to the power grid in order to handle the associated uncertainty of supply.

Risk is an intrinsic parameter in characterizing a particular option. Using nuclear power presents a risk. Risk also presents if opting to abandon deployment of nuclear energy. We may consider recent developments in Germany by way of example. Shortly after the Fukushima disaster, Germany announced the abandonment of its nuclear programs, with immediate shutdown of eight reactors and the phased shutdown of the remaining nine reactors by 2022. The share of nuclear generated electricity dropped from twenty-three to seventeen percent while renewables increased from twenty to twenty-five percent. A consequence of this motion is that consumers will be required to pay €250 per household to sponsor cleaner energy. Over thirty billion euros of investment is also needed over the next two decades to build more power lines to transmit and deliver the electricity generated from renewables, such as offshore wind turbines. However, wind is an intermittent resource, and thus the construction of a new supplementary power plant will be required to meet the power deficit requirement. This will present challenges to the country in its ability to cut carbon emissions by forty percent of 1990 levels by 2020, however Germany has shown innovation and has been an exemplar country in renewable technology development and grid connection. Neighboring countries to Germany, in particular France and the Czech Republic generate a major portion of their electricity from nuclear power, hence buying and selling between countries may require import of nuclear sourced energy. This fact introduces some complex dynamics into energy and climate policy with respect to national vs. economic trading bloc priorities.

7.9. Future for Nuclear Fusion

The atomic reaction in nuclear fusion is different from nuclear fission; in fusion two light atomic nuclei fuse to form a heavier nucleus and in so doing a large amount of energy is released. Research in fusion is the domain of plasma physics and is sometimes considered a holy grail in the search for future energy provision. From early twentieth century scientific experimentations to the present day, research has continued to endeavor to perfect the capture of energy transfer via nuclear fusion. A *Tokamak*, devised by

Russian physicists Igor Tamm and Andrei Sakharov in the 1950s, is a magnetic device that contains a *plasma* in the shape of a torus, and which remains central to research in nuclear plasma physics to this day. Today, the International Thermonuclear Experimental Reactor (ITER) project is comprised of an international research consortium with objective to build the world's largest experimental tokamak nuclear fusion reactor. In the process fusion between deuterium and tritium (D-T) will produce one helium nucleus, one neutron and excess energy. The aim is to complete the transition from experimental studies in plasma physics to electricity generating fusion power plants.

The timeline toward achieving these goals however remains relatively long term; the first demonstration of electricity production is not expected for thirty years. In addition, as with current day fission power generation, safety and waste management will be of paramount importance to nuclear fusion power generation. It is important that support is maintained for ITER and related research projects in the expectation that experimental physics, material science and engineering practice will combine to enable development of a mature and reliable fusion energy future.[16]

Notes

1. Ian Hore-Lacy, *The World Nuclear University Primer: Nuclear Energy in the 21st Century*, 2nd ed. (London: World Nuclear University Press, 2006).
2. Ibid.
3. Jefferson W. Tester et al., *Sustainable Energy: Choosing Among Options* (Cambridge, MA: MIT Press, 2005), 362; and "The Economics of Nuclear Power," World Nuclear Association, last modified October, 2013, http://www.world-nuclear.org/info/Economic-Aspects/Economics-of-Nuclear-Power/.
4. Byron King, "Nuclear Power: The King of All Energies," *Daily Reckoning*, last modified August 11, 2010, http://dailyreckoning.com/nuclear-power-the-king-of-all-energies/.
5. Energy Information Administration, *Annual Energy Outlook 2011* (Washington, D.C.: Department of Energy, 2011).
6. Energy Information Administration, "Levelized Cost of New Generation Resources in the Annual Energy Outlook 2013," *Annual Energy Outlook 2013*, last modified January 28, 2013, http://www.eia.gov/forecasts/aeo/electricity_generation.cfm.
7. One exception is the Chernobyl reactor, which used graphite as moderator and water as coolant. This unique feature was one of the contributors to the Chernobyl accident.
8. World Nuclear Association, "Nuclear Power in the World Today," last modified November 2013, http://www.world-nuclear.org/info/Current-and-Future-Generation/Nuclear-Power-in-the-World-Today/.
9. See Nuclear Waste Repository Act, Pub. L. No. 97–425, 96 Stat. 2201 (1983) and subsequent amendments. The Act was extensively amended in identical form by Pub. L. 100–202 (101 Stat. 1329–121) and Pub. L. 100–203 (101 Stat. 1330–243) on Dec. 22, 1987. The Act appears in the United States Code at 42 U.S.C. 10101 et seq. Bracketed notes at the end of each section indicate the United States Code citation for the reader's convenience.
10. James Murray and David King, "Climate policy: Oil's tipping point has passed," *Nature* 481, no. 7382 (2012): 433–435, http://dx.doi.org/10.1038/481433a.
11. Energy Watch Group, *Coal: Resources and Future Production* (Berlin: Energy Watch Group, 2007).
12. EIA, *Annual Energy Outlook 2011*.
13. "Natural Gas Extraction: Hydraulic Fracking," Environmental Protection Agency, last modified November 18, 2013, http://www.epa.gov/hydraulicfracturing.

14. J. David Hughes, *Will Natural Gas Fuel America in the 21st Century?* (Santa Rosa, CA: Post Carbon Institute, 2011).
15. Life extension from forty to sixty years is possible and subject to regulatory approval. In the US, nearly fifty percent of the cases are approved. Possible extension to eighty years' operation has also been discussed but not likely to happen in the current political atmosphere.
16. ITER and the Environment: "Fusion has the potential to play an important role as part of a future energy mix for our planet." http://www.iter.org/environment.

Bibliography

Energy Information Administration. *Annual Energy Outlook 2011*. Washington, D.C.: Department of Energy, 2011.

Energy Information Administration. "Levelized Cost of New Generation Resources in the Annual Energy Outlook 2013." *Annual Energy Outlook 2013*. Last modified January 28, 2013. http://www.eia.gov/forecasts/aeo/electricity_generation.cfm.

Energy Watch Group. *Coal: Resources and Future Production*. Berlin: Energy Watch Group, 2007.

Environmental Protection Agency. "Natural Gas Extraction: Hydraulic Fracking." Last modified November 18, 2013. http://www.epa.gov/hydraulicfracturing.

Hore-Lacy, Ian. *The World Nuclear University Primer: Nuclear Energy in the 21st Century*. 2nd ed. London: World Nuclear University Press, 2006.

Hughes, J. David. *Will Natural Gas Fuel America in the 21st Century?* Santa Rosa, CA: Post Carbon Institute, 2011.

King, Byron. "Nuclear Power: The King of All Energies." *Daily Reckoning*. Last modified August 11, 2010. http://dailyreckoning.com/nuclear-power-the-king-of-all-energies/.

Murray, James, and David King. "Climate policy: Oil's tipping point has passed." *Nature* 481, no. 7382 (2012): 433–435. http://dx.doi.org/10.1038/481433a.

Organization for Economic Cooperation and Development and the Nuclear Energy Agency. *Uranium 2011: Resources, Production and Demand*. Paris: OECD Publishing, 2012.

Tester, Jefferson W., Elisabeth M. Drake, Michael J. Driscoll, Michael W. Golay, and William A. Peters. *Sustainable Energy: Choosing Among Options*. Cambridge, MA: MIT Press, 2005.

World Nuclear Association. "The Economics of Nuclear Power." Last modified October, 2013. http://www.world-nuclear.org/info/Economic-Aspects/Economics-of-Nuclear-Power/.

World Nuclear Association. "Nuclear Power in the World Today." Last modified November 2013, http://www.world-nuclear.org/info/Current-and-Future-Generation/Nuclear-Power-in-the-World-Today/.

World Nuclear Association. "Outline History of Nuclear Energy." Last modified June, 2010. http://www.world-nuclear.org/info/Current-and-Future-Generation/Outline-History-of-Nuclear-Energy/.

PART 3

ENERGY DISTRIBUTION AND USE

Chapter 8

Taking Emerging Renewable Technologies to Market

MELISSA DARK, JENNY DAUGHERTY,
PETER CAMPBELL, AND WILLIAM GRIMSON

Abstract

This chapter sets out to demonstrate that introducing renewable energy sources into an existing market is a complex socio-techno-economic challenge. The important role that competition plays is reviewed in the context of demand and the number of suppliers: too few can result in the market being difficult to penetrate. The cost of switching technologies is very relevant to taking advantage of renewable energy sources; for example the electrical supply networks would need considerable augmentation. The role of regulatory frameworks, the distinction between inducements and mandates are discussed and consideration is given to the type of political systems in place where renewable technologies are being adopted. The nature of the social dimension is stressed where it is noted that different countries react in different ways to the promotion of new technologies. A key concept, levelized cost, is used to compare the economics of a range of energy sources. Levelized costs take into account the cost of a kilowatt-hour (kWhr) in terms of both initial construction of the system and the recurring operating costs over its financial life. The question of reliability and maintainability is stressed, in particular with respect to offshore wind energy systems where costs are currently relatively high. However the technical learning that comes with exploiting offshore wave energy is, over time, expected to moderate the levelized costs. In turn, this is seen as a driver to various initiatives by which governments seek to promote such technologies and encourage much needed venture capital.

8.1. Introduction

This chapter provides a brief overview of various economic, political, social, and maintainability factors that influence bringing technologies to market. In order to illustrate the interactions among social, technical, and economic systems, a few salient aspects of these systems are elaborated as a way of laying a foundation for a more detailed account focused on wave energy. Furthermore, all factors are endogenous, meaning that 1) these factors are all a part of the environment or system in which a technology is brought to market, and 2) the value of these factors is determined by the states of other variables in the system. The multiple interactions and impacts of these factors on bringing a technology to market are beyond the scope of this chapter. Instead of an exhaustive treatment, we offer a few critical ideas with illustrations in hopes that we provide sufficient grounding for readers to extend their thinking with regard to the topic.

8.2. Economic Factors

All technologies encounter prevailing market conditions, which influence whether or not the technology comes to market, how, when, by whom, and where. Prevailing market conditions are factors such as the number of competitors, the nature and intensity of competitiveness, the market's growth rate, and so on. The competitors within a market are the persons or firms that offer a similar product or service. Competition in a market is generally viewed as socially desirable because it requires companies to make efficient use of their resources in order to reduce costs.

All markets can be said to have a competitive structure that affects bringing a new technology to market. The competitive structure of a market refers to the current state of the market with regard to several interrelated factors: 1) the number of competitors in the market; 2) the relative strength of these competitors; 3) the level of demand for the technology and the existing supply; and 4) the ease of entry into the market. There are several known obstacles that can make it difficult to enter a market with a new technology, and the more barriers to entry there are, the stronger the position for the incumbents. For example, a market that is occupied by a couple of dominant firms that have economy of scale advantages can be difficult to penetrate. Incumbent firms usually have favorable access to resources, existing supply and distribution chains, and "know how" that positions them strongly for holding onto and growing market share. Another barrier to entry is capital (equipment, buildings, or raw materials) investment. Generally speaking, as the needed investment in capital increases, it becomes more difficult to take a new technology to market. A third barrier to entry is strong brand recognition and customer loyalty among the existing competitors. A product or service that is established in the market and known to provide a given value for the price paid has a distinct advantage over a new product whose price-value proposition is uncertain.

A fourth possible barrier to entering a market with a new technology is switching costs. Switching costs are the costs incurred with switching from one product to the next. For instance, the potential costs to switch from gasoline powered to electric vehicles (EVs) would include producing an infrastructure of charging stations to support the fleet of EVs. As with many new technologies, there is the familiar chicken-or-the-egg phenomenon. The market needs to produce more charging stations in order to encourage more drivers to drive EVs. At the same time, the market needs to see more EVs on the road in order to build more charging stations. Switching costs can go beyond installation and startup costs to include search costs, that is, the costs associated with searching for and learning about the new technology, as well as fees to exit a market, for instance, the costs to disassemble the infrastructure built to sell gasoline for cars. A fifth potential barrier to entering a market is network effects. Network effects arise in cases where the value of a product increases as the number of users increase. This is often the case with information technology products and services where the value an individual derives from a product arises both from their own personal use as well as from the usage of others. In cases of network effects, consumers are reluctant to switch one product because of the consequent effects on the other products/services in the network that they enjoy and rely on. Finally, tariffs and government regulations can prevent or delay entry into a market by protecting the existing technology or failing to incentivize the development and/or adoption of the new technology, which we discuss again in the section on social factors.

The competitive intensity of a market has been characterized by Porter as including five factors.[1] While many of these factors are similar to the barriers to entry already mentioned, Porter offers elaboration that is useful and worth repeating. These are: 1) the threat of substitute products or services; 2) the threat of established rivals; 3) the threat for new entrants; 4) the bargaining power of suppliers; and 5) the bargaining power of consumers. The threats to substitute products that prohibit or constrain bringing a new technology to market arise from buyers' propensity to substitute products, the relative price of the substitute to the incumbent product/service, the number of substitute products, the ease of substitution, associated switching costs, and perceived differentiation. The threat of established rivals refers to the competitive struggle for market share among firms in an industry where increased rivalry among established firms leads to stronger threats to profitability. The strength of rivalry among established firms within an industry stems from the extent of exit barriers, the amount of fixed costs, the presence of global customers, the growth rate of the industry, the demand for the new product, and the possible absence of switching costs. The threat for new entries comprises the risk to the entrant to get into a market and includes patents (which essentially block new entries for a time period), brand equity and customer loyalty, switching costs, sunk costs and capital requirements, access to distribution, and absolute cost. The bargaining power of consumers is the ability of consumers to exert pressure, often through price, on the firm(s) attempting to enter with new technology. Several factors influence consumer bargaining power including the ratio of buyer concentration to firm concentration, buyer switching costs relative to firm switching costs, buyer information availability, price sensitivity among buyers, and the availability of substitute products with differential advantage. The fifth and final threat category in the Porter model is the bargaining power of the suppliers. Suppliers can increase the cost of inputs to the product/service; the presence of substitute inputs will affect the bargaining power of suppliers, as will the differentiation of the inputs. If the inputs from a given supplier are more costly, but result in a differential product that consumers will buy, such that switching costs are also justified, entry of the technology into the market is more likely.

8.3. Political Factors

Intersecting the myriad of economic factors, political factors also influence bringing a new technology to market and the deployment of technology in the market. We can think about these political factors as having direct and/or enabling (or disabling as the case may be) effects. A full treatment of how political factors affect bringing a technology to market is beyond the scope of this chapter and book; we will focus on describing two factors with examples that illustrate how the factor affects marketization. The two political factors we address are 1) issues that deal with the structure or affairs of government, and 2) particular policies, laws, or regulations. Both political structure and the policies and laws enacted by a government can lead to either an enabling or constraining environment. And political structures have implications for the types of policies and laws that may be enacted.

What do we mean by a political structure? Political structure refers to groups, such as political parties, lobbying groups, and institutions; it includes the presence or absence of, and nature of: the judicial, legislative, and executive systems. Political structure also refers to the relation of these groups to each other, and their patterns of interaction within the political system. These factors will vary across different political structures, their relation to each other, and their patterns of interaction over time influence the laws, regulations, and norms present in the political system. These variations matter in bringing technologies to market and in enacting policies and laws that enable or constrain bringing technologies to market.

One very clear example of different groups and their relation to each other is non-federalist versus federalist governments. The two primary types of non-federalist political structures are unitary political systems and confederate political systems. A unitary system is a system of political organization in which most or all of the governing power resides in a centralized government, which delegates authority to subnational units and channels policy decisions down to them for implementation. There are many examples of unitary political systems, including Bolivia, Chile, Egypt, Ireland, France, the People's Republic of China, and Vietnam. And while there are several nation states that have unitary political systems, they vary greatly.

A confederate government is the type of government where the power rests with the local entities, which dictates directives to the national government; the confederate government can only do what the confederation allows it to do. In this system, there is little central political control and power is very diffuse. Confederate government powers are often in the areas of defense and foreign commerce. Two current day examples of confederate political systems are Canada and the European Union. Energy policy in Canada's confederation is interesting and illustrative. Jurisdiction over energy resources is divided between the federal, provincial, and territorial governments. Canada's national government has authority regarding the regulation of inter-provincial and international trade and commerce, which has implications for energy, as well as the management of non-renewable resources on federal lands. The provinces control the exploration, development, conservation, and management of non-renewable resources, as well as the generation and production of electricity. As a result, Canada's national government must coordinate its energy policies with those of the provincial governments without any guarantee of success and little power to issue mandates. A case in point is Canada's participation in the Kyoto Protocol. While the national government of Canada had the authority to sign the Kyoto Protocol, the brunt of implementing the legislation fell to the provinces given their control over energy resources and consumption. The greenhouse gas reduction targets for Canada in the treaty remained an unrealized dream due in large part to Canada's confederate political structure. The national government in Canada reduced funding for Canada's climate change plan and cut most of Canada's climate change programs, including successful programs like the Wind Power Production Incentive, which subsidizes the installation of wind power, and Energuide for Houses, which gives incentives for Canadians to make their homes more energy efficient.

In a federalist political structure, the powers of government are divided between the national (federal) government and state and local governments. Certain powers are delegated to the national government, and all other powers are reserved by the states or the people. A case in point is the US system, which is federalist; a fact that has significant implications on our views toward and use of energy both regionally and nationally. In the US, the states rather than the federal government hold important authority for planning energy system expansions, siting energy facilities and regulating energy facilities and transmission. States control important policy decisions that affect bringing energy technologies to markets. A supplier looking to bring a new technology to market may have to deal with fifty different state positions. The complexity of this for a supplier should be fairly obvious. Furthermore, it should be noted that such a company is facing several other possible barriers to entry described in the previous section. The states' role in energy policy in the United States is particularly important for wind power. According to Wilson and Stephens, state-level authority has had huge implications when it comes to bringing wind technology to market. Sixty-five percent of all turbines installed in 2008 were in just six states (TX, CA, IA, MN, WA, and OR), yet the Great Plains states (ND, SD, and NE) have both some of the nation's greatest wind resource potential and installed wind power capacity of only four percent of the national total.[2] In a unitary system, the policies would be enacted at the national level and passed down to the local level for implementation.

Within similar political systems, policy development, implementation and outcomes can vary greatly for numerous reasons. Some of the factors that shape the policy agenda, which in turn affects bringing technologies to market include: the nature of relationships and the patterns of interaction among the institutions (federal, state, local), government stability, public trust in government, consumerism, and the political agenda toward the market economy. What is important to keep in mind is that while societies are complex and dynamic systems, and no two are identical, the structure of the government will impact policy, which in turn will impact if and how energy technologies are brought to market. In a recent analysis, Shobe and Burtraw found that "the design and implementation of climate policy in a federal union will diverge in important ways from policy design in a unitary government. National climate policies built on the assumption of a unitary model of governance are unlikely to achieve the expected outcome because of interactions with policy choices made at the sub-national level."[3] These interactions could be many. For example, whereas a unitary government is better positioned to enact international agreements, a federalist system is argued to be closer to people, the locus of their needs, and therefore more responsive and unique in solutions for attacking social, economic, and political problems.

All political structures, regardless of type, create policy instruments (laws and regulations) to tackle social problems. The political structure impacts the types of policy instruments that are realized, which in turn impacts if and how new technologies are brought to market. Successful policies are successful in a context, that is, a political structure. Policies that lead to increased energy security, growth of domestic economic activity, and environmental benefits in one context, may lead to high energy prices, perverse incentives, and public dissatisfaction with renewable energy technologies in another context. This satisfaction and dissatisfaction impacts bringing the technology to market. While there are several types of policy instruments, we are only addressing the two most common policy instruments: inducements and mandates.

Inducements are some sort of incentive that helps bring about an action or a desired result, such as tax breaks, subsidies, and rebates. Inducements are conditional grants of money and are often accompanied by rules to ensure that the money is used consistently with policymakers' intent. Any inducement, regardless of its nature or intended objective, is comprised of three main parts: (1) the inducement giver, or the person or party offering the incentive; (2) the inducement receiver, who is the target, or the individual or entity being offered the inducement; and (3) the inducement or incentive itself. Inducements rely on the power of persuasion as opposed to force. And while inducements are frequently used policy instruments, their complexity must be acknowledged. As policy instruments, the efficacy of inducements is dependent on the following factors: 1) the extent that the inducer is willing and able to make good on its promise; 2) the degree to which the inducement that is actually provided reflects what was offered; and 3) the extent to which the inducement receiver complies with responsibilities and obligations present in the agreement made with the inducer. Inducements are not really suited to modify behavior in a single episode. Instead, the goal of most inducements is to secure ongoing compliance with overarching, long-range policy goals, and this assumes that the inducement giver is willing and able to monitor the inducement receiver over time. Inducements to develop innovations often come in the form of a research grant to create the new technology and supply the needed resources to do so. Other forms of inducements aimed at stimulating supply are loans, loan guarantees, tax breaks, free land, etc. Because innovations rarely find ready-made markets, it is often necessary to stimulate or create a market, that is, stimulate demand. For example, in the United States, HUD (Housing and Urban Development) has several multi-faceted programs that offer incentives to citizens to utilize energy-savings techniques. Using a tax break as the inducement, a citizen may get a tax break for implementing an energy saving technique, thereby aiming to create demand in the market for the technology.

Inducements can be negative as well as positive. Sanctions or fines can be used to deter undesired behavior. When the inducement receiver fails to modify the undesired behavior, the inducement giver must be willing and able to impose the sanction or punishment, else the inducement will be ineffective. The cost to enforce the inducement has to be carefully considered; it is not socially advantageous when it costs more to enforce a sanction than is gained by the diffusion of the technology. However, it is often very difficult to quantify these costs and trade-offs.

Mandates are official rules that specify actions to be taken in specific situations or contexts, as opposed to incentives (or disincentives) that attempt to stimulate behavior. Mandates state that one will do something, whereas incentives ask whether one would like to. Regulatory mandates and incentives can produce similar results, but mandates generally require no loss of revenue by the government. Generally speaking, mandates can be efficient in terms of expenditures; the primary costs of mandates are enforcement costs. Mandates can operate on individual people, organizations, collectives, or governments. A few examples of mandates that affect bringing energy technologies to market in the United States include: 1) state renewable portfolio standards (RPS) for renewable electricity, which require the increased production of energy from renewable energy sources, such as wind, solar, biomass, and geothermal; 2) the federal Renewable Fuel Standard (RFS), which is a program that requires transportation fuel sold in the US to contain a minimum volume of renewable fuels; and 3) California's Low Carbon Fuel Standard regulation that mandates use of an increasing amount of fuels with lowered GHG emissions each year in the state. In much the same way as incentives, mandates aim to affect supply and demand in bringing a new technology to market.

8.4. Social Factors

Technology can be seen as being influenced and shaped by the societal context within which it resides.[4] A given technology being adopted by society is not necessarily a predictable occurrence; the result of a linear process that moves from research and development to a commercialized product. The best or most inventive technology does not necessarily translate into a technology that is adopted and used by society. A technology's success in the market is not based solely on the technology's merits or capabilities, but is contingent on several factors including the preferences and choices of individuals, as well as societal values and norms. Often there is a right time and right place for a technology to be picked up by the majority of people within a society. For example, in 2004, David Cohen's idea for a social network using mobile phones seemed innovative and viable enough for investors so that he could found his company iContact. However, after eighteen months of trying, he was unable to convince phone carriers that his software was the next best thing that consumers would be clamoring for; not until Apple's iPhone's app store hit the market in 2008 would this trend be manifested.[5]

The social context (the time and place) in which a technology emerges is important to its adoption and diffusion. Adoption and diffusion are two concepts that are often used to examine the social factors involved in emerging technologies moving from concept to market. The adoption and diffusion of technology is an explanation of how, why, and at what time a technology spreads. These terms stem from Rogers' book entitled *The Diffusion of Innovations*, in which he described diffusion as the process by which an innovation develops over time and spreads through a social system.[6] Adoption is the process an individual undergoes from first hearing about a technology to the point of deciding to accept it. The diffusion process, on the other hand, explains group behavior, in how an innovation spreads among consumers; how a technology ends up in factories, homes, schools, offices, and so on. Essentially, the diffusion process encompasses the adoption process of several individuals over time.

The factors that affect adoption and diffusion often reside within a society's culture. A way to understand a society's culture is through its social institutions. A society structures and reproduces itself overtime through institutions such as its government, family, language, and legal systems. These are developed by people and fulfil certain roles over generations. For example, the family as a social institution has become the primary site of reproduction and initial socialization; it is where individuals first learn about the norms and values of their larger social group. Because of this socialization process, it is often difficult to see social institutions as being constructed by people as they have typically existed for generations and generations. They are taken for granted by the individuals within a society and become essentially invisible to them. Technology's development is subject to these social institutions and, in particular, the political and economic arrangements of the society. The power dynamics within these institutions impact the priorities that are established, the investments that are made, and the projects that are funded that lead to technological innovations.

The importance of social institutions on emerging technologies is evident when comparing nations.[7] Differences exist in how technologies are adopted even among countries with similar technological and economic infrastructures. For example, the United States and Great Britain could be said to be similar in many regards, but the adoption and diffusion of certain technologies has varied greatly. Take, for example, the adoption of genetically modified food; Great Britain and other European countries have instituted stricter regulations than the United States thus slowing the diffusion of these types of foods in the market. This has been attributed to differences in their cultures in regulating risk; the United States has become less restrictive in this regard.[8] The societal institutions and individuals acting within those societies affect how a technology emerges. In other words, technological adoption results from the choices of individuals who operate within social institutions. Social institutions, including governmental policies and societal values, in a sense, define the rules of the game or define the right time and right place for a specific technology within a specific society.

Diffusion, or how individuals learn about a new technology, is a function of the social system and the interconnected nature of its institutions. Rogers provided a framework for understanding the role of individuals within a society in spreading the knowledge and behaviors associated with an innovation. Generally, change agents bring the innovations to a society, typically through the gatekeepers and opinion leaders of that society. The gatekeepers and opinion leaders are those individuals who have the expertise and power to impact societal behaviors and values. In terms of their influence, individuals within that society can be categorized across five different rates of adoption. These categories are innovators, early adopters, early majority, late majority, and laggards. Innovators are the risk-takers who are willing to be first to adopt an innovation. Early adopters are more discreet in their adoption choices but are typically quick to try new innovations. Their opinions carry the most weight among the other adopter categories. The early majority tends to adopt after some time and once the success of the innovation is more predictable. The late majority adopts after the average member of the society does and are typically skeptical about innovation. Laggards are last to adopt, if they ever do.

Beyond diffusion, however, the acceptance or resistance of a new technology can be impacted by how individuals make decisions and calculate their own costs and benefits. Do the benefits outweigh the costs (not only financial costs but also other potential costs such as to their health, time, and so on) for the individual? This cost-benefit analysis changes depending on its impact over time, such as long-term environmental impacts and impacts on others. Some argue that individuals are willing to assume more risk or costs for others than for themselves. The acronym *NIMBY*, which stands for the phrase "not in my back yard," is used to describe this resistance to a particular technology. If the individual incurs disproportionate costs or experiences adverse impacts, then resistance can be higher than otherwise. For example, even though proponents note the environmental and economic benefits of wind energy, an often-cited opposition to wind turbines is from individuals in communities where projects have been proposed because their large structures negatively impact the landscape. This occurred on the east coast of the United States where several residents opposed the construction of Cape Wind, an offshore wind farm in Nantucket Sound, on the grounds that the wind turbines would obstruct their oceanfront views. Wind farms are a good idea in theory, but perhaps less so when they are in one's own back yard.

Another complaint about wind turbines is the noise they produce. Some have argued that the sounds produced by the turbines can cause insomnia, dizziness, and headaches. For example, Jim Cummings of the Acoustic Ecology Institute, an online clearinghouse for sound-related environmental issues, has said that about a dozen or so of the 250 new wind farms have generated significant noise complaints.[9] This issue has been researched by several groups and the findings indicate that there is no evidence that wind turbine noise poses any health risks. While there is no evidence to support such claims, this issue is dependent on context. For example, a previously quiet setting "is more likely to produce irritated neighbors than, say, a mixed-use suburban setting where ambient noise is already the norm."[10] Again, right time, right place.

8.5. Maintainability Factors

The final set of factors is more narrowly focused on the technology itself and its design. Maintainability factors are considered during the design and installation of technological products and systems. *Maintainability* is defined as the probability that when maintenance is necessary, a failed technology will be restored to its operational effectiveness within a specified time.[11] *Operational effectiveness* is the ability of a technology to perform as expected when operated. Maintainability is related to the reliability-failure propensity of a given technology. Reliability is a measure of the ability of a technology to avoid failure. Failure includes lost performance, compromised safety, and the need for restorative actions such as diagnosis, repair, spare part replenishment, or maintenance. The ease and economy of restorative actions that are necessary to restore a failed product is a function of its maintainability. Restoration involves isolating the source of failure, correcting the problem, and testing. For a technology to be maintainable, the design should not be too complex; equipment should be easy to access, remove, and replace; component parts should be uniform or standardized; and there should be minimal specialized parts or tools.

Maintainability is an important factor when considering offshore wind turbines as a solution to wind energy technology. Offshore wind farms are increasing due to the demand for renewable and environmentally friendly energy and the social and political concerns over onshore wind farms as already discussed. However, offshore wind energy production faces other issues including the design, installation, and maintenance of the turbines. Research has been undertaken to inform the design of wind turbines in a marine environment and the related logistical and accessibility issues. One of the primary challenges is the operation and maintenance of the offshore turbines including their accessibility, corrosion, and related costs. For example, the type of vessels needed to install and maintain the turbines is but one factor that must be determined. One approach is to develop ways to ensure the reliability of the wind farm as a whole by implementing a more standardized reliability system within the industry.[12] Many industries have implemented similar standardized reliability and maintenance approaches to improve cost-effectiveness, control, and safety. If offshore wind turbines are to emerge as a primary energy sources, the reliability and maintainability factors will have to be resolved.

8.6. Economics of Energy

When you consider the worst or near-worst climate change scenarios there seems to be little point in arguing about the relative costs of alternative sources of energy versus fossil fuel based energy, such as oil, gas, and coal. In a variation of Pascal's wager, the best bet appears to be to assume that climate change is real and that without appropriate action highly undesirable and very costly outcomes will occur. One such action is to drastically reduce our dependence on fossil fuels and in turn control or even reduce carbon dioxide levels in the atmosphere, and in such a course of action the high cost would be more than offset by the avoidance of a highly changed climate. If the assumption turns out to be invalid the consequence is that money will have been wasted on the search for alternative energy sources. It is, really, in a rational world, a choice of how two very different consequences should guide or even dictate our actions with respect to energy. But we live in a world of competing rationalities and undoubtedly one ground on which there are multiple views concerns the economics of energy. The following section of this chapter deals with some of these economic data.

The economics of alternative energy, just like any other energy source, are difficult to establish to the point that all interested parties accept the conclusions. Furthermore, when it comes to comparing, say, the cost of wind with nuclear power generation various conclusions can be generated depending on the assumptions made. For example, is the cost of an additional standby plant factored in to the wind side of the equation to account for downtime when there is no wind? Is there a figure allocated to environmental costs associated with uranium ore extraction and nuclear waste disposal, or the costs of related security aspects? How might the sequestering of CO_2 on very large scale be factored into estimates of the cost of energy in the event that such a course of action is required? In spite of such problems, some overview of the relative costs associated with diverse energy sources is at the least informative.

For tidal and wave energy the first piece of data that needs to be gathered is how much energy is there and how much of it is available at an affordable cost. A State of the Industry Report for the UK reports that "globally, it is estimated that there are 180 TWh of economically accessible tidal energy and over 500 TWh of economically accessible wave energy available annually."[13] Now, both tidal and wave energies depend on local physical characteristics of the surrounding sea and seabed, and it turns out the UK, despite its small size, has a disproportionate share of this globally accessible energy. "It has been estimated that UK waters contain around 15 per cent of the world's economically accessible tidal resource and over 10 per cent of the world's economically accessible wave resource. This tidal resource is estimated at 29 TWh, and wave is 50 TWh."[14] To put these figures in context, in 2011 the total UK overall primary energy consumption in primary energy terms was approximately two hundred million tonnes of oil equivalent. In units of watt-hour this amounts to 2.36 TWh. In terms of electricity consumption the wave energy could contribute up to about fifteen percent of the UK's needs. Clearly then there is an case on the basis of availability, especially in the UK, that marine energy should be exploited.

The second piece of economic data that needs to be considered is the cost of producing energy and making measured comparisons across a sweep of technologies. Two sets of estimates are presented, with the first set comparing levelized costs for a range of both dispatchable and non-dispatchable sources in the United States. The second set of estimates deal with tidal and wave generation based on data assembled in the UK with specific reference to Scotland. A close reading of both reports is required if conclusions are to be drawn as between the United States and the UK; but that is not the intention here, rather conclusions are only reached with respect to each jurisdiction but which are nevertheless possibly valid globally with respect to the various technologies. A final point—the actual amount of costs estimated are of less importance than the relative costs across technology domains, so the year on which costs are estimated is of no great significance with respect to the conclusions drawn later.

The US Energy Information Administration (EIA) in a recent report presents the average levelized costs for a diverse range of generating technologies.[15] The report defines *levelized cost* as the per-kilowatthour cost of both building and operating a generating plant over an assumed financial life and duty cycle. Factors in calculating the estimated levelized costs include "fuel costs, fixed and variable operations and maintenance (O&M) costs, financing costs, and an assumed utilization rate for each plant type."[16] It is worth noting that financial incentives are not factored into the estimated levelised costs. A few points can be made in reference to Table 8.1 in the EIA report, presented here for convenience.

Some of the points that emerge are:
- The network investment for non-dispatchable technologies is approximately three hundred percent above that of dispatchable technologies.[17]
- The capital cost of offshore wind is 275% above that of on-land wind.
- The levelized capital cost of Natural Gas-fired Conventional Combined Cycle is 17.4 (near the lowest among dispatchable technologies), compared to 83.4 for Advanced Nuclear and 193.4 for Offshore Wind.

What is immediately clear is that the economic data favor, in terms of cost, the use of dispatchable technologies over the non-dispatchable ones. The issue that the electricity network or grid is unsuitable currently, say, for the large scale introduction of alternative energy systems is simply a reflection that high capacity transmission does not exist to and from much of the coastal regions where wave, tidal and wind generation could be sited. But addressing this problem would not represent a recurring cost, so in that sense the investment cost is not overly serious. The acceptance of large transmission systems in areas often of outstanding beauty would certainly be considered problematic. However the issue of the emissions downside to the use of fossil fuel remains.

Focusing now on the relative costs of wave and tidal where the data in the following table is taken from a report prepared for the Scottish Government.[18]

The report asserts that the diminishing costs in time are due to global deployment of the technologies and, as a consequence, the expected impact on learning and knowledge accumulation in the marine business; , the advantage that would come with an expected increased level of deployment (scale); and a declining rate of increase in the underlying costs. Overall, Tidal Stream appears to have the initial advantage but Wave technology is predicted to have, in the longer term, the lowest levelized cost.

8.7. Some Challenges for Emerging Wave Energy Technologies

This section provides an illustration of how economic, political, social, and maintainability factors influence wave energy technologies' route to market in the UK. It opens with an exploration of an oil embargo in the Middle East, an area not renowned for its abundance of water or waves.

Table 8.1. Estimated Levelized Cost of New Generation Resources, 2018.

US average levelized costs (2011 $/megawatthour) for plants entering service in 2018						
Plant type	Capacity factor (%)	Levelized capital cost	Fixed O&M	Variable O&M (w/ fuel)	Transmission investment	Total system levelized cost
Dispatchable Technologies						
Conventional Coal	85	65.7	4.1	29.2	1.2	100.1
Advanced Coal	85	84.4	6.8	30.7	1.2	123.0
Advanced Coal with CCS	85	88.4	8.8	37.2	1.2	135.5
Natural Gas-Fired						
Conventional Combined Cycle	87	15.8	1.7	48.4	1.2	67.1
Advanced Combined Cycle	87	17.4	2.0	45.0	1.2	65.6
Advanced CC with CCS	87	34.0	4.1	54.1	1.2	93.4
Conventional Combustion Turbine	30	44.2	2.7	80.0	3.4	130.3
Advanced Combustion Turbine	30	30.4	2.6	68.2	3.4	104.6
Advanced Nuclear	90	83.4	11.6	12.3	1.1	108.4
Geothermal	92	76.2	12.0	0.0	1.4	89.6
Biomass	83	53.2	14.3	42.3	1.2	111.0
Non-Dispatchable Technologies						
Wind	34	70.3	13.1	0.0	3.2	86.6
Wind-Offshore	37	193.4	22.4	0.0	5.7	221.5
Solar PV1	25	130.4	9.9	0.0	4.0	144.3
Solar Thermal	20	214.2	41.4	0.0	5.9	261.5
Hydro2	52	78.1	4.1	6.1	2.0	90.3

The Organization of Petroleum Exporting Countries (OPEC) was formed in 1960 in order to create a system that stabilized the price of oil and in turn, the world's energy markets. This aim was achieved successfully for many years by coordinating the

Table 8.2. Summary of Levelized Costs (in January 2010 Terms). Case (£/MWhr).

Technology	Cost scenario	2020	2035	2050
Wave	High	253	142	105
	Medium	214	118	86
	Low	177	97	71
Tidal Range	High	349	323	286
	Medium	279	258	229
	Low	205	190	168
Tidal Stream (shallow)	High	211	199	166
	Medium	173	166	138
	Low	141	134	111
Tidal Stream (deep)	High	250	159	129
	Medium	203	126	102
	Low	166	102	82

policies of the member states, thereby protecting their mutual economic interest. In its early days OPEC was not a political organization. In 1973, however, some OPEC members sent a strong political signal to the West by using a tool of economic warfare, the embargo.[19]

In 1972 the price for a barrel of crude oil was around three dollars. This price had quadrupled to over twelve dollars by the end of 1974. Western economies were given the rare chance of a ride in a time machine and saw what the world would be like when there was no longer an endless supply of cheap oil.[20] The UK government began to assess a wide range of possible energy options to ensure security of energy supplies.

One possible energy option was to develop the indigenous oil reserves held under the UK's North Sea. With extreme waves of up to thirty meters high, not to mention long term effects of erosion and corrosion, it was clear that any development in this harsh environment would not be without its technological and economic challenges.

It was estimated that overall, the development and production investment required for an oilfield in the North Sea was in the region of £1200–£1500 per barrel per day.[21] This significant capital investment posed a potential barrier. However, following the events of the oil embargo, the oil was becoming so expensive that the North Sea was viewed as being economically viable to exploit on a large scale, and consequently a high level of investment took place. Investment in oil and gas extraction accounted for six to eight percent of all UK fixed investment from 1975 to 1983.[22] What made these prospects even more attractive was the fact that the North Sea was surrounded by politically stable states, and possession of an indigenous source of oil and gas increased energy security of supply.

In 1975, the required infrastructure of connecting pipelines and terminal facilities were completed and the first oil from the Argyle field was brought ashore. This was followed soon after by oil from BP's massive Forties field. Discoveries and production from the region grew as the UK sold leases on sectors in the North Sea to British, European, and American companies.

Throughout this period of early oil industry development, the UK was still assessing a wide range of energy options. Aware of the available resource lapping on the shores of the UK and perhaps buoyed by offshore development possibilities being demonstrated in the North Sea, government advisers advocated an urgent research program into the potential of harnessing the UK's wave resource. As a result, the Department of Energy provided inducements to support innovation, funding wave energy research from 1974 to 1983 under its Wave Energy Programme. The program objectives were twofold: first to establish the feasibility of extracting energy from ocean waves and second, to estimate the cost of energy if used on a large scale to supply UK requirements.

A large number of devices were considered during the program but were found to be uneconomic. With hindsight, industry experts felt that the program's objectives were over ambitious resulting in massive devices, with corresponding high capital and generating costs.[23]

By the early 1980s the UK was a net exporter of oil and as the energy crisis subsided, so did the interest in wave energy. A period of consolidated research followed before a resurgence of interest during the mid 1990s, concurrent with a widespread public recognition of both climate change issues and the finite nature of fossil and nuclear energy sources.

As part of its commitment to tackle climate change, the UK is legally committed to delivering fifteen percent of its energy demand from renewable sources by 2020, contributing to its energy security and decarbonization objectives.[24] A policy framework has been put in place to ensure that the wave energy industry can grow and help to meet this clear mandate and encourage wave energy technology development.

When considering the progress of the wind energy industry since the 1970s, it has been shown that the turbine industries have established themselves in countries such as Denmark, where there have been suitable framework conditions in place providing a stable domestic market.[25] The innovators and early adopters now hold a significant market share of the multi-billion-pound wind turbine market; in 2008 the combined global market share of Denmark's two largest wind turbine manufacturers was just over twenty-seven percent.[26] To date the UK's policy framework is working to promote a stable domestic market for the wave industry. As a result a large proportion of the world's marine energy device developers are either based in the UK or conducting tests in UK waters.[27]

An important element of the UK's framework is funding. To support the deployment of early wave energy technologies, short term capital funding is required. In the recent economic downturn, a realization of technology risk and the long time to market of early-stage wave energy technologies has led to a shift away from reliance on venture capital money to fund the early stage of the industry. In order to fill the gap, public funding has increased via initiatives such as the Marine Energy Array Demonstrator (MEAD) fund, and some major industrial companies and utilities have taken equity stakes in technology companies.

Longer term funding streams are also essential. These provide certainty to investors and an enhanced revenue stream for the first commercial projects. They are also vital to encourage long-term investments by supply chain companies and offshore operations and maintenance contractors, key when considering maintainability. At present this funding is provided through the Renewables Obligation (RO), the main market support mechanism for renewable electricity projects in the UK. It places an obligation on licensed electricity suppliers in the UK to source a proportion of their supply to customers from eligible renewable sources. It is monitored through Renewables Obligation Certificates (ROCs); with suppliers required to acquire a certain number of ROCs per MWh of electricity supplied to customers.[28] Through a system banding within the RO, there are incentives in place to create demand in the market for nascent technologies. Under current banding, a wave energy installation receives five ROCs for each MWh produced while an onshore wind installation receives one.[29]

While the funding streams available provide confidence, social factors require a decrease in reliance on such policy instruments, paid for through levies on consumers. The industry needs to demonstrate that considerable wave resource can be accessed, converted to a useful form and delivered at a cost effectively in comparison with other methods of energy generation (shown in Table 8.1).

Future reductions in wave energy's levelized costs of energy (LCoE) lie through programmes of Research and Development (R&D) combined with learning through knowledge accumulation with deployment of technologies. As stated earlier, inducements to develop innovations often come in the form of research grants. In 2011, around forty universities were identified globally which were focused on research into marine renewable devices, with over a quarter of these based in the UK.[30] This research base is supported by high capital infrastructure required for testing. The National Renewable Energy Centre (NaREC) and the European Marine Energy Centre (EMEC) currently provide facilities in the UK to test components and deploy full scale devices respectively.

When considering significant offshore operations, such as the deployment of a full scale wave device, the nascent wave industry in the UK is able to leverage the learning and knowledge that has been accumulated in the oil industry over the previous four decades. The rapid development of the North Sea's oil industry was aided by the strong competition and investment, which lead to the development of new technology that was at the forefront of the industry's major achievements. Technologies such as dynamically positioned vessels and geophysical survey tools used to develop oil reserves are now being effectively used in the development of offshore renewables.

As stated in the Economics of Energy section, decreased levelized costs are expected to come as a result of an increased scale of deployment of technology. To date, forty-one wave and tidal sites have been leased from the Crown Estate, which owns or has vested interests in almost the entirety of the seabed to twelve nautical miles around the UK. With a total potential installed capacity of approximately two GW, this is the largest development pipeline of wave and tidal projects in the world, effectively creating demand by providing a route to market for developing technologies.[31]

The majority of these leased areas are in an early development phase, seeking regulatory permissions from the consenting authorities before the implementation of the project. As part of any consent application, and depending on the scale of the proposed project, an Environmental Impact Assessment (EIA) may be required under legislation. The International Association for Impact Assessment (IAIA) defines an EIA as "the process of identifying, predicting, evaluating and mitigating the biophysical, social, and other relevant effects of development proposals prior to major decisions being taken and commitments made."[32] The process ensures that all likely impacts, both positive and negative, are fully appraised before the project is allowed to proceed. For any proposed wave development this appraisal would include consideration of the effects of proposed project infrastructure upon the landscape and seascape.

The process ensures that all likely impacts, both positive and negative, are fully appraised before the project is allowed to proceed. For any proposed wave development, this appraisal would include consideration of the visual impact on the landscape and seascape, as well as a range of other environmental factors. Grid infrastructure upgrades required to meet any increased generation capacity would also be appraised through this process.

As with most forms of renewable resources, wave energy is unevenly distributed. With exposure to the prevailing westerly wind direction and the long Atlantic fetch, the far north and west (and least populous) areas of the UK, tend to have the most energetic wind and wave climates. In order to help to meet renewable energy targets, the UK has an opportunity to generate energy from its far-flung periphery.

The grid network was designed to transmit and distribute electricity from big, central power stations to our cities, towns and onwards, through ever thinner wires, to remote regions of the UK. Akin to the infrastructure of connecting pipelines and terminal facilities required for the successful development of the North Sea oil industry, appropriate grid infrastructure is essential for wave technologies to succeed in the UK.

8.8. Conclusion

Introducing renewable energy sources into an existing market is a complex socio-techno-economic challenge, which requires many agencies to align their objectives if a successful outcome is to be achieved. Because of the economics of delivering energy to the consumer, which favors traditional sources, the commitment to embark on a heavy cycle of investment in renewable energy systems is contingent on governments accepting the need to manage environmental impacts and address climate change (for example by controlling CO_2 and other gases in the atmosphere). Governments are the agents of society; tl is individuals as part of the community have a clear role to play. To date a balanced debate between all stakeholders has been marked by its absence and special interest groups with their particular agenda have dominated the debate. As ever with such complex situations a well-informed understanding of what is involved is critical if sound decisions on the future of energy are to be made. And that process of being informed and informing should not be confined to technologists, economists, and engineers. Society in general has a role especially bearing in mind that the choices involved are difficult ones to make. And perhaps we do not have the tools available to make such choices. The introduction of additional power transmission networks (grids) and devices in areas of often great beauty has to be balanced against the advantages of curtailing climate change. How are citizens to make such choices with such diverse and competing factors at play?

The debate is not simply terrestrial. Andre Bryans has noted that "the installation of Tidal Energy Devices will result in the earth moon distance increasing at a rate of approximately 1 cm per year per 1 TW year extracted."[33] The point here is not so much the increase in orbit of the moon but rather the complexity of systems by which the unexpected, to some, can result. We are undoubtedly at the beginning of a long learning curve.

Notes

1. Michael. E. Porter, "The Five Competitive Forces That Shape Strategy," *Harvard Business Review* 86, no. 1 (2008): 78–93.
2. Elizabeth Wilson and Jennie C. Stephens, "Wind Deployment in the United States: States, Resources, Policy and Discourse," *Environmental Science and Technology* 43, no. 24 (2009): 9063–9070, http://dx.doi.org/10.1021/es900802s.
3. William M. Shobe and Dallas Burtraw, *Rethinking Environmental Federalism in a Warming World* (Washington D.C.: Resources for the Future, 2012), 2, http://www.rff.org/documents/RFF-DP-12-04.pdf.
4. Rudi Volti, *Society and Technological Change*, 6th ed. (New York: Worth, 2008).
5. John Tozzi, "Think Twice about Being First to Market," *Bloomberg Businessweek* (May 19, 2009), http://www.businessweek.com/smallbiz/content/may2009/sb20090519_306313.htm.
6. Everett M. Rogers, *Diffusion of Innovations* (New York: Free Press, 1962).
7. Naubahar Sharif, "Contributions from the Sociology of Technology to the Study of Innovation Systems," *Knowledge, Technology, & Policy 17*, no. 3–4, (2004): 83–105, http://dx.doi.org/10.1007/s12130-004-1005-4.
8. Diahanna Lynch and David Vogel, "The Regulation of GMOs in Europe and the United States: A Case-Study of Contemporary European Regulatory Politics," *Council of Foreign Relations Press* (April 5, 2001), http://www.cfr.org/agricultural-policy/regulation-gmos-europe-united-states-case-study-contemporary-european-regulatory-politics/p8688.
9. Tom Zellar, "For Those Near, the Miserable Hum of Clean Energy," *The New York Times* (October 5, 2010), http://www.nytimes.com/2010/10/06/business/energy-environment/06noise.html?_r=1&.
10. Ibid.
11. Harold S. Balaban, Ned Criscimagna, and Michael Pecht, "Product effectiveness and worth," in *Product Reliability, Maintainability, and Supportability Handbook*, 2nd ed., ed. Michael Pecht, 1–18 (Boca Raton, FL: CRC Press, 2010).
12. Z. Hameed, J. Vatn, and J. Heggset, "Challenges in the Reliability and Maintainability Data Collection for Offshore Wind Turbines," *Renewable Energy* 36, no. 8, (2011): 2154–2165, http://dx.doi.org/10.1016/j.renene.2011.01.008.
13. RenewableUK, *Marine Energy in the UK: State of the Industry Report 2012* (London: RenewableUK, 2012), p. 39. http://www.renewableuk.com/en/publications/reports.cfm/Marine-SOI-2012.
14. Ibid.
15. Energy Information Administration, "Levelized Cost of New Generation Resources in the Annual Energy Outlook 2013," Report Number: DOE/EIA-0383ER(2013). *Energy Information Administration* (January 28, 2013), http://www.eia.gov/forecasts/aeo/er/electricity_generation.cfm.
16. Ibid.

17. Non-dispatchable technologies are those whose operation is tied to the availability of an intermittent resource.
18. Ernst & Young, LLP, *Cost of and Financial Support for Wave, Tidal Stream and Tidal Range Generation in the UK* (London: Ernst & Young, LLP, 2010), http://webarchive.nationalarchives.gov.uk/20121205174605/http:/decc.gov.uk/assets/decc/what%20we%20do/uk%20energy%20supply/energy%20mix/renewable%20energy/explained/wave_tidal/798-cost-of-and-finacial-support-for-wave-tidal-strea.pdf.
19. "Brief History," *Organization of the Petroleum Exporting Countries*, last modified 2013, http://www.opec.org/opec_web/en/about_us/24.htm.
20. James L. Williams, "Oil Price History and Analysis," *WRTG Economics*, last modified 2011, http://www.wtrg.com/prices.htm.
21. Hugo Manson, "Lives in the Oil Industry: Oral History of the UK North Sea Oil and Gas Industry," *Department of History at the University of Aberdeen*, last modified June 12, 2006, http://www.abdn.ac.uk/oillives/about/nsoghist.shtml.
22. Roger Backhouse, *Applied UK Macroeconomics* (London: Blackwell, 1991). See especially chapter 9, "North Sea Oil."
23. Thomas Thorpe, "An Overview of Wave Energy Technologies: Status, Performance and Costs," *Waveberg.com* (November 30 1999), http://waveberg.com/pdfs/overview.pdf.
24. Department of Energy and Climate Change, *UK Renewable Energy Roadmap* (London: Department of Energy and Climate Change, 2011), https://www.gov.uk/government/uploads/system/uploads/attachment_data/file/48128/2167-uk-renewable-energy-roadmap.pdf.
25. Sustainable Energy Ireland, *Offshore Wind Energy and Industrial Development in the Republic of Ireland* (Dublin, Ireland: Sustainable Energy Ireland, 2004), http://www.seai.ie/Publications/Renewables_Publications_/Wind_Power/Offshore_Wind_Energy.pdf.
26. Center for Politiske Studier, *Wind Energy: The Case of Denmark* (Copenhagen, Denmark: CEPOS, 2009). http://www.cepos.dk/fileadmin/user_upload/Arkiv/PDF/Wind_energy_-_the_case_of_Denmark.pdf.
27. The Carbon Trust, *Accelerating Marine Energy: The Potential for Cost Reduction; Insights from the Carbon Trust Marine Energy Accelerator* (London: The Carbon Trust, 2011), http://www.carbontrust.com/media/5675/ctc797.pdf.
28. Ofgem, *Renewables Obligation Annual Report 2011–12* (London: Ofgem, 2012), https://www.ofgem.gov.uk/ofgem-publications/58133/ro-annual-report-2011-12web.pdf.
29. RenewableUK, *Marine Energy in the UK*.
30. The Carbon Trust, *Marine Renewables Green Growth Paper* (London: The Carbon Trust, 2011).
31. "Further wave and tidal leasing to accelerate technology development," *The Crown Estate* (May 22, 2013), http://www.thecrownestate.co.uk/news-media/news/2013/further-wave-and-tidal-leasing-to-accelerate-technology-development/.
32. Directive 97/11/EC of 3 March 1997 amending DIRECTIVE 85/337/EEC of 27 June 1985 on the assessment of the effects of certain public and private projects on the environment, http://www.energy-community.org/pls/portal/docs/36281.PDF; International Association for Impact Assessment, *What is Impact Assessment?* (Fargo, ND: IAIA, 2009), http://www.iaia.org/publicdocuments/special-publications/What%20is%20IA_web.pdf.
33. Andre G. Bryans, " Impacts of Tidal Stream Devices on Electrical Power Systems," Unpublished doctoral dissertation, The Queen's University Belfast, 2006.

Bibliography

Backhouse, Roger. *Applied UK Macroeconomics*. London: Blackwell, 1991.

Balaban, Harold S., Ned Criscimagna, and Michael Pecht. "Product effectiveness and worth." In *Product Reliability, Maintainability, and Supportability Handbook*. 2nd ed., edited by Michael Pecht, 1–18. Boca Raton, FL: CRC Press, 2010.

Bryans, Andre G. "Impacts of Tidal Stream Devices on Electrical Power Systems." Unpublished doctoral dissertation, The Queen's University Belfast, 2006.

Carbon Trust. The. *Accelerating Marine Energy: The Potential for Cost Reduction; Insights from the Carbon Trust Marine Energy Accelerator.* London: The Carbon Trust, 2011. http://www.carbontrust.com/media/5675/ctc797.pdf.

Carbon Trust. The. *Marine Renewables Green Growth Paper.* London: The Carbon Trust, 2011.

Center for Politiske Studier. *Wind Energy: The Case of Denmark* (Copenhagen, Denmark: CEPOS, 2009). http://www.cepos.dk/fileadmin/user_upload/Arkiv/PDF/Wind_energy_-_the_case_of_Denmark.pdf.

Crown Estate, The. "Further wave and tidal leasing to accelerate technology development." *The Crown Estate* (May 22, 2013). http://www.thecrownestate.co.uk/news-media/news/2013/further-wave-and-tidal-leasing-to-accelerate-technology-development/.

Department of Energy and Climate Change. *UK Renewable Energy Roadmap.* London: Department of Energy and Climate Change, 2011. https://www.gov.uk/government/uploads/system/uploads/attachment_data/file/48128/2167-uk-renewable-energy-roadmap.pdf.

Energy Information Administration. "Levelized Cost of New Generation Resources in the Annual Energy Outlook 2013." Report Number: DOE/EIA-0383ER(2013). *Energy Information Administration* (January 28, 2013). http://www.eia.gov/forecasts/aeo/er/electricity_generation.cfm.

Ernst & Young. LLP. *Cost of and Financial Support for Wave, Tidal Stream and Tidal Range Generation in the UK.* London: Ernst & Young, LLP, 2010. http://webarchive.nationalarchives.gov.uk/20121205174605/http:/decc.gov.uk/assets/decc/what%20we%20do/uk%20energy%20supply/energy%20mix/renewable%20energy/explained/wave_tidal/798-cost-of-and-finacial-support-for-wave-tidal-strea.pdf.

Hameed, Z., J. Vatn, and J. Heggset. "Challenges in the Reliability and Maintainability Data Collection for Offshore Wind Turbines." *Renewable Energy* 36, no. 8 (2011): 2154–2165. http://dx.doi.org/10.1016/j.renene.2011.01.008.

International Association for Impact Assessment. *What is Impact Assessment?* Fargo, ND: IAIA, 2009. http://www.iaia.org/publicdocuments/special-publications/What%20is%20IA_web.pdf.

Lynch, Diahanna, and David Vogel. "The Regulation of GMOs in Europe and the United States: A Case-Study of Contemporary European Regulatory Politics." *Council of Foreign Relations Press* (April 5, 2001). http://www.cfr.org/agricultural-policy/regulation-gmos-europe-united-states-case-study-contemporary-european-regulatory-politics/p8688.

Manson, Hugo. "Lives in the Oil Industry: Oral History of the UK North Sea Oil and Gas Industry." *Department of History at the University of Aberdeen.* Last modified June 12, 2006. http://www.abdn.ac.uk/oillives/about/nsoghist.shtml.

Ofgem. *Renewables Obligation Annual Report 2011–12.* London: Ofgem, 2012. https://www.ofgem.gov.uk/ofgem-publications/58133/ro-annual-report-2011-12web.pdf.

Organization of the Petroleum Exporting Countries. "Brief History." *Organization of the Petroleum Exporting Countries.* Last modified 2013. http://www.opec.org/opec_web/en/about_us/24.htm.

Porter, Michael. "E. "The Five Competitive Forces That Shape Strategy." *Harvard Business Review* 86, no. 1 (2008): 78–93.

RenewableUK. *Marine Energy in the UK: State of the Industry Report 2012.* London: RenewableUK, 2012. http://www.renewableuk.com/en/publications/reports.cfm/Marine-SOI-2012.

Rogers, Everett M. *Diffusion of Innovations.* New York: Free Press, 1962.

Sharif, Naubahar. "Contributions from the Sociology of Technology to the Study of Innovation Systems." *Knowledge, Technology & Policy* 17, no. 3–4 (2004): 83–105. http://dx.doi.org/10.1007/s12130-004-1005-4.

Shobe, William M., and Dallas Burtraw. *Rethinking Environmental Federalism in a Warming World.* Washington, D.C.: Resources for the Future, 2012. http://www.rff.org/documents/RFF-DP-12-04.pdf.

Sustainable Energy Ireland. *Offshore Wind Energy and Industrial Development in the Republic of Ireland.* Dublin, Ireland: Sustainable Energy Ireland, 2004. http://www.seai.ie/Publications/Renewables_Publications_/Wind_Power/Offshore_Wind_Energy.pdf.

Thorpe, Thomas. "An Overview of Wave Energy Technologies: Status, Performance and Costs." *Waveberg.com* (November 30 1999). http://waveberg.com/pdfs/overview.pdf.

Tozzi, John. "Think Twice about Being First to Market." *Bloomberg Businessweek* (May 19, 2009), http://www.businessweek.com/smallbiz/content/may2009/sb20090519_306313.htm.

Volti, Rudi. *Society and Technological Change*. 6th ed. New York: Worth, 2008.

Williams, James L. "Oil Price History and Analysis." *WRTG Economics*. Last modified 2011. http://www.wtrg.com/prices.htm.

Wilson, Elizabeth, and Jennie C. Stephens. "Wind Deployment in the United States: States, Resources, Policy and Discourse." *Environmental Science and Technology* 43, no. 24 (2009): 9063–9070 http://dx.doi.org/10.1021/es900802s.

Zellar, Tom. "For Those Near, the Miserable Hum of Clean Energy." *The New York Times* (October 5, 2010). http://www.nytimes.com/2010/10/06/business/energy-environment/06noise.html?_r=1&.

Chapter 9
Transportation and Energy

RICHARD A. SIMMONS, SHAUN MCFADDEN,
DAVID KENNEDY, AND MARY JOHNSON

Abstract

The role of transportation in society and commerce today is undoubtedly more far-reaching than at any prior time in history. This chapter explores both the literal and figurative prime movers that have defined more than a century of innovation in transportation. In section one we commence with an overview of transportation energy and its links with the environment, consumers and related policies. Particular attention to the implications of the sector's reliance on petroleum is given. Gasoline and diesel fuel have proven extremely well suited in providing ample energy, in a dense, portable and low cost manner. We review basic thermochemistry behind the combustion of liquid petroleum fuels in setting the stage to compare strategies aimed at reducing environmental impacts sector-wide.

Sections two and three explore recent developments and case studies focused on electric and hybrid vehicles, and aviation respectively. Historical trends, key policies and global interactions are also discussed along with noteworthy actions taken by lead nations. For both ground transport and aviation, priority is being given to efficiency, as economic paybacks are attractive and environmental benefits can be felt immediately. Next and equally critical, are focused efforts to develop and commercialize alternative fuels, advanced vehicles and technologies over the mid to longer term. A brief overview of fuel cells is also provided. Environmental impacts and sector emissions are discussed as increasingly imperative, but not exclusive, inputs to energy transitions in transport.

A more coordinated overall approach is suggested to help achieve secure and sustainable economic, environmental and social objectives in the coming decades. Prompt execution of definitive actions, largely known today, can be critical in stimulating what many experts believe are necessary reduction trends. Through the overviews, case studies, and policy analysis, the authors suggest that near term steps will heavily influence the composition of automotive and aviation fleets and their fuel supplies by 2030.

9.1. Transportation Energy Overview

9.1.1. Introduction

As we shift gears to consider energy for transportation, it quickly becomes apparent that the sector's Achilles heel is its disproportionate reliance on oil. From one's daily commute, to the family vacation, and even the laptop shipped from overseas—chances are that transportation, fueled by petroleum, has made it possible. This has significant geopolitical, economic, and environmental consequences, none of which are easily navigated in the near term. While the oil dependency of major consuming countries varies, the United States is not unique with a transportation sector that accounts for nearly thirty percent of total domestic energy consumption and a similar percentage of greenhouse gas (GHG) emissions.[1] Strides have been made in the United States and elsewhere to ensure that electricity needs for stationary power can be met by a diverse suite of domestically sourced coal, natural gas, nuclear, hydro and a host of emerging renewable resources. By contrast, the US remains reliant on petroleum for about ninety percent of transportation needs.[2] And the US is not alone.

For many countries, including the US and many European states, oil is often the single most significant imported item, accounting for ever-increasing shares of GDP and driving trade deficits. Even oil rich nations face concerns including inefficient domestic use, wasteful national subsidies, air quality, and capacity management in the face of volatile global markets. It is estimated that there are nearly one billion vehicles on the road today, with expectations that the global fleet will exceed 1.3 billion by 2020.[3] In China alone, the vehicle fleet grew ten-fold from about two million in 1980 to twenty million by 2005.[4] In 2012, more than eighteen million new light duty vehicles entered the Chinese market, a tally that led the world in units produced and surpassed sales in the US (fourteen million) and Europe (sixteen million).[5] Any discussion seeking to advance an understanding of the global energy crisis must include an analysis of the transportation sector and the central role played by oil. It must also consider some of the challenges, opportunities, technologies, and policies associated with replacing petroleum and developing more sustainable forms of energy for the transit of people and goods.

Experts agree that substantial reductions in oil consumption and emissions can be achieved through a combination of measures over near, intermediate, and long term horizons. With regard to ground transportation, improvements to vehicle efficiency, diversification of alternative fuel supplies, and realistic scale up of hybrid, electric, and advanced vehicles over the long term constitute key strategic developments. An eventual transition to hydrogen as an alternative fuel may also become technically and economically viable in the long term, though many significant challenges loom. In addition to passenger cars, transportation segments such as aviation, rail and maritime represent areas of considerable opportunity for reducing reliance on traditional petroleum fuels and reducing the sector's carbon footprint. Technology deployment will be the driving force behind this strategy, but this will take time and impose significant costs. Successful implementation would help mitigate national security, environmental and trade concerns, but expectations should remain realistic. Resource diversification can pay double dividends by not only curbing demand, but also by redirecting significant quantities of oil to higher value purposes with fewer alternatives such as aviation, fertilizer, plastic, and chemical production. And yet even successful conservation and optimization initiatives in developed regions like the US and Europe, may do little to change global oil consumption as the developing world grows and competes for scarce supplies.

9.1.2. Efficiency and Vehicle Technology

Efficiency enabled by vehicle technology is the single most effective tool for reducing oil use in the near term. Higher oil prices combined with aggressive new rules by the US Environmental Protection Agency (EPA) for increasing Corporate Average Fuel Economy (CAFE) standards are expected to result in higher efficiency internal combustion engines and reductions in vehicle weight. In 1973 during the oil embargo, cars in the US averaged fourteen miles per gallon (mpg). Five years later, CAFE standards were introduced for passenger vehicles at a level of eighteen mpg. This was increased gradually to 27.5 mpg by 1990, where it remained unchanged until 2011. The recent EPA rulemaking sets a goal of 35.5 mpg to be met by 2016, followed by a five percent annual increase for nine years to an equivalent fuel economy of 54.5 mpg by 2025. Many experts indicate that fuel economies in the 40 mpg range are achievable within ten years from technologies under development today. To achieve interim CAFE targets within this decade, recent estimates predict that technology upgrades will result in cost premiums in the range of $1500–$4500 per vehicle at production scales.[6] Many believe that these investments are justified, given that they will be offset or exceeded by fuel savings; though given the variability of gasoline and diesel prices, it proves difficult to predict benefit/cost ratios with certainty. Despite this, there is reason to have confidence in selected near-term estimates, as the technology under development is largely identified. Though not exhaustive, key technological innovations include the following:

- Weight Savings (while maintaining safety and crash worthiness)
    - Advanced materials (composites and alloys)
    - Removal of excess payload
    - Downsizing within classification footprints
- Improvements to Internal Combustion Engine (ICE)
    - Higher thermal efficiencies
    - Increased use and optimization of turbochargers, reduction in displacement at constant performance
    - Improved combustion (higher compression, variable valve actuation, advanced sensors, optimized control)
    - Selective cylinder de-activation
- Improvements to Vehicle
    - Regenerative braking and idle-off (stop-start) modes
    - Improved transmissions (increased number of discrete gears)

- Higher voltage systems (reduced resistive losses)
- Reduced aerodynamic drag, rolling resistance, and friction
- Improved air conditioning systems, including lower impact refrigerants

Looking further down the road, a marked point of diminishing economic returns on increased fuel economy at affordable premiums may potentially occur in the 2020–2030 timeframe and above fifty mpg. Actual viability of ultra-efficient vehicles clearly depends on the future price of oil, a very difficult commodity to predict even one year ahead, let alone ten to fifteen. Reaching 54.5 mpg for the fleet target *average*, meaning essentially half of new vehicles sold (including light duty vehicles and light trucks) should exceed this value, will only be possible by significantly growing the market share of hybrid/electric, fuel-cell, and advanced alternative vehicle technologies. However, this technology remains extremely expensive and may require additional infrastructure. These technologies, and their associated cost estimates, are less evolutionary by nature, and risk factors are therefore higher. In addition, for grid-recharged electric vehicles, the impact on GHG emissions is highly dependent on the source of electricity used. Studies indicate that the lifecycle environmental impact and emissions of an electric vehicle recharged with coal-derived electricity may not, in fact, be better than an equivalent vehicle with a gasoline powered internal combustion engine. If natural gas, renewables, or nuclear power is used for recharging, then the net lifecycle emissions impact is superior. To quantify this, one study projects that by 2015, a fully electrified Nissan Leaf would emit twenty grams of CO_2 per kilometer in France, where much of the electricity comes from nuclear power, but 114 g/km in the UK, where there is a greater reliance on coal. In heavily coal-dependent countries such as Poland or Luxembourg, estimated emissions are 135 g/km.[7] Thus, while one overarching objective is to reduce oil consumption, a parallel objective is to reduce the environmental footprint by increasing the use of lower carbon sources. Experts urge caution against unrealistic expectations for the growth rate of purely electric vehicles, noting that market share in a decade could remain in the single digits.

Challenges notwithstanding, consumer demand for vehicles with a reduced impact on energy resources and the environment has increased, and products showcasing many of the aforementioned technologies are beginning to satisfy that market trend. Vehicle efficiency standards (such as CAFE) which encourage the introduction of such advanced technologies, taken along with advanced alternative fuels could together help reduce oil consumption on the order of thirteen to forty percent by 2035.[8] These levels of reduction can only be realized if efficiency gains are cashed in directly to reduce oil consumption, rather than being traded for increased performance, weight, or additional use (known as the rebound effect). This trade-out phenomenon characterized the Sport Utility Vehicles (SUVs) and sports cars of the 1990s and early 2000s, as engine power, vehicle capacity, and weight increased dramatically, while fuel economy remained relatively flat. Looking forward, oil reductions in transportation that are not displaced by coal-electricity should reduce the associated GHG emissions proportionately.

9.1.3. Alternative Fuels and Diversity at the Pump: Biofuels, Natural Gas, and Hydrogen

Alongside aggressive efficiency improvements, alternative fuels have the potential to offer considerable near-term promise in campaigns to reduce oil and emissions. It should be noted that alternative fuel availability and adoption are highly regionally dependent. Conventional biofuels currently account for an impressive ten percent of US gasoline supply (or about seven percent of US liquid transport fuels at nearly thirteen billion gallons per year or 600,000 boe/d), primarily in the form of corn-based ethanol blended gasoline. By 2022, US policy calls for twenty-one of the predicted thirty-six billion gallons of renewable fuels to come from advanced non-food sources such as cellulosic biomass and algae, more than doubling current biofuels penetration. There is increasing concern that renewable fuel targets will be missed, as significant technical and economic challenges have delayed the commercialization of advanced fuels. Other infrastructure, storage and delivery aspects of higher alternative fuel blend percentages will also require attention. Despite this, research indicates that ethanol can reduce GHG emissions as compared to gasoline. Reductions vary depending upon the process and feedstock, for example: corn ethanol reduces emissions by twenty percent; sugarcane ethanol by sixty percent; and cellulosic ethanol by sixty to ninety percent. Several US companies are bringing pilot facilities online in 2013 and 2014, with projected full-scale production of cellulosic ethanol within five years. On January 1, 2012, Congress allowed fiscal support and import tariffs for conventional ethanol to expire. Legislation and support policies that would re-direct future tax incentives or loan guarantees toward advanced biofuels and infrastructure upgrades have been introduced, but have not been adopted as law.

The United States and Brazil (with bioethanol from corn and sugarcane) and Europe (with biodiesel from rapeseed and waste cooking oil) account for more than ninety percent of the world's biofuel production. The commercial availability of flex-fuel vehicles, beginning in the early 2000s, has enabled higher blends of biofuels with conventional petroleum fuels. So-called first generation biofuels dominate the biofuels market, owing to their simple conversions from established crops, such as corn, sugarcane, rapeseed, and soybeans. Increasing attention to land-use and environmental impacts has resulted in constructive national and international dialogues over ways to ensure the sustainable production of biofuels. Scale-up of advanced biofuels will require the utilization of low cost, non-food feedstocks, such as agricultural residue, municipal solid waste, woody and cellulosic biomass, and potentially algae.

Natural gas currently comprises less than one percent of transport fuel, but demonstrates remarkable potential for growth—particularly with urban fleets of buses, delivery vehicles, taxis, and other mass transit vehicles. Heavy duty and long haul transport offer additional segments where natural gas could serve as a key transport fuel. Abundant supplies of low cost natural gas and developed infrastructure in both centralized urban and residential locations could make this fuel a competitive alternative in the medium term. Gaseous fuels include Liquefied Petroleum Gas (LPG), Compressed Natural Gas (CNG), Dimethyl Ether (DME), Bio-Gas, and Hydrogen; all have the potential to emit cleaner air than liquid petroleum transport fuels. Atomised gas fueled (AG-F)-powered vehicles for instance perform significantly better than their conventional counterparts in terms of NO_x and PM emissions. They have reduced CO_2 emissions over gasoline and diesel, and uptake of AGF could constitute a step in the transition toward a hydrogen-based road transport system. Propane (LPG) has a higher energy density than gasoline, burns cleaner, and results in less fouling of plugs and contamination of oil.[9] As a result, it is broadly used in industry for factory applications, as a fuel for forklifts and maintenance vehicles.

Natural gas engine technology and infrastructure upgrades are currently expensive, and few consumers are willing to pay an estimated $2000–$4500, or more, to switch a light duty vehicle over from gasoline. However, a sustained fuel price spread with oil due to the shale gas revolution improves the prospects, reducing payback periods and resulting in a more compelling overall value proposition for natural gas. While US natural gas prices have recently decoupled from global crude oil, they are predicted to rise modestly in the coming years, particularly if the United States embarks on an export strategy or substantially accelerates the transition from coal powered electricity. Studies suggest that CNG has the potential to displace 180 million barrels of crude oil per day between 2025 and 2040, particularly if it can be phased in for fleet applications that rely heavily on diesel. A gradual conversion and replacement of transit, delivery and heavy-duty vehicles could result in a five to ten percent reduction in diesel demand, and yield significant benefits by improving energy security and fuel price stability while reducing carbon emissions.[10]

Public and private research in alternative fuels continues at an aggressive pace. Many vehicles can be converted to run on hydrogen as a green alternative to gasoline, however research is addressing concerns related to storage, safety, infrastructure and sourcing. These, and other economic constraints, will have to be resolved before hydrogen and hydrogen fuel cells can achieve appreciable market share. Meanwhile, the Vehicle Technologies Program of the Department of Energy funds and coordinates a portfolio of vehicle research into such areas as hybrids, energy storage, advanced combustion engines, and advanced materials. Similarly, the European Union has launched the European Green Cars Initiative, involving "research on a broad range of technologies and smart energy infrastructures essential to achieve a breakthrough in the use of renewable and non-polluting energy sources."[11] Both US and European initiatives focus on mid to long-term R&D and emphasize the importance of aligned approaches between industry and policy makers.

9.1.4. Combustion Overview and Primary Transport Emissions

When fossil fuels burn, hydrocarbons combine with oxygen (or air) to produce carbon dioxide, water and heat. The amount of heat produced is a function of the chemical energy potential of the fuel and the efficiency of the process. The energy content, or so-called *higher-heating value* or *calorific value* of a fuel is the quantity of heat produced by its combustion under standard conditions (that is, at a temperature of 0°C and a constant pressure of 1,013 mbar). On a mass basis, the calorific values of diesel and gasoline (petrol) are comparable at 45.5 MJ/kg and 45.8 MJ/kg, respectively. Diesel fuel is, however, more dense than gasoline and contains approximately ten percent more energy by volume (roughly 36.9 MJ/liter compared to 33.7 MJ/liter). Diesel engines operate at a higher thermodynamic efficiency than gasoline counterparts. These facts help explain why diesel powered vehicles can generally achieve better efficiencies than gasoline powered vehicles.

Just as energy content determines heat output, the amount of carbon produced by a combustion reaction depends largely on the carbon intensity of the fuel. A rough rule of thumb suggests that for equivalent units of energy produced, natural gas emits about half and petroleum fuels about three-quarters of the carbon dioxide produced by coal. The emissions from a four-stroke Internal Combustion Engine (ICE) depend on the ratio of air to fuel as they enter the cylinder during the intake stroke. The combustion of octane, a key component of gasoline, with the theoretical or "stoichiometric" amount of air is shown below.[12]

$$C_8H_{18} + 12.5(O_2 + 3.76\ N_2) \rightarrow 8CO_2 + 9H_2O + 47N_2$$

With less air than the theoretical amount, the mixture is rich, forming some Carbon monoxide (CO) and unburned hydrocarbons (HC). This may inhibit the production of nitrous oxides (NO_x) due to lower combustion temperatures. With a higher proportion of air, fuel becomes the limiting reactant, and this lean mixture generally results in the formation of fewer unwanted products of combustion. If, however, the mixture becomes excessively lean, misfire can occur, resulting in emissions of unburned hydrocarbons (HC). Diesel engines operate with a lean mixture and emissions of HCs and COs are low, but NO_x emissions are high due to high operating temperatures. Diesel engines also emit soot particles (or black carbon), and as a result have been known to pose health and respiratory risks. Led by a core group of national governments, the multi-phase/multi-year implementation of stringent environmental regulations on gaseous and particulate matter (PM) from diesel engines has helped mitigate such risks considerably.

Catalytic converters serve the purpose of reducing emissions of the main pollutants (CO, HC, and NO_x) that result from the incomplete combustion of transport fuels. Operating at elevated temperatures with specialized materials, the converters oxidize HCs and COs to CO_2; as well as reducing NO_x to N_2, nitrogen gas, a benign agent comprising seventy-eight percent of the earth's atmosphere.

As noted, fuels differ in the amount of carbon and energy they contain and this has implications for fuel economy and greenhouse emissions. To quantify one potential comparison, a car fueled by gasoline emitting 148 g CO_2/km would emit 130 g CO_2/km if fueled on LPG. Despite its lower energy density by volume, a twelve percent net reduction in emissions on a per kilometer basis would result. It should be noted that the proportion of tailpipe versus upstream GHG emmissions comprising the total CO_2eq/km for gasoline, LPG, and other fuels can vary substantially. Table 9.1 shows typical amounts of CO_2 emitted from the combustion of three primary transport fuels.

Table 9.1 Typical CO_2 Tailpipe Emissions Per Volume of Fuel Consumed.[13]

Fuel Type	CO_2 Emissions Kg/Liter of Fuel	CO_2 Emissions Lbs/Gallon of Fuel
Gasoline/Petrol	2.3 kg/L	19.4 lbs/gal
Liquid Petroleum Gas (LPG)	1.6 kg/L	12.7 lbs/gal
Diesel	2.7 kg/L	22.2 lbs/gal

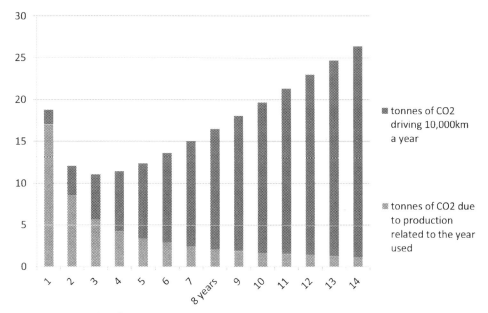

Figure 9.1. Example of CO_2 Emissions (Tonnes) from Production and Use Over Vehicle Lifespan.

In the production process of a vehicle it is estimated that 720 kg of CO_2e is produced for every €1,000 of purchase price (about 1220 lbs CO_2e per $1000). The production of a medium car therefore can generate up to 17 tonnes of CO_2e and a typical gasoline fueled car produces approximately 1.8 metric tonnes of CO_2 for every 10,000 km driven. Figure 9.1 charts the combined CO_2 emissions profile for an average vehicle including production and use, projected for a vehicle's operating life.[14]

9.1.5. Regulating Emissions from the Transportation Sector

As noted earlier, the Kyoto Protocol is generally seen as an important first step toward a global emission reduction regime aimed at stabilizing GHG emissions, and providing essential architecture for future international agreement on climate change. This has significant implications for transport emissions, which account for about twenty-five percent of the total from fossil fuel combustion worldwide, and grew forty-five percent between 1990 and 2007.[15] Regulation on vehicle fuel economy (as in the United States) or fuel consumption (such as in Europe) for new vehicles has been the primary policy mechanism aimed at curbing emissions in the sector. As noted above, aggressive fuel economy targets have been established via CAFE standards through 2025, with similar initiatives in place for Europe. The prevalence of gasoline and diesel in transport means that there is, at present, a tight correlation between fuel consumption (or fuel economy) and vehicle emissions. For example, 54.5 mpg is roughly equivalent to about seventy to eighty grams of CO_2 equivalent per kilometer driven. The existing CAFE policies combine fuel economy and emission regulations into one standard. More precisely, while the Department of Transportation's National Highway Traffic Safety Administration (NHTSA) sets fuel economy specifications directly, EPA regulations specify emission targets which can be correlated to effective fuel economy targets based upon vehicle size (or *footprint*) classification.

The European Commission imposes standards on GHG emissions for new vehicles explicitly. In 2010, the approximate level of CO_2 equivalent emitted per km from new cars sold in Europe was 143 grams, having fallen from 167 grams in 2002. Currently, the EU has in place a target of 130 grams by 2015, and a proposed target of 95 grams for 2020. By 2030, a range of 50 grams CO_2/km and 70 grams CO_2/km (depending on vehicle technology) is targeted for cars, and a range between 75 grams CO_2/km and 105 grams CO_2/km applies to vans. These targets are seen as credible but challenging by industry, and they are consistent with the EU goal of ensuring that average emissions of new car and vans are near-zero at the tailpipe by 2040.

9.1.6. Consumer Behavior and Intelligent Transit

Clearly, consumers play a critical role in long term strategies to reduce oil use. In the absence of a price on carbon, persuading consumers to reduce their energy footprint will remain challenging. However, higher gasoline prices have already initiated a market-induced behavioral shift. Increasing availability of affordable lower carbon fuels and efficient vehicle options will continue this trend. Reduction in vehicle miles traveled (VMT) via carpooling, teleworking, real-time route optimizations, and expanded use of mass transit will further reduce oil consumption. At the vehicle and system level, it will be increasingly possible to leverage powerful data acquisition and network resources to inform driving behavior and enhance overall efficiency. As transportation represents the second highest expense in most American household budgets, consumers stand to directly benefit from reduced dependence on and volatility of oil. Complemented by coordinated policies at the state and municipal levels, consumer behavioral change has the potential to contribute to double digit reductions in oil used for transport.

An important aspect of consumer behavior is clearly product selection. An individual's decision to purchase a given vehicle is a complex matter of personal need and preference, value judgment, budget, and a critical assessment of major attributes. Yet, it is certainly a topic to which a wide audience can relate, and potentially, an opportunity to examine a host of theories and research findings. Much has been written on this by a range of experts in business, academia, and government; and a thorough discussion of this topic is beyond the scope of this book. However, energy transitions in the context of transportation technology and policy can be illuminated by a brief glance at fundamental operating costs of existing and emerging vehicles. Consider a hypothetical comparison of 6 selected vehicles, using a variety of fuel sources and technologies, all subjected to the same real world conditions. Table 9.2 summarizes first order results from one such thought experiment. It is by no means exhaustive, and while many assumptions must be made, such an exercise can begin to help quantify relative operating costs and impact on emissions for various vehicle architectures.[16]

The purpose of such a thought experiment is not to "advocate" for any particular option, nor to suggest that any specific vehicle, technology or company is best. Instead, it is meant to introduce the notion that numerous factors can be significant in a vehicle decision, and economic and environmental implications must be balanced against other

Table 9.2 Comparison of Vehicle Operating Cost and Emissions for Various Vehicle Architectures.[16]

Vehicle Energy Source	Vehicle Make/Model and Driving Mode	2013 MSRP	Operating Cost Scenario 1 ($/mile)	Operating Cost Scenario 2 ($/mile)	Total Est. Emissions (t CO_2)	Fuel Economy (mpg or mpge)
Gasoline	Ford Focus (All Gasoline)	$16,200	$0.365	$0.462	30.0	31
Diesel	VW Jetta TDI (All Diesel)	$22,990	$0.462	$0.538	31.8	34
Hybrid Gasoline	Toyota Prius (Hybrid Gasoline)	$24,200	$0.427	$0.487	18.6	50
Electric Vehicle (EV)	Nissan Leaf (Electricity Mix A)	$28,800	$0.445	$0.455	16.8	115
	Nissan Leaf (Electricity Mix B)	$28,800	$0.445	$0.455	6.7	115
	Nissan Leaf (With Subsidy)	$21,300	$0.339	$0.349	6.7–19.3	115
Plug-in Hybrid EV (PHEV)	Chevy Volt (All Electric, Mix A)	$39,145	$0.600	$0.611	18.9	98
	Chevy Volt (All Gasoline)	$39,145	$0.668	$0.749	23.5	37
	Chevy Volt (80/20 EV Gas, Mix A)	$39,145	$0.612	$0.639	19.2	74
	Chevy Volt (80/20 EV Gas, Mix B)	$39,145	$0.613	$0.639	9.7	74
	Chevy Volt (All Electric, with Subsidy)	$31,645	$0.494	$0.504	9.7–20.4	98
	Chevy Volt (All Gasoline, with Subsidy)	$31,645	$0.561	$0.634	23.5	37
Compressed Natural Gas (CNG)	Honda Civic NG	$27,255	$0.488	$0.569	29.1	31

criteria and preferences, including safety and styling. That said, a more analytical assessment of comparative options may help inform the interdependent development of technology and policy, while meeting long term consumer and social objectives. The reader is thus encouraged to consider how the rubber literally meets the road as personal values and preferences are put increasingly into a greater social and global energy/climate context.

9.1.7. Policy Overview

The enactment of robust policies is essential to ensuring successful outcomes, as policies experience rapid iteration to varying degrees of success. Transport policy in particular is essential to achieve successful reduction of CO_2 emissions. Some policies are aimed at taxing fuel or creating market-based mechanisms for trading emissions and/or carbon credits. Others are encouraging, by subsidy, regulation or mandate, greater adoption of cleaner fuels, energy efficiency, and alternative energy sources. Economic uncertainty and broad differences of opinion among voters and their elected officials have made consensus on energy and climate policy difficult to achieve. State and local authorities often have more flexibility to execute energy policy measures than large national governments. In the United States as of 2013, for example, a clean energy standard (CES) has been controversial at the Federal level, yet at least half of the fifty states have passed legislation such as a renewable portfolio standard (RPS) specifying certain amounts of renewable energy.[17] In addition, many US states mandate the use of ethanol blended fuel (E10), and some impose additional requirements on renewable fuels, as in the case of the California Low Carbon Fuel Standard. The implementation of local, state, or regional policies may benefit from coordination with neighboring efforts, as policy effectiveness can be sensitive to scale. In addition, automakers generally prefer a consistent policy and regulatory context to facilitate standardization in design and manufacturing. Flexible and comprehensive policies that accommodate technological, economic, and social considerations can help relieve geopolitical and global economic stress and reduce emissions while helping to normalize trade balances. Thus, rectifying an overdependence on crude oil becomes a tremendous opportunity, with triple-bottom-line benefits. As the industry adapts and pursues a more sustainable future, the visionary spirit of Henry Ford, Thomas Edison, and the Wright brothers is alive and well today. This is welcome indeed, because a new journey of a thousand miles has begun.

9.2. Electric and Hybrid Vehicles

9.2.1. Introduction

Today's ground vehicles represent over a century of sustained technological progress and provide mobility and access that quite literally open doors to new worlds. They are seemingly ubiquitous; some eighty-one million were produced in 2011 alone. Figure 9.2 shows global vehicle production between 1997 and 2011.[18]

It is well-known that the Internal Combustion Engine (ICE) Vehicle has dominated the industry over the last century, and effectively powers the vast majority of the world's estimated one billion vehicles. As noted in the previous section, ensuring a sustainable

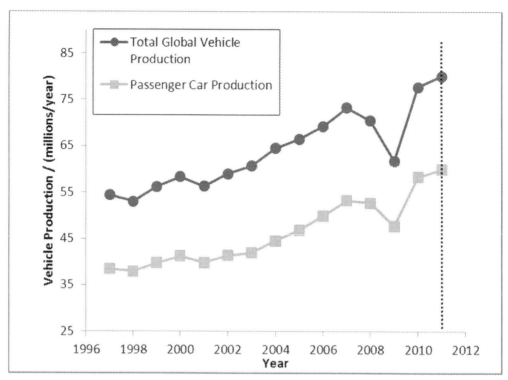

Figure 9.2: Global Vehicle Production Figures for Recent Years.

future for ground transportation will require solutions that help wean the world from petroleum-derived fuels and reduce emissions. In 2011, seventy-one percent of the petroleum supplied to all sectors in the United States found its way to the fuel tanks of the nation's boats, planes, trains, and predominantly, ground vehicles.[19] The market is not yet diverse, as Hybrid Electric Vehicles (HEVs) accounted for 3% of total vehicles sales in the US in 2012, Electric Vehicles (EVs) just 0.3%, and Compressed Natural Gas, 0.008%.[20] The remainder of new car sales in the US (96.7%) are safely assumed to be conventional ICE vehicles operating with gasoline or diesel fuel. Recent annual growth trends for HEVs are noteworthy, and the US has more HEVs than any other county having recently surpassed the one million mark, but change progresses slowly. While the market share of many alternatives is growing, diversification is still a relatively recent development. Despite that newer vehicles account for a greater share of the miles driven, the estimated fleet turnover ratio in the US is about fifteen to twenty years, implying the overall US fleet remains comprised of nearly ninety-nine percent gasoline or diesel powered vehicles.[21] The composition of national fleets in other countries is not substantially different, though the ratio of gasoline to diesel engines can vary considerably.

9.2.2. Challenges Faced by Transportation

The challenges to the transportation sector are as clear as they are significant, and while they may not be unique to this sector, they are certainly felt more acutely. The most important challenges are the threat of oil depletion, greenhouse gas emissions, and other tailpipe pollutants.

Predictions about peak oil abound, but since 1983, global production of oil has exceeded global oil discoveries. Unconventional and deep offshore resources are coming online, but shortfalls and gaps between production and reserves have been known to result in higher prices. In turn, this may justify the use of new and unproven technologies and greater risks to extract oil that has previously been deemed uneconomic. Even with potentially expensive and unconventional supplies, crude oil is clearly a finite resource and its depletion is a legitimate concern.

The products of combustion of a petroleum derived fuel burned in oxygen include water (H_2O) and carbon dioxide (CO_2). Both emissions were once thought of as benign, non-polluting substances that could be safely released into the atmosphere. However, it has become clear that increasing levels of greenhouse gases, notably CO_2, contribute to global warming and climate change. In 2009, transportation accounted for twenty-three percent of global CO_2 emissions, ranking second among all sectors behind electricity.[22]

In the past, vehicles burned lower quality fuels in less efficient engines, resulting in high levels of air pollution concentrated in urban centers. Modern vehicles have significantly reduced air pollutants such as NO_x and SO_x (gases comprising either nitrogen and oxygen or sulfur and oxygen). However, tailpipe emissions such as particulate matter from diesel exhaust can still cause health and respiratory conditions. With increasing urbanization and vehicle use, localized air pollution poses a significant problem. This is particularly true for developing nations where vehicle fleets may be older or non-compliant with modern emissions standards.

9.2.3. The Role of Technology

From a global and social perspective we may consider a quantity of interest to be dependent on multiple related factors namely: population size (P), affluence (A), technology (T), and end-user usage (U). A dimensional analysis illustrates how the resultant quantity of interest can be expressed as the product of the disaggregated factors:

$$Quantity\ of\ Interest = P \times A \times T \times U$$

For example, consider global CO_2 emissions from transport in terms of appropriate factors.

$$CO_2\ emissions\ per\ year = (Population) \times \left(\frac{Vehicles}{People}\right) \times \left(\frac{CO_2}{km}\right) \times \left(\frac{km/year}{Vehicle}\right)$$

The CO_2 emissions per year (on a mass basis) depends on population size, the ratio of vehicles to people (affluence factor), the average mass of CO_2 produced per kilometer of travel (technology factor), and the average number of kilometers traveled per year per vehicle (usage factor).

It may be difficult to calculate the factors individually but the effects of a change in any given factor are evident. The global population is about seven billion, and growing steadily. Similarly, vehicle ownership per capita is on an increasing trajectory worldwide.

Therefore, in order to reduce the level of CO_2 emissions, we must look at reducing the remaining disaggregated factors: improvements in the technology to reduce the average mass of CO_2 emitted per km, or reductions in vehicle usage in a given year.

Direct comparisons of CO_2 emissions for ICE and EVs are complicated. CO_2 estimates for ICE vehicles are based on combustion chemistry of standardized and well-understood fuels. As noted in the previous section, the ideal combustion of one liter of octane produces approximately 2.3 kg of CO_2 (Please see Table 9.1). The technology factor is estimated by multiplying the carbon intensity by the fuel economy: $2.3 \times (L/km)$. Slight variances may apply for different fuel types or blends, but the approach and order of magnitude remain valid. This methodology is called a tank-to-wheel (T2W) analysis. On the other hand, EVs have no direct tailpipe emissions and we may assume that the corresponding technology factor is zero CO_2/Km. However, if we extend the analysis to include the primary electricity source we find that the CO_2 emissions and CO_2 intensity per kWh vary greatly and are dependent upon the source of electricity. For EVs this approach for calculating CO_2 emissions is called a well-to-wheel (W2W) analysis.

Similar thought exercises, as just described, may be performed on petroleum consumption or tailpipe emissions in transportation by changing the technology factor. For petroleum consumption the technology factor is changed to fuel economy (L/km). For tailpipe emissions (such as NO_x) the technology factor is changed to units of mass of unwanted emissions per km (for example, NO_x/km). The role of technology is clear—improvements in technology can reduce petroleum consumption, CO_2 emissions, and other tailpipe emissions.

Recently, much discussion has revolved around the electrification of vehicle drivelines and the benefits conferred by these new technologies. Two distinct modes of electric vehicles have come to the fore, namely, the Electric Vehicle (EV) and the Hybrid Electric Vehicle (HEV). New electric mobility technologies are emerging as both viable competitors and complementary systems to the internal combustion engine.

9.2.4. A Brief Automotive Journey Through Time

It is a mistake to consider vehicle electrification as a new technology, and the historical record is instructive as we look forward to a new era in transportation. The electric vehicle has its own long-established history, enjoying a significant share of the vehicle market during the industry's early years (1895–1905). In the year 1900, for example, 4200 vehicles were sold in the United States, of which forty percent were steam driven, thirty-eight percent were EV, and twenty-two percent were ICE vehicles.[23] Following early steam-powered vehicles by Cugnot (1769) and Trevithick (1801), Thomas Davenport developed one of the first DC electric motors and demonstrated its use on a small model vehicle in the 1830s. Shortly afterward, Robert Anderson developed a non-rechargeable Battery Electric Vehicle (BEV) which was followed by the invention of a rechargeable battery by Gaston Planté in the 1860s. Camille Faure (1881) improved this technology for use in BEVs.

Most of the initial EV designs were little more than a battery box and motor on a 4-wheel frame, using simple chain drive transmission systems. On a full charge, the range from the lead acid battery cell was typically fifty miles with a top speed of thirty mph. The battery was charged in between uses by a stationary generator. In the early days of distributed electricity, slow and frequent recharging plagued the growth of the EV. However, 1897 would bring some milestone developments: in the US, Oldsmobile was formed to manufacture electric vehicles; in the UK, the London Electric Cab Company was launched; and in France, M. A. Darracq demonstrated regenerative braking technology.[24] Regenerative braking converts a vehicle's inertia back into stored electrical potential by using the electric motor as a generator. Despite this, the expense, short life, slow speed and limited range coupled with the large mass of chemical required to store electrical energy made early electric vehicles obsolete.

Conversely, gasoline/petrol/diesel vehicles overcame most of these shortcomings very early and grew quickly in appeal. Karl Benz is credited with developing the first ICE vehicle in 1885. By the early 20th century ICE vehicles were beginning to replace BEVs. Three major factors were responsible for the domination of the ICE vehicle over the BEV in the early years of the automobile:

1. BEV's limited driving range compared to the ICE vehicle;
2. Advent of the Ford Model T production assembly line; and
3. Invention of the starter motor for the ICE.[25]

The BEV's limited range was largely due to the low energy density of the battery technology of the day, which failed to mature sufficiently to compete with the energy stored in liquid fuels. While only gasoline, diesel, and some alcohols were available as liquid fuels in the early years, early batteries were far inferior to today's NiMH and LiIon variants.

Henry Ford introduced the ICE-based Model T Ford in 1909, successfully pioneering the principles of mass production to place motor vehicles within financial reach of many in society. Unable to capitalize on economies of scale for reasons of both supply and demand, BEVs remained relatively expensive.

The early ICE vehicles (including the Model T) had one major drawback: the need for a crank start. This meant that the vehicle operator had to step out of the vehicle and manually turn the engine crank shaft via an external handle until the engine started. This operation required great effort and could be dangerous. In 1911, Charles Kettering solved this problem by connecting an electric starter motor to the engine's crank shaft enabling the operator to easily start the ICE without exiting the vehicle. It seems ironic that batteries and electric motors played a major role in the early conquest of the ICE over the BEV!

The first hybrid electric vehicles (HEV) came about in the early years too. Beginning in 1897, hybrid electric vehicles operating with natural gas or gasoline were introduced by Entz, Porsche, Jenatzy, the Electric Vehicle Company, Baker, and Owen, the latest of which was in production from 1915 to 1922.

It was not until the 1970s and the Arab oil embargo that concerns about oil supply resurfaced, renewing interest in electric vehicles. In the US, the Electric Vehicle Act of 1976 was introduced while Victor Wouk, the so-called "godfather of the hybrid," developed the concept of the HEV based on a Buick Skylark from General Motors. After testing by the EPA, Wouk's prototype was verified to consume half the fuel and emit just nine percent of the emissions of the stock version.[26]

Meanwhile, significant improvements in ICEs and reformulated gasoline took place as well, constantly raising the bar for competitive technologies. Many had a positive and lasting impact, such as the introduction of the catalytic converter, the phase-out of lead, the commercialization of biofuels, fuel injection, and sophisticated combustion controls.

More recently, modern EV and HEV have enjoyed a surge in attention through invigorated research and development. A pivotal moment came when General Motors launched the EV1 and, between 1996 and 1999, became the first to mass produce electric vehicles in the modern era. Ford quickly followed in 1998 with an all-electric pickup truck, the Ranger Electric, marketed primarily as a limited-use service vehicle to parks, couriers and utilities. The HEV entered the modern era to stay when Toyota released the Prius to the Japanese market in 1997 and later worldwide, making it the first mass-produced HEV. Since these pivotal launches, HEVs have enjoyed greater market share than their EV counterparts. In 2012, the Society of Automotive Engineers listed twelve major original equipment manufacturers (OEMs) with production hybrid vehicles on the market. A few OEMs have brought EVs to market, such as Nissan with its Leaf. The Chevy Volt is classified as an extended range EV, and if driven less than forty miles per day and recharged every night, could theoretically never require gasoline. By combining power plants and energy sources, the Volt functions as a plug-in hybrid electric vehicle (PHEV). Though difficult to generalize and sensitive to electricity sources, one study suggests that an electric car can actually have a higher W2W carbon footprint than a gasoline vehicle until it has exceeded 130,000 km.[27] This is partly due to the carbon emissions generated in the mining of materials such as lithium, copper and rare earth metals, and in the production of batteries for electric cars. Though not yet mainstream, fuel cell vehicles are also entering the mix, suggesting that our roads could be characterized by a great deal more technical diversity in the future, as noted in a market simulation performed by IEA shown in Figure 9.3.

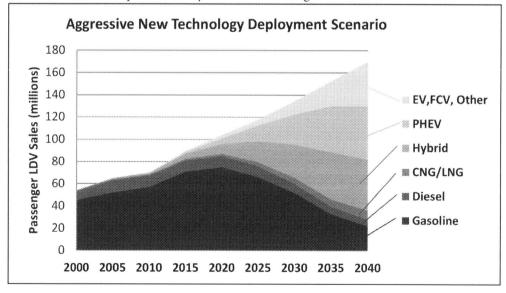

Figure 9.3. IEA Scenarios of Light Duty Vehicle (LDV) Market Share Through 2050.[28]

9.2.5. Modern Vehicle Configurations

A vehicle may be classified by the technology used in its propulsion system. From an energy perspective, we consider how the energy is stored within the vehicle and how it is converted to kinetic energy. Hence, there are two distinct subsystems to consider: the Energy Storage (ES) system and the Driveline (DL) configuration.

From a thermodynamic viewpoint, we consider the system boundary around the energy system (dashed lines in Figure 9.4). We categorize vehicle energy systems as open or closed. In an open system, matter flows into and out of the system boundaries, such as air, fuel and exhaust. In closed systems, no matter crosses the thermodynamic system boundary. Figure 9.4 shows the thermodynamic principles of open and closed systems applied to the automotive context.

In both open and closed systems, energy may pass the system boundary in the form of electrical energy, mechanical work, chemical potential, or heat. The law of conservation of energy applies to both systems. Note that in some systems (such as *b*) regenerative energy from the braking system may pass from the DL to ES device. In the open system, the law of conservation of matter applies; the matter that crosses into the system boundary must either be stored within or passed through the system boundary. The overall efficiency of the propulsion system is evaluated by dividing the useful kinetic energy by the total stored energy in the system. The effectiveness of a vehicle propulsion system can be assessed in view of three competencies:

1. The energy stored and made available at the wheel (total capacity to do work).
2. The power available at the wheel through the propulsion system (work performed in a given time).
3. The rate and capacity to replenish the energy stored (ease of refueling or recharging).

Competency 1, the energy at the wheels, determines the potential range of the vehicle. It is a function of the capacity of the ES system and the overall efficiency of the propulsion system.

Competency 2, the power available at the wheels, is effectively the rate of conversion of energy from the ES to the drive train. It is a function of the overall efficiency of the propulsion system, and the capacity of the ES to deliver power to the DL system over time. Competency 2 differs from Competency 1 in that it represents the ability to extract stored energy at a sufficient rate to meet driving demands. Hence, the power available at the wheels determines vehicle driving performance.

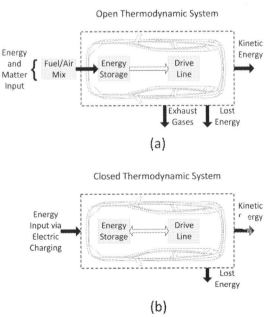

Figure 9.4. Open and Closed Thermodynamic Systems.

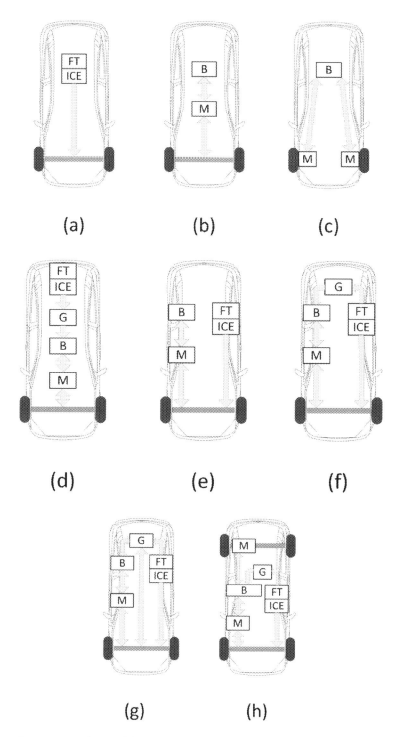

Figure 9.5. Series Drivelines: (a) ICE Driveline, (b) BEV Driveline with Transaxle (c) BEV with In-Hub Motors; Hybrid Electric Drivelines: (d) Series Hybrid, (e) Parallel Hybrid, (f) Series-Parallel Hybrid; Complex Hybrid Drivelines (g) Two Wheel Drive, (h) Four Wheel Drive. FT: Fuel Tank; B: Battery; M: Motor; G: Generator.

Competency 3, the capacity to replenish the ES, is a measure of how readily energy removed from the vehicle can be replaced, for example by refueling or recharging. It is purely a matter of end-user convenience and includes considerations of time and availability of energy infrastructure, such as fueling or charging stations, to replenish the ES device.

The following sections highlight key points of ICE, EV, and HEV technology, but are not meant to be exhaustive.[29]

9.2.5.1. Internal Combustion Engine Vehicles

In an ICE vehicle, the ES system is a fuel tank where liquid fuel is stored in chemical form. The fuel is pumped to the engine on demand where it is mixed with air from the atmosphere and combusted within the engine. As noted in the preceding section, this process converts chemical potential to thermal energy which, in turn, imparts kinetic energy through the engine mechanism to a rotating shaft. The mechanical DL transmits this rotational kinetic energy via a system of gears, couplings, and driveshafts and delivers energy to the wheels, thus imparting kinetic energy to the vehicle. All of the energy conversion processes in this propulsion system occur in series. Diagram *a* in Figure 9.5 shows a schematic of this driveline with the direction of energy flow.

The ICE vehicle is an open thermodynamic system. Fuel and air are the system inputs that enter the system boundary at the fuel tank throat and air intakes respectively; and the products of combustion, such as exhaust gases, exit the system boundary at the tailpipe. From this, it follows that a vehicle's overall system efficiency can be defined as the useful work performed divided by the total value of the energy input.

Consider gasoline, which has a specific higher calorific value of approximately 46 MJ/kg or in volumetric terms 34 MJ/L. For a full 50 liter (13.2 gal) tank of gasoline, the chemical potential (ES) is 1700 MJ (475 kWh). Since the energy available at the wheels is dependent on the overall efficiency, then stored energy available at the wheels, given a typical overall efficiency of twenty percent, is 340 MJ (or 95 kWh).

The distance a vehicle can travel on 50 L of fuel (or 340 MJ of energy at the wheel) depends on vehicle performance, driver inputs, and driving conditions. Using average fuel economy values, a typical modern gasoline ICE vehicle with an engine size of 1.6 L is 6.6 L/100 km and a fifty-liter fuel tank would, on average, permit 757 kilometers of driving. This exercise demonstrates that an ICE vehicle, even with its low overall efficiency, converts a modest amount of fuel into ample energy at the wheels, and by extension, range to the vehicle.

Next, consider the rate of conversion of energy, or power, for such a vehicle. Imagine that a particular driving condition requires 20 kW of power at the wheels. Again assuming a twenty percent overall efficiency, 100 kW of energy are required through the propulsion system from the tank; this is referred to as *fuel power*. Knowing the energy content is 46 MJ/kg, a fuel power of 100 kW would require 2.17 g/s of fuel to be pumped into the engine and injected into the cylinders for combustion. Again, we note that even in this low efficiency system, small amounts of fuel (on the order to grams per second) are required to deliver the specified instantaneous power.

A refill of a fifty-liter fuel tank typically takes less than five minutes to complete. Hence, with respect to all three competencies, we see that gasoline is a dense, powerful, and convenient source of energy. Its overall suitability as a source of energy to an open thermodynamic system helps to explain the dominance of the ICE vehicle in everyday use.

That said, ICE vehicles are actually quite inefficient in their use of energy. Because of the open thermodynamic system, undesirable exhaust gases are steadily emitted. The required range of engine performance must match a wide range of driving conditions: take off from a standstill, repeated stop and go, high speed cruising, accelerating at high speed, and hill climbing. To achieve this, the DL must have different gear ratios to help match the engine's torque-speed characteristic to those required where the wheels meet the road. Hence, the DL, with its gearboxes and drive shafts, is bulky and adds considerable weight to the vehicle. A great deal of the energy input for many vehicles is consumed in simply conveying the vehicle itself. In addition, the engine cannot provide a stall torque, or a torque without rotation, which means the engine has to be running as the vehicle takes off from a standing start. Historically, this has meant that in urban driving when the vehicle is waiting in traffic, the engine is consuming fuel to keep it idling, but produces no useful kinetic energy. Under this condition, the overall efficiency is effectively zero percent. Even though OEMs have begun addressing such inefficiencies (for example by applying selective cylinder shut-off or idle-off control schemes), the incredible convenience and suitability of petroleum fuels in the ICE has created inherent shortcomings that leave much room for improvement.

9.2.5.2. Electric Vehicles

The Electric Vehicle (EV), or Full Electric Vehicle (FEV), is one where the only energy source on the vehicle is an electrical one. The Battery Electric Vehicle (BEV) is an example of a Full Electric Vehicle. Since the battery must be charged by an external source, the term Plug-In Electric Vehicle (PEV) is sometimes used.

Figures 9.5(b) and 9.5(c) show two possible architectures for a BEV. Both versions of EV drivelines shown are examples of series architecture. Figure 9.5(b) shows a version with a single motor, which means that a transaxle with a differential gear system is required to divide and direct the drive torque to the wheels, not unlike a conventional ICE drivetrain. Figure 9.5(c) demonstrates the flexibility of the EV architecture, where two motors may be used: one to drive each wheel. The electric motor can act as a generator in reverse, thus regenerative braking is readily achievable with little additional cost. For this reason the power flow from the wheels to the motor to the battery is shown as being bi-directional.

EV performance under Competency 1 is determined by the capacity of the ES and the overall efficiency of the propulsion system. Typically, EV drivelines exhibit very high overall efficiencies, on the order of ninety percent. However, the energy densities of modern batteries are a significant constraint on the ES capacity and hence vehicle range. State-of-the-art battery technologies are based on Nickel-Metal Hydride (Ni-MH) and

Lithium Ion (Li-ion) chemistry. Practical values for the specific energies of these technologies are 270 kJ/kg for Ni-MH and 648 kJ/kg for Li-ion.[30] Lithium-ion technology, popularized in cell phones and laptop applications, is becoming the preferred choice for EV architectures. However, the specific energy values should be compared to the calorific value of petroleum fuels, 46 MJ/kg, up to two orders of magnitude greater. At the system level, we can compare the ICE vehicle's 95 kWh of energy content from its 50L fuel tank with the EV and PHEV shown in Table 9.2, which have battery capacities of 24 kWh and 16.5 kWh respectively.

EV performance under Competency 2 is also limited by the ES and the battery technology. Typical specific power values for modern batteries vary widely and range from 5 to 400 W/kg,[31] or about 25 kWh per charge for the EV and PHEV vehicles shown in Table 9.2. Furthermore, a technology tradeoff exists in modern batteries between specific power and specific energy. In other words, most batteries are optimized to either deliver low levels of continuous power over extended periods (like a laptop), or surges of peak power for brief periods (like a cordless drill). Today's vehicles frequently require both in sufficient quantity on any given trip. Studies of various battery technologies demonstrate that maximum specific power is achieved with a significant reduction in specific energy and vice versa.[32]

Also under Competency 2, acceleration performance can be excellent for vehicles propelled with an electric motor, as full torque is available throughout the speed range. This provides for a quicker response from start and smoother transitions between speeds at greatly enhanced efficiencies, as compared to ICE vehicles that require discrete gears and sub-optimal engine power matching.

EV performance under Competency 3 has three major considerations: the time to fully recharge at one sitting, the availability of charging points in a recharging infrastructure, and the number of recharges permitted during the lifetime of the battery. Typically recharging a battery takes much longer than filling a fuel tank (even in socalled "quick charge" high voltage modes). Vehicle charging stations are increasing in number in many urban centers, but are not inexpensive and lag significantly behind the number of conventional refueling stations. A related consideration is that in many urban centers where EV uptake is targeted, many consumers lack a garage or a reserved parking space to reliably recharge overnight. Batteries have lifecycles in the range of 150–1500 recharge cycles.[33] Li-ion batteries in particular have issues with electrolyte decomposition and the formation of oxide films on the battery terminals that affect the life of the battery.

9.2.5.3. Hybrid Electric Vehicles

The Hybrid Electric Vehicle, or HEV, is used to describe a vehicle that has at least two onboard Energy Storage systems, where the energy from each storage systems is used to propel the vehicle either directly or indirectly. The purpose of the HEV architecture is to exploit the benefits of two energy storage technologies. Possible combinations include, but are not limited to ICE with an electric motor powered by a battery, capacitor or flywheel, or an electric motor powered by a fuel cell and a battery or a capacitor. Our focus will be on the most common hybrid combination on the market, namely, ICE paired with a battery-powered electric motor.

Figure 9.5(d) shows the simplest hybrid driveline, called a *series hybrid*. As with the previous series architectures the energy conversion processes may be logically considered to occur one after another. This architecture can enjoy the benefits of regenerative braking. Two particular advantages of this architecture include reduced demand for the ICE and optimization of operating speed for the ICE as its sole purpose is to run the generator. Figure 9.5(e) shows an alternative architecture called the *parallel hybrid*. Here, two power streams interface at a suitable point in the driveline, delivering their combined effort through a customized mechanism. Figure 9.5(f) shows a *series-parallel hybrid driveline*. This driveline leverages two control methodologies- one that exploits the presence of the generator to improve engine efficiency and a second that provides for operation in a more conventional direct gasoline mode. The drive from the ICE may be used to directly drive the axle or generate electricity to recharge the battery. The trade-off for this type of architecture is that it requires a sophisticated Engine Control Unit (ECU).

Even more sophisticated *complex hybrid drivelines* have reached the market. Figure 9.5(g) shows a two-wheel drive version of a complex hybrid driveline which is found on the latest Toyota Prius. In this architecture the generator may be used in motoring mode to add additional drive to the axle. Figure 9.5(h) shows a four-wheel drive version of a complex hybrid driveline, which is found on the Lexus RX450h. Parallel-series drive is provided to one axle via the combined efforts of the ICE and motor. A second motor supplies drive to the second axle in isolation. Coordinated control for such architectures is critical.

Each HEV has specific advantages and disadvantages. However, in general HEV performance under Competency 1 is an improvement over EV and ICE vehicles. Overall efficiencies are on the order of thirty-five percent, which is not as good as EVs that can reach sixty percent, but considerably better than the twenty to twenty-five percent range typical of ICE vehicles. Caution is advised here when comparing vehicle system efficiencies that are purely tank-to-wheels (T2W). A well-to-wheels measurement would capture electricity efficiency, oil mining, and refining, which could narrow significantly the margin in overall efficiency between EV and gasoline vehicles. Because petroleum fuel is used as the primary on-board energy source, the ES storage capacity is high. Ni-Mh storage capacity is typically sufficient for HEV. HEV vehicles currently have the greatest stored energy available to the wheels and hence the longest driving ranges.

HEV performance under Competency 2 is generally comparable to other vehicle technologies. The combination of stored energy in the fuel tank and the battery can be made available to the wheels at the required power level by using either by the motor, the engine, or both. One drawback impacting performance is the additional mass (due the larger battery pack and sophisticated DL) as compared to the ICE vehicle's conventional DL.

HEV performance under Competency 3 is directly comparable to the ICE Vehicle. This is because petroleum-based fuel is the primary energy source to be replenished.

Some HEV architectures, such as plug-in HEV or PHEV, have been designed to tap into cheaper electrical energy available from the grid. As mentioned, the Chevy Volt is an example of PHEV architecture.

As discussed, the power density of electric batteries is much lower than that of gasoline, meaning electric cars require more space to store that power in batteries, adding energy-sapping weight. The engines of electric cars are generally smaller than gasoline or diesel engines, but the size and mass of the batteries offsets any net advantage. In addition, the user is effectively paying for multiple drivetrains which indeed improve efficiency, but at a cost premium. The main performance advantage of the electric car is its low energy consumption when accelerating from start and at lower speeds, enabled by precision control. Hybrid cars take advantage of the electric drivetrain in acceleration at low speeds and then switch automatically to gasoline for higher speeds requiring sustained power. This enables HEVs to get the best of both worlds. When the engine runs on gasoline, it automatically charges the battery of the electric engine.

9.2.6. Government Policies and Initiatives for Change in the Automobile Market

It is clear that ICE is the dominant and mature technology in the automotive sector. Because of evolving longer term sustainability concerns with ICE (high CO_2, unwanted gaseous and particulate emissions, low efficiency, and oil dependence) competing technologies will increasingly replace the ICE over time. In the short and medium terms, improvements in ICE technologies are expected to further improve the performance of the ICE, as noted in the previous section.

Presently, the technology closest to matching the performance of the ICE is the electric motor. However, just as the Achilles heel of ICE is oil dependence, the Achilles heel of the EV sector is low energy density and the high cost of battery technology. The EV revolution will not happen overnight, nor will it happen in the short term. It could also incur large system costs, beyond the traditional purchase price of the vehicle itself. Some cost premiums may be justified, but more data will be required in the coming years.

Currently, there is a major political drive to deploy a diverse array of new vehicle technologies including hybrid, electric, and future alternatives that substitute batteries, fuel cells, or even hydrogen for traditional fuels.

Though not exhaustive, Table 9.3 provides a few examples of areas where specific policies are being enacted across themes and regions.

Governments could use a *feebate* system (McKinsey Report), where revenue from high taxes and penalties on poor performers could be used to provide the fiscal incentives to support a growing market share for electric vehicles.[34]

A real objective of policymakers is to reduce oil consumption and emissions, but achieving market share for EV and HEV technologies has become a proxy for these goals. Economists warn that solutions be market-based and realistic. Table 9.4 provides a snapshot of targets worldwide. This list is not exhaustive and is subject to change in policy.

At the municipal level, many cities and large urban centers have introduced local initiatives and policies for promoting the use of EV technologies. Short commuting

Table 9.3. Policy Themes Being Enacted Worldwide.

Policy Theme	Embodiment	Example Country or Region
Fuel Economy	Numeric standards and targets in the form of mpg or L/100 km	United States, Japan, Canada, Australia, Taiwan, South Korea
Emission Standards	Emission targets for new vehicles to market	European Union, United States (California & Federal level)
Fuel Taxes	Tax fuel sales to reduce fuel consumption.	European Union, Japan
Fiscal Incentives	Give tax relief based on engine size, efficiency, CO2 emissions. Provide subsidies for new cleaner technologies	European Union, Japan
Fiscal Penalties	Penalise poor performers: low efficiency vehicles operators and high polluters	Paris (Ban on SUVs)
Research and Development	Promote private and public sector research for cleaner energy technology and alternative fuels	European Union, Japan, United States, others
Traffic Control Measures	Allow lane privileges to EV and HEV vehicles	United States, Norway
Change Consumer Behavior	Educational programs on EV technology	
Government Fleet Procurement	EV technology in public service fleets via public and public-private procurement	
Market Share Targets	Set mandates to introduce new technology targets	See Table 9.4.

distances within urban centers will suit the current performance characteristics of EV. The OECD and IEA have jointly published the *EV City Casebook*, providing a review of the global electric vehicle movement in sixteen urban centers across the United States, Europe, China and Japan.[36] IEA has also launched an international effort, the Electric Vehicle Initiative (EVI), in which more than a dozen countries are collaborating to remove hurdles and speed adoption via policies, standards, infrastructure, and information sharing. EVI has recently published the "Global EV Outlook," with a detailed summary of the EV landscape through 2020.[37]

Table 9.4 Electric or Advanced Vehicle Targets by Region.[35]

Country/Region	Target	Target Date
United States	1,000,000	2015
China	5,000,000	2020
Japan	20% "Next-generation" autos	2020
	100% "Next-generation" autos	2050
Canada	500,000	2018
Europe	15,000,000	2025
Denmark	200,000	2020
France	2,000,000	2020
Germany	1,000,000	2020
Ireland	350,000	2020
	40% market share	2030
The Netherlands	10,000 in Amsterdam	2015
	200,000 in Amsterdam	2040
Spain	1,000,000	2014
Sweden	600,000	2020
United Kingdom	1,200,000 BEV; 350,000 PHEV	2020
	3,300,000 BEV; 7,900,000 PHEV	2030
India	100,000	2020

Policymakers will play an important role in the gradual adoption of EV technology to the market. A joint approach with a full range of stakeholder views should inform policy decisions. Stakeholders include auto manufacturers and suppliers, fuel and energy providers, government agency and public sector authorities, and consumers. Policymakers must balance the need for quick action on climate change and energy security with the pitfalls of economic hardship arising from bad policy decisions. It is important that policy works at the pace of technology maturation. Sound policy will in turn expedite and encourage technology development and maturation in key areas, thus, providing consumer confidence and an increased market share for clean technologies.

9.3. Aviation Fuels and Regulation

A predominant reliance on petroleum may be a significant issue for cars, trucks, and ground transportation, but it is felt even more acutely in aviation; as yet, little compares in delivering the necessary energy density, availability, and cost. A century-old industry, air transport has been built around fossil fuels- arguably been made possible by them; and the technological and financial barriers to entry are among the greatest of any industry. Debates ensue about whether such demanding specifications justify the complex conversion of limited biomass to jet fuel. And yet, an interesting journey has

Fuel Cells

Much basic research has been allocated to the development of hydrogen fuel cells. While this technology is by no means commercially viable as yet, it may eventually help resolve some of the range, power, and emission concerns prevalent today. In addition to potential cost barriers, hydrogen sourcing and infrastructure loom large as key challenges. That said, it is important that a robust vehicle diversification strategy include a range of options. Fuel cell energy storage for future generation vehicles may equal or indeed surpass advanced battery systems. If so, they will certainly help satisfy some of the longer term objectives in the drive toward sustainable transportation.

A basic fuel cell operates by conversion rather than combustion and is based on an electrochemical cell. Fuel cells contain an anode and cathode, separated by an electrolyte. The anode receives hydrogen (H_2) from a supply at low pressure while the cathode receives oxygen or filtered air. At the anode, hydrogen molecules split into protons and electrons. If the anode and cathode are then connected via an electrical conductor, the protons from the anode move through the electrolyte toward the cathode. The electrons move through the conductor to the cathode and can supply electric curent to a moter or other appliance load. At the cathode, protons and electrons react to form water with the supplied oxygen. In the long term, hydrogen fuel cells may become suitable green car technology with the potential to offer the power and range of a conventional gasoline engine with water as the only byproduct.

Fuel cells have relatively few moving parts, so recurring maintenance costs are lower, helping reduce overall lifetime cost. They are comparatively efficient—some technologies can convert more than fifty per cent of the energy content of the hydrogen to electrical power. Notwithstanding the tank-to-wheel (T2W) uncertainty of the fuel supply, however, fuel cells are more efficient than diesel generators. Due to their modular design, additional cells can be added to the system, increasing the amount of power produced without replacing the whole system. Fuel cells are generally distinguished by their electrolyte and brief description of major technologies follows.

Alkaline fuel cells were among the first to be developed, and use an alkaline electrolyte such as potassium hydroxide. Catalysts include a variety of non-precious metals, but their main drawback is poisoning of the cell by carbon dioxide.

Direct methanol fuel cells use a polymer membrane as the electrolyte but are differentiated in their fuel type, which is pure methanol.

Proton exchange membrane or *polymer electrolyte membrane* (PEM) fuel cells use a polymeric membrane as the electrolyte, with carbon electrodes and a platinum catalyst. Their high power density and fast start-up time make them the most suitable fuel cell for transport applications.

Phosphoric acid fuel cells use liquid phosphoric acid as an electrolyte and porous carbon electrodes containing a platinum catalyst. This is one of the most mature fuel cell technologies and is mainly deployed in stationary power applications.

> *Molten carbonate* fuel cells (MCFCs) are high temperature fuel cells in which a molten salt mixture is suspended in a ceramic matrix. They can reach much higher efficiencies and are currently being developed for electrical applications in natural gas and coal power plants. Durability is a primary disadvantage due to corrosion resulting from high temperature operation.
>
> *Solid oxide* fuel cells (SOFCs) operate at extremely high temperatures and as a result are projected to have extremely high efficiencies. SOFCs can utilize less expensive catalyst materials and a greater variety of fuel sources. Slow start up and durability may limit their use in mobile applications, yet development is underway to optimize performance at lower temperatures.

begun as leading aviation stakeholders are engaged in the pursuit of sustainable economic and environmental solutions.

9.3.1. The Global Aviation Industry

The goal of the aviation industry is to provide quick, safe, and affordable transportation of passengers and cargo to destinations all over the globe. Air transportation is the only viable means of transport between many areas of the world, as other means of travel such as by ship, automobile or train are impractical or are unavailable. Aviation provides social benefits as well, for example by providing access to distant or remote locales and relief during times of disaster. The Air Transport Action Group projects that aviation accounts for 56.6 million jobs worldwide and $2.2 trillion in GDP.[38] Yet the economic engine that is aviation does not run without burning fuel and producing emissions. Therefore, efforts to reduce energy consumption and emissions in aviation are vital to the long-term sustainability of the industry. As noted in Chapter 2, a pivotal US policy developed to improve energy independence while reducing emissions is the Energy Independence and Security Act (EISA) of 2007. This law identifies specific goals for improved efficiency standards for fuel economy, appliances, lighting and buildings; as well as increased production of biofuels.[39]

The transportation sector accounts for about 28.1% of US energy consumption.[40] In 2012, US jet fuel consumption was 22 billion gallons (512 million barrels), or about eight percent of total US petroleum use. Commercial air carrier operations accounted for the majority of this total, with domestic flights out-consuming US-originated international flights by nearly two to one.[41] Aviation gasoline accounts for about one percent of US aviation fuels, with a consumption of about 221 million gallons (5 million barrels). In comparison, 192 billion gallons (4.6 billion barrels) of gasoline and diesel were used in ground transportation, a sector that accounts for more than seventy percent of total US petroleum consumption.[42] Annual worldwide jet fuel consumption for the aviation industry is around 63 billion gallons (1500 million barrels) as estimated by the Air Transport Action Group.[43]

In 2012, 2.9 billion passengers flew over 5.3 trillion kilometers. The International Civil Aviation Organization (ICAO) is projecting an annual growth rate of 4.9% through

2030, meaning that passenger traffic and total flight kilometers would more than double in the next twenty years. Such growth will put extraordinary demands on fuel supply with significant environmental implications. Of particular importance is the impact of air transportation on CO_2 emissions, which are expected to increase at an annual rate of two to three percent, up to three percent of the global total from all sources by 2050.[44] Numerous initiatives are aimed at reducing the carbon footprint of aviation on both national and international levels. As we delve further into this, it becomes apparent how the future of the aviation industry is inextricably linked to its energy source.

9.3.2. Aviation Fuels Overview

In addition to creating propulsive power, fuel serves several other important functions in aircraft. In jet aircraft, heat is generated by the combustion of fuel and by friction between aircraft surfaces and surrounding air at operating speeds. Aircraft designers use the heat absorbing capacity of fuel to dissipate heat and help cool the aircraft. The lubricity of fuel is a critical property in the engine and fuel system of the aircraft. Fuel can also be used as the working fluid in hydraulic systems and for load balancing.

Most commercial and private aircraft use one of three fuels: aviation gasoline (avgas), jet fuel of various types (such as Jet A, Jet A-1, Jet B, or JP-8), and diesel. Diesel is used in small quantities for ground operations and in a limited number of flight applications; hence the principal focus here will be on the other two primary aviation fuels. Avgas represents the last leaded transportation fuel permitted in the US, and current efforts are focused on replacing lead in this fuel by 2018. For jet fuel, the current focus is on petroleum alternatives and managing the emissions of carbon dioxide (CO_2) and other GHGs. Particulate matter emissions are gaining in importance as well, particularly because some alternative jet fuels emit smaller particulate matter and health impacts are believed to increase with decreasing particle size.

In 2012, fuel costs comprised thirty-three percent of airline operating expenses, and represented the largest portion of airline costs. This is in dramatic contrast to their fourteen percent share in 2003. Due to a strong correlation between fuel and operating expenses, the industry is eager to reduce fuel consumption on economic grounds. Enduring technical focus on efficiency, weight reductions, and flight optimization techniques have resulted in significant fuel savings in recent decades. Figure 9.6 documents these trends in aircraft energy intensity during the past fifty years.

Air carriers have significantly increased overal net efficiency (the number of aircraft miles flown per gallon of fuel). This metric incorporates both operational and technological efficiency gains. Table 9.5 uses data from the US Department of Transportation to calculate percent improvement, indicating that aircraft miles flown per gallon for domestic air operations has increased sixty-seven percent from 0.32 in 1990 to 0.54 in 2011. A forty-one percent increase has been estimated for international operations. It should be noted that "aircraft miles per gallon" are lower for international flights owing to the use of larger aircraft with increased weight. However, such flights typically carry more passengers than domestic flights, meaning the energy efficiency per passenger mile is actually comparable to shorter flights.

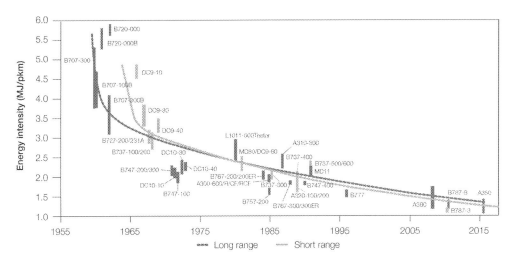

Figure 9.6. Energy Intensity in Aviation.[45] Note: Energy Intensity (vertical-axis) is Defined as Megajoules Per Passenger Kilometer Traveled (MJ/p-km); 1.4 MJ/p-km Equals Approximately 49 Passenger-miles Per US Gallon; 1.0 MJ/p-km ~ = 70mpg.

Despite these promising trends linking sustained innovation to reduced energy use, the recent volatility of fuel prices has made it difficult for airlines to forecast key operating costs and has impacted profitability. During the past decade, many global air carriers have been required to restructure their businesses, and many have faced bankruptcy. While a variety of factors may have contributed to this situation, financial stress in aviation has been exacerbated by vulnerability and risk associated with fuel costs. Continuing to reduce fuel usage is therefore crucial in order to help the sector achieve economic, environmental and social sustainability. Accordingly, the Federal Aviation Administration (FAA) has set a goal to "improve the National Airspace System (NAS) energy efficiency by at least two percent per year, and develop and deploy alternative jet fuels for commercial aviation."[47]

Today's aviation fuels are almost exclusively petroleum derived. To an even greater extent than ground transport, the capital-intensive aviation sector developed around petroleum fuels due to their unique combination of operating characteristics, energy density, availability, and reasonable cost. Engines and aircraft systems are extremely expensive

Table 9.5. Aircraft Miles Flown Per Gallon.[46]

Year	Domestic Operation	Improvement from 1990	International Operations	Improvement from 1990
1990	0.32		0.19	
1995	0.36	11%	0.22	15%
2000	0.38	17%	0.23	21%
2005	0.48	48%	0.26	33%
2010	0.53	64%	0.28	45%
2011	0.54	67%	0.27	41%

and therefore designed for long life through a rigorous product development schedule where ten years is not atypical. Fleet turnovers are commensurately slow, on the order of twenty-five years. This is easily understood when one considers that a Boeing B737–800 can cost $89 million; a B787–8, $206 million; and an Airbus A380 reportedly costs $375 million.[48] Taken along with entrenched support infrastructure, it is important to note that major changes of any kind can be costly, long term propositions in the aviation business.

9.3.3. Alternative Aviation Fuels and Emissions

Despite the complexity of introducing alternatives to petroleum jet fuel, the potential benefits of improved price stability, supply diversity, and reduced environmental impacts are broadly supported. Renewable fuels are alone among viable petroleum fuel replacements in aviation, as neither batteries, fuel cells, nor any other low carbon technologies can yet deliver on the overall requirements. An important constraint on potential new fuels is the requirement they be "drop-in," meaning operationally indistinguishable from petroleum-based fuels. If fuels are not drop-in, then the required modifications to aircraft, engines, and aviation infrastructure for the transportation, storage and delivery of aviation fuels would make alternative fuels cost-prohibitive and operationally difficult. Should an alternative fuel necessitate changes to the aircraft system, the aircraft would be required to undergo extensive testing and evaluation and be certified (or *type-certificated*) to use that fuel.

Under the drop-in caveat, significant strides have been made including the certification of two alternative fuel specifications which have been tested on a myriad of military and commercial test flights. Blends of petroleum-based Jet A/A-1 with up to fifty percent alternative fuels are accepted for aviation turbine engine use if compliant with jet fuel specifications ASTM D1655 and ASTM D7566. Because neat (one hundred percent) biofuels have certain characteristics that fall outside of jet fuel requirements such as low aromatic content or higher freezing points, the fifty percent blending allowance permits mixtures to comply with specifications. There is no currently approved alternative fuel for aviation gasoline, but research efforts are underway to produce an unleaded aviation gasoline that performs as well as 100-octane leaded avgas (100LL) and meets the same specifications (ASTM D910 and ASTM D6227).

Liquid drop-in alternative jet fuels can be produced by the conversion of feedstocks (such as coal, natural gas, or biomass) through a variety of technical pathways. The two ASTM-approved methods are the Fischer-Tropsch method, which converts solids to synthesis gases and then liquid fuels, and the Hydrotreated Esters and Fatty Acids (HEFA) method, which converts oil-based biomass into hydrocarbon molecules similar to kerosene. Pathways that utilize non-food feedstocks are preferable, as they do not compete for conventional agricultural land. Certified fuels must satisfy alternative jet fuel production methods and jet fuel specifications. A summary of primary conversion pathways is provided in Table 9.6.

In addition to alternative fuel technologies, research in aircraft propulsion is also a growing area of development. While the energy density of electric power is a barrier for transport category electric aircraft, significant advances in unmanned and personal-use aircraft have been achieved. Comparative Aircraft Flight Efficiency's CAFE Foundation conducts annual electric aircraft symposia to usher in the "age of electric flight." The CAFE Foundation hosts a Green Flight Challenge Program incentivizing the development of sustainable, ultra-quiet and emission-free flight.[50] Solar powered aircraft and hybrid aircraft are being developed for selected applications as well. In addition to reducing fuel consumption, the aviation industry is endeavoring to reduce greenhouse gas and particulate emissions; yet without successfully introducing some quantity of renewable alternatives, the aviation sector has few options to meaningfully reduce its carbon footprint. The primary emissions of jet fuel combustion are carbon dioxide, water vapor, nitrogen oxides, and particulate matter. Petroleum-based fuels contain sulfur and emit trace quantities of sulfur oxides, whereas bio-derived jet fuels have little or no sulfur. An important difference between aviation and other industries is that emissions are delivered to the atmosphere not only locally, but also in the atmosphere at altitude.

Comparison of alternative to petroleum-based fuels are made on a life cycle basis from well-to-wake, or from the origin of the inputs through point of use in the aircraft. The well-to-wake method offers a more complete environmental and cost impact analysis than that of a wheel-to-wake approach. A new fuel may appear environmentally attractive based on the wheel-to-wake process, but may be characterized by greater environmental impacts during its conversion than petroleum-based fuels. Improper land use change or inefficient collection/conversion methods might explain why total life cycle analysis (LCA) could become less favorable for a biofuel. Agreement on metrics

Table 9.6 Summary of Alternative Jet Fuel Conversion Pathways, 2013.[49]

ASTM Specification	Status	Description	Typical Feedstocks
D1655–99	Approved, June 1999	Petroleum Jet Fuel	Crude Oil
D7566-Annex A1	Approved, Sept. 2009	Fischer Tropsch (FT)	Coal, Natural Gas, Biomass
D7566-Annex A2	Approved, July 2011	Hydroprocessed Esters & Fatty Acids (HEFA)	Plant oils, animal fats, algae
D7566-Future Annexes	Under Review	Alcohol-to-Jet (ATJ) Fuel	Ethanol, Butanol, Methanol
D7566-Future Annexes	Under Review	Direct Sugar to Hydrocarbons (DSHC)	Sugar cane, Cellosic Biomass
D7566-Future Annexes	Under Review	Hydrotreated Depolymerized Cellulosic Jet (HDCJ)	Cellulosic Biomass, Ag. Residue
D7566-Future Annexes	Under Review	Catalytic Hydrothermolysis (CH) or Pyrolysis	Plant oils, other renewable oils
D7566-Future Annexes	Under Review	Catalytic Conversion of Sugar	Sugar cane, Corn Stover, Biomass

and methods for LCA are inherently complicated, yet are being discussed within governments, organizations, and ICAO. Thankfully, the aviation community can draw from a wide body of existing literature and research regarding environmental impacts of biofuels in general.

9.3.4. Policy Frameworks and Initiatives for Greening the Aviation Industry

The simultaneous challenges of expanding aviation while reducing its environmental impact clearly requires the implementation of practical and coordinated policies on both national and international levels. International aviation is excluded from the Kyoto Protocol, and instead the International Civil Aviation Organization (ICAO) is charged with the development of policies and standards that will reduce GHG emissions in aviation.[51]

ICAO is a United Nations specialized agency that provides a forum for UN nations to collaborate on standards and regulations, and is charged with promoting global aviation safety, security, environmental protection, and sustainable development of air transportation. ICAO supports policies for market-based emission reduction and sustainable alternative fuel.[52] Due to the expected climatic impact of aviation, ICAO alternative fuels efforts fall under its environmental protection mission. ICAO created the Global Framework for Alternative Aviation Fuels (GFAAF) to highlight and communicate goals and progress in the development and use of alternative aviation fuels. GFAAF facilitates fora to harmonize definitions, standards, methodologies, and financial incentives.

The ICAO Committee on Aviation and Environmental Protection (CAEP) seeks to develop an international carbon dioxide emissions standard that will include life cycle and emissions impacts of alternative fuels and technologies. The ICAO Aircraft Engine Emissions Databank contains information on exhaust emissions for production aircraft engines with rated output exceeding 26.7kN (6000 lbs). Emissions data is provided by engine manufacturers using ICAO approved procedures and is currently available for Jet A/A-1 petroleum-based fuels. The standards limit the emissions of smoke, unburned hydrocarbons, carbon monoxide, and oxides of nitrogen from turbojet and turbofan aircraft engines.[53] Studies by NASA, including the Alternative Aviation Fuel Experiment (AAFEX II) and the Alternative Fuel Effects on Contrails and Cruise Emissions (ACCESS), are being conducted on the ground and at altitude to quantify the effects of alternative fuels on aircraft and exhaust emissions.[54]

To promote awareness and aid passengers and airlines in carbon offset efforts, ICAO has developed a Carbon Calculator which is accessible on their website.[55] The Carbon Calculator estimates the carbon dioxide and fuel consumed per passenger between destinations worldwide using route specific data. The calculator takes in to account cargo carried, number of passengers on board, aircraft type and other route data. For example, an economy class passenger traveling the 11,822 km roundtrip from Dublin to Dubai is estimated to have a carbon footprint of approximately 846 kg of CO_2; an economy class passenger traveling the 3,186 km roundtrip from Birmingham, England to Budapest is estimated to have a carbon footprint of approximately 323 kg of CO_2.[56]

The Federal Aviation Administration (FAA) and the European Aviation Safety Agency (EASA) are the US and EU government agencies charged with regulatory oversight for alternative fuels and emissions reduction in aviation. The FAA's strategic planning document, *Destination 2025*, sets several energy and environment-related performance metrics to be achieved by 2018, including:

- one billion gallons of alternative jet fuel consumed by 2018,
- a replacement for leaded aviation gasoline,
- a two percent annual improvement in energy efficiency (fuel consumed per miles flown),
- a fifty percent reduction in health impacts from emissions, and
- the establishment of a trajectory for carbon neutral commercial aviation growth based on a 2005 baseline.[57]

Other FAA initiatives have set similarly ambitious goals by 2018 such as reductions in overall aviation fuel consumption by thirty-three percent, noise by thirty-two decibels, and NO_x emissions by sixty percent, as described in the Continuous Lower Energy, Emissions and Noise (CLEEN) program.[58]

In order to provide demand assurance for industry and FAA aspirational goals, the service branches of the US Department of Defense have announced a range of goals to increase the utilization of renewable energy over the next decade. According to the US Department of Defense, "any alternative fuels for DOD operational use must: be 'drop-in'; that is, requiring no modification to existing engines; be cost-competitive with conventional petroleum fuels; be derived from a non-food crop feedstock; and have lifecycle greenhouse gas emissions less than or equal to conventional petroleum fuels."[59] The Air Force has more specific goals "to test and certify all aircraft and systems on a 50:50 alternative fuel blend by 2012, and to be prepared to acquire 50% of the Air Force's domestic aviation fuel as an alternative fuel blend by 2016."[60] While the Air Force has met the 2012 goal, it may be worth noting that "being prepared to acquire" and "actually acquiring commercially competitive quantities" by 2016 are quite different matters. Cost remains a major challenge.

The European Aviation Safety Agency's *Flightpath 2050* also seeks carbon neutral aviation growth using a 2005 baseline, and a fifty percent reduction in emissions by 2050.[61] Sustainable Way for Alternative Fuels and Energy for Aviation (SWAFEA) is a study commissioned by the European Commission's Directorate General for Transport and Energy. Covering multiple disciplines and interests in the well-to-wake lifecycle, the SWAFEA team includes representation from twenty European and international organizations that are studying the feasibility and impact of the use of alternative fuels in aviation.[62] International collaborators within the International Civil Aviation Organization (ICAO) have set goals aimed at improving global fuel efficiency by two percent annually and seeking funding for the broader development of sustainable aviation fuels.[63]

While the aviation industry and regulatory agencies agree on the need to reduce emissions, there is disagreement over the best approaches to accomplish such far-reaching goals. Market-based incentives, fuel taxes, and voluntary offsets are just a few of the ways that use of alternative fuels may be promoted. Using a market-based mechanism known as cap-and-trade, the European Commission via its Emissions Trading System (ETS) sets a limit on CO_2 emissions for certain emission-intensive industries. In January 2012, the Commission included the aviation industry in the ETS and initially set an emissions cap for flights taking off or landing at European Union airports. For the purposes of fiscal and environmental accounting, the EU ETS estimates the CO_2 based on the amount of fuel consumed during flights to or from an EU city. The ETS is intended to incentivize and accelerate the reduction of CO_2 through reduced fuel consumption and improvements in aircraft operation. Non-EU countries have protested the implementation of the EU ETS as a unilateral tax. The EU does not view the ETS as a fuel tax or charge that violates any international agreement. However, it does have some unintended consequences. For instance, some cargo carriers responded by re-routing long haul cargo flights around the EU. The cargo still found its way to the EU destination or departed from EU airports, but the cargo airlines added stops to previously non-stop long haul flights to significantly reduce the impact of the flight on their ETS assessments. Adding a stop actually adds to the net emissions, but saves on the ETS assessments. As a result of controversy over the ETS policy and its implementation for aviation, the ETS was postponed in 2013 for international flights in anticipation of a pending multi-lateral agreement through ICAO.[64]

The Commercial Alternative Aviation Fuels Initiative (CAAFI) is a multi-stakeholder endeavor which aims to promote the development of alternative jet fuel in the context of the industry's energy and sustainability challenges. CAAFI includes members from the aviation and aerospace industries, academia, and government.[65]

IATA's policy is to reduce the environmental impact of aviation, and is supporting sustainable biofuels as a means of achieving carbon dioxide reductions throughout the well-to-wake lifecycle. While IATA obviously does not control the price of fuel, IATA policies are aimed at indirect measures of managing fuel costs, for example by increasing the availability and diversity of supply, and the reliability of certified fuels.[66]

9.3.5. Conclusion

While aggressive national and international policies and performance goals can be useful in developing the assurances that industry stakeholders require, it should be reiterated that major transformations in technology and infrastructure are costly, time-consuming, and complex. While fuel efficiency in air transit has increased dramatically, the implementation of market-based carbon emissions regulatory programs at both national and international levels appears non-trivial. History has often demonstrated that policy and technology must progress in a coordinated manner in order to deliver optimal results. Given its involved regulatory framework, expensive and technically advanced capital equipment, and international reach, this may be even more relevant with respect to the aviation sector. At the same time, it is clear that alternative fuels and reduced emissions are enabling priorities for the future of aviation. A robust policy discourse and international cooperation can be essential in facilitating the deployment and commercialization of emerging technologies.

Notes

1. Energy Information Administration, "Total Energy Flow, 2011," *Annual Energy Review*, September 2012, http://www.eia.gov/totalenergy/data/annual/diagram1.cfm; and Energy Information Administration, "Table 11.1 Carbon Dioxide Emissions From Energy Consumption by Source, 1949–2011," *Annual Energy Review*, September 2012, http://www.eia.gov/totalenergy/data/annual/showtext.cfm?t=ptb1101.
2. Energy Information Administration, "Primary Energy Consumption by Source and Sector, 2011," *Annual Energy Review*, September 2012, http://www.eia.gov/totalenergy/data/annual/pecss_diagram.cfm.
3. Daniel Sperling and Deborah Gordon, *Two Billion Cars: Driving Toward Sustainability* (New York: Oxford University Press, 2009), 1–5.
4. H. Cai, and S. D. Xie, "Temporal Variations and Spatial Distribution of Greenhouse Gas Emissions from On-road Vehicles in China," paper presented at the 2009 Sustainable Energy and Environmental Protection Conference, 12–15 August 2009, Dublin, Ireland.
5. Carlos Gomes, "Global Auto Report," *Scotia Bank* (December 5, 2013), 2, http://www.gbm.scotiabank.com/English/bns_econ/bns_auto.pdf.
6. Anup Bandivadekar et al., *On the Road in 2035: Reducing Transportation's Petroleum Consumption and GHG Emissions* (Cambridge: Massachusetts Institute of Technology, 2008), http://web.mit.edu/sloan-auto-lab/research/beforeh2/otr2035/On%20the%20Road%20in%202035_MIT_July%202008.pdf.
7. Ed Pike, *Calculating Electric Drive Vehicle GHG Emissions* (Washington, D.C.: The International Council on Clean Transportation, 2012), http://www.theicct.org/calculating-electric-drive-vehicle-ghg-emissions; and Martin Campestrini and Peter Mock, *European Vehicle Market Statistics* (Washington, D.C.: The International Council on Clean Transportation, 2011), http://www.theicct.org/sites/default/files/publications/Pocketbook_LowRes_withNotes-1.pdf.
8. Bandivadekar et al., *On the Road in 2035*, and Anup Bandivadekar, *Transportation Energy Challenge* (Washington, D.C.: The International Council on Clean Transportation, 2008), slide 6. http://csis.org/files/attachments/090608_bandivadekar.pdf.
9. Ministry of Agriculture and Food, *Energy Factsheet: Propane as a Fuel for Vehicles and Tractors*, Order No. 210.510–2 (Victoria, BC: Ministry of Agriculture and Food, British Columbia, 1982), http://www.agf.gov.bc.ca/resmgmt/publist/200Series/210510-2.pdf.
10. EIA, *Annual Energy Outlook 2013* (Washington, D.C.: US Energy Information Administration, 2013), http://www.eia.gov/forecasts/aeo/; and American Clean Skies Foundation, "Driving on Natural Gas: Fuel Price and Demand Scenarios for Natural Gas Vehicles to 2025" (Washington, D.C.: American Clean Skies Foundation, 2013), 20.
11. European Commission. *European Commissions' and Member States' R&D Programmes for the Electric Vehicle*. Report on a European Commission Workshop (November 15, 2009), 1, http://www.green-cars-initiative.eu/workshops/Report_WS_EC-MS_Electric_Vehicle_R-D.pdf.
12. Michael J. Moran and Howard M. Shapiro. *Fundamentals of Engineering Thermodynamics*, 5th ed. (New York: John Wiley and Sons, 2004), 660.
13. Environmental Protection Agency, *Unit Conversions, Emissions Factors, and Other Reference Data* (Washington, D.C.: Environmental Protection Agency, 2004), 2, www.epa.gov/cpd/pdf/brochure.pdf.
14. "What's the Carbon Footprint of Building Your Car, and How does that Compare to Tailpipe Emissions?" *Global Change Blog* (September 24, 2010), http://www.globalchangeblog.com/2010/09/whats-the-carbon-footprint-of-building-your-car-and-how-does-that-compare-to-tailpipe-emissions/.
15. International Transport Forum, *Reducing Transport Greenhouse Gas Emissions: Trends & Data 2010* (Paris: OECD Publishing, 2010), 5.
16. Table assumes comparable driving habits and conditions. It also assumes a seven year life (84,000 miles) and residual value at end of life equal to twenty percent of initial capitalized cost. Operating cost calculations assume and include a discount (interest) rate of eight percent; inflation rate of two percent; annual price increase for gasoline, diesel, and natural gas of three percent; annual price increase for electricity of one-half percent. MSRP for new 2013 model year vehicles for sale to the US market. Scenario 1 sets initial

(year 1) fuel/energy prices as follows: gasoline at $3.50/gal; diesel at $3.85/gal; Electricity at $0.12/kWh; CNG at $2.10/GGE (gallon of gasoline equivalent); Scenario 2 sets initial (year 1) fuel/energy prices as follows: gasoline at $6/gal; diesel at $6/gal; Electricity at $0.15/kWh; CNG at $4.20/GGE (gallon of gasoline equivalent). Total lifetime emissions based upon US EPA published CO_2e rates per mile for specified vehicles, carbon intensities of known petroleum fuels (including tailpipe and upstream GHGs associated with their extraction, refining, transport, etc.), CO2 emissions of US electricity generation from primary sources, modes of driving, and electricity mix; then projected for 84,000 mile vehicle life. Fuel economy from EPA published mpg or mpge. Electricity Mix A represents the US national average blend, approximately: 50% coal, 30% natural gas, 20% nuclear and renewables; Electricity Mix B represents clean power (available only in limited areas): 50% natural gas, 50% nuclear and renewables. A third Electricity mix representing 100% coal electricity is shown as the upper limit of lifetime emissions in the "with subsidy" simulation case for Leaf and Volt. As of 2013, a US federal tax credit (subsidy) of up to $7500 applies to certain EV and PHEV up to a maximum of 200,000 units per manufacturer. EPA fuel economy and emissions information can be found at http://www.fueleconomy.gov/feg/label/calculations-information.shtml.

17. Center for Climate and Energy Solutions, "Renewable and Alternative Energy Portfolio Standards," www.c2es.org, last updated 2013, http://www.c2es.org/us-states-regions/policy-maps/renewable-energy-standards.
18. Organisation International de Construceurs d'Automobiles (OICA), "Production Statistics," last modified 2013, http://www.oica.net/category/production-statistics/.
19. EIA, *Annual Energy Outlook*.
20. http://www.hybridcars.com.
21. http://www.calculatedriskblog.com/2011/09/vehicle-sales-fleet-turnover-ratio.html.
22. EIA, *Annual Energy Outlook*.
23. João Vitor Fernandes Serra, *Electric Vehicles: Technology, Policy and Commercial Development* (New York: Earthscan, 2012).
24. James J. Eberhardt, "Fuels of the Future for Cars and Trucks," presentation made ate the 2002 Diesel Engine Emissions Reductions (DEER) Workshop, San Diego, August 25–29, 2002, http://www1.eere.energy.gov/vehiclesandfuels/pdfs/deer_2002/session1/2002_deer_eberhardt.pdf.
25. Serra, *Electric Vehicles*.
26. Brad Berman, "Victor Wouk and the Great Hybrid Car Cover-up of 1974," *hybridcars.com*, last modified on March 28, 2006, http://www.hybridcars.com/the-great-hybrid-car-cover-up-of-74/.
27. Michael H. Westbrook, *The Electric Car: Development and Future of Battery, Hybrid, and Fuel-cell Cars* (London: The Institution of Electrical Engineers, 2005).
28. International Energy Agency, *Energy Technology Perspectives 2010* (Paris: International Energy Agency, 2010). See especially chapter 7, 270–276.
29. Readers seeking a more detailed review are directed to Mehrdad Ehsani, Yimin Gao, and Ali Emadi, *Modern Electric, Hybrid Electric, and Fuel Cell Vehicles: Fundamentals, Theory, and Design* (2nd Ed.) (Boca Raton, FL: CRC Press, 2010).
30. Bruno G. Pollet, Iain Staffell, and Jin Lei Shang, "Current Status of Hybrid, Battery, and Fuel Cell Electric Vehicles: From Electrochemistry to Market Prospects," *Electrochemica Acta* 84 (2012): 237. See Table 2 and conversion from $Whkg^{-1}$ to kJ/kg.
31. Ibid., 237, Table 2.
32. International Energy Agency, *Technology Roadmap: Electric and Plug-in Hybrid Electric Vehicles* (Paris: IEA, 2009).
33. Pollet, "Current Status," 237, Table 1.
34. McKinsey & Company, *Roads Toward a Low-carbon Future: Reducing CO_2 Emissions from Passenger Vehicles in the Global Road Transportation System* (New York: McKinsey & Co., 2009).
35. Guzay Pasaoglu, Michel Honselaar, and Christian Thiel, "Potential Vehicle Fleet CO_2 Reductions and Cost Implications for Various Vehicle Techology Deployment Scenarios in Europe," *Energy Policy* 40 (2012): 404–21; Fabien Leurent and Elisabeth Windisch, "Triggering the Development of Electric Mobil-

ity: A Review of Public Policies," *European Transport Research Review* 3, no. 4 (2012): 221–35; and *European Roadmap: Electrification of Road Transport* (2nd Ed.) (Paris: European Commission, 2012), http://www.smart-systems-integration.org/public/documents/content-elements/files/Electrification_Roadmap_2_Web.pdf.

36. International Energy Agency, *EV City Casebook* (Paris: IEA, 2012).
37. International Energy Agency, *Global EV Outlook: Understanding the Electric Vehicle Landscape to 2020* (Paris: IEA, 2013).
38. Air Transport Action Group, "Facts and Figures," ATAG.org, last modified March 2012, http://www.atag.org/facts-and-figures.html.
39. Office of the Press Secretary, "Fact Sheet: Energy Independence and Security Act of 2007," *Whitehouse.gov* (December 19, 2007), http://georgewbush-whitehouse.archives.gov/news/releases/2007/12/20071219-1.html.
40. S. C. Davis, S. W. Diegel, and R. G. Boundy, *Transportation Energy Data Book, ORNL-6989* (Edition 32 of ORNL-5198) (Oak Ridge, TN: Oak Ridge National Laboratory, 2013). See Chapter 2, Table 2.1, "U.S. Consumption of Total Energy by End-Use Sector, 1973-2012."
41. Research and Innovative Technology Administration (RITA), "Airline Fuel Cost and Consumption (U.S. Carriers)," *US Department of Transportation*, 2012, http://www.transtats.bts.gov/fuel.asp?pn=1.
42. Energy Information Administration, "Petroleum & Other Liquids: Product Supplied," *EIA.gov*, last updated September 27, 2013, http://www.eia.gov/dnav/pet/pet_cons_psup_dc_nus_mbbl_a.htm.
43. ATAG, "Facts and Figures."
44. International Civil Aviation Organization, *ICAO Environmental Report, 2013: Aviation and Climate Change*" (Montreal: ICAO, 2013). http://cfapp.icao.int/Environmental-Report-2013/#210/z.
45. International Energy Agency, *Transport, Energy and CO2: Moving Toward Sustainability* (Paris: OECD/IEA, 2009), 317.
46. Bureau of Transportation Statistics, "National Transportation Statistics," *U.S. Department of Transportation*, last updated October 2013. http://www.rita.dot.gov/bts/publications/national_transportation_statistics/index.html.
47. Nate Brown, "Sustainable Alternative Jet Fuels: Update on ASTM Approval," *Federal Aviation Administration* (May 18, 2012), http://www1.eere.energy.gov/bioenergy/pdfs/10_brown_roundtable.pdf.
48. Boeing Corporate website: http://www.boeing.com/boeing/commercial/prices/; and Howard Mustoe, "Airbus Raises A380 Price by 8.4% as Model Beats Airline Cost-Savings Goal," *Bloomberg* (January 18, 2011), http://www.bloomberg.com/news/2011-01-18/airbus-raises-a380-price-by-8-4-as-model-beats-airline-cost-savings-goal.html.
49. Brown, "Sustainable Alternative Jet Fuels."
50. CAFE Foundation, "The 2011 Green Flight Challenge Sponsored by Google," *Cafefoundation.org*, last modified 2013, http://cafefoundation.org/v2/gfc_main.php.
51. Jane A. Leggett, Bart Elias, and Daniel T. Shield, *Aviation and the European Union's Emission Trading Scheme* (Washington, D.C.: Congressional Research Service, 2012).
52. International Civil Aviation Organization, *Annual Report of the Council: 2011* (Montreal: ICAO, 2012), http://www.icao.int/publications/Documents/9975_en.pdf.
53. International Civil Aviation Administration, *ICAO Annex 16: Environmental Protection, Volume II; Aircraft Engine Emissions* (Montreal: ICAO, 2008).
54. B. E. Anderson et al., *Alternative Aviation Fuel Experiment (AAFEX)* (Hampton, VA: NASA Langley Research Center, 2011), http://nix.nasa.gov/search.jsp?R=20110007202&qs=N%3D4294966788%2B4294724624%26No%3D10.
55. http://www.icao.net/Pages/default.aspx.
56. International Civil Aviation Organization, "Carbon Emissions Calculator," *ICAO.int* (2013), http://www.icao.int/environmental-protection/CarbonOffset/Pages/default.aspx.
57. Federal Aviation Administration, *Destination 2025* (Washington, D.C.: Federal Aviation Administration, 2011), 9–10, http://www.faa.gov/about/plans_reports/media/Destination2025.pdf.

58. Federal Aviation Administration, "Continuous Lower Emissions, Energy, and Noise (CLEEN) Program," *FAA.gov*, last modified January 9, 2013, http://www.faa.gov/about/office_org/headquarters_offices/apl/research/aircraft_technology/cleen.
59. Ibid.
60. Katherine Blakeley, *DOD Alternative Fuels: Policy, Initiatives and Legislative Activity* (Washington, D.C.: Congressional Research Service, 2012), ii.
61. European Commisson, *Flightpath 2050: Europe's Vision for Aviation* (Luxembourg: Publications Office of the European Union, 2011).
62. Sustainable Way for Alternative Fuels and Energy in Aviation, "State of the Art on Alternative Fuels and Energy," http://edepot.wur.nl/180370.
63. ICAO, *ICAO Environmental Report 2013*, 18–26.
64. "Decision of the EEA Joint Committee No 50/2013 amending Annex XX (Environment) to the EEA Agreement," *Official Journal of the European Union* L 231/24 (2013), http://www.efta.int/sites/default/files/documents/legal-texts/eea/other-legal-documents/adopted-joint-committee-decisions/2013%20-%20English/050-2013.pdf
65. See the Commercial Alternative Aviation Fuels Initiative (CAAFI) website at http://www.caafi.org/.
66. International Air Transport Association, *IATA 2012 Report on Alternative Fuels* (Montreal: IATA, 2012), http://www.iata.org/publications/Documents/2012-report-alternative-fuels.pdf.

Bibliography

Air Transport Action Group. "Facts and Figures." *ATAG.org*. Last modified March 2012. http://www.atag.org/facts-and-figures.html.

American Clean Skies Foundation. "Driving on Natural Gas: Fuel Price and Demand Scenarios for Natural Gas Vehicles to 2025." Washington, D.C.: American Clean Skies Foundation, 2013.

Anderson, B. E., A. J. Beyersdorf, C. H. Hudgins, J. V. Plant, K. L. Thornhill, E. L. Winstead, L. D. Ziemba et al. *Alternative Aviation Fuel Experiment (AAFEX)*. Hampton, VA: NASA Langley Research Center, 2011. http://nix.nasa.gov/search.jsp?R=20110007202&qs=N%3D4294966788%2B4294724624%26No%3D10.

Bandivadekar, Anup. *Transportation Energy Challenge*. Washington, D.C.: The International Council on Clean Transportation, 2008. http://csis.org/files/attachments/090608_bandivadekar.pdf.

Bandivadekar, Anup, Kristian Bodek, Lynette Cheah, Christopher Evans, Tiffany Groode, John Heywood, Emmanuel Kasseris, Matthew Kromer, and Malcolm Weiss. *On the Road in 2035: Reducing Transportation's Petroleum Consumption and GHG Emissions*. Cambridge: Massachusetts Institute of Technology, 2008. http://web.mit.edu/sloan-auto-lab/research/beforeh2/otr2035/On%20the%20Road%20in%202035_MIT_July%202008.pdf.

Berman, Brad. "Victor Wouk and the Great Hybrid Car Cover-up of 1974." *Hybridcars.com*. Last modified on March 28, 2006. http://www.hybridcars.com/the-great-hybrid-car-cover-up-of-74/.

Blakeley, Katherine. *DOD Alternative Fuels: Policy, Initiatives and Legislative Activity*. Washington, D.C.: Congressional Research Service, 2012.

Brown, Nate. "Sustainable Alternative Jet Fuels: Update on ASTM Approval." *Federal Aviation Administration* (May 18, 2012). http://www1.eere.energy.gov/bioenergy/pdfs/10_brown_roundtable.pdf.

Bureau of Transportation Statistics. "National Transportation Statistics." *U.S. Department of Transportation*. Last updated October 2013. http://www.rita.dot.gov/bts/publications/national_transportation_statistics/index.html.

CAFE Foundation. "The 2011 Green Flight Challenge Sponsored by Google." *Cafefoundation.org*. Last modified 2013. http://cafefoundation.org/v2/gfc_main.php.

Cai, H., and S. D. Xie. "Temporal Variations and Spatial Distribution of Greenhouse Gas Emissions from On-road Vehicles in China." Paper presented at the 2009 Sustainable Energy and Environmental Protection Conference, 12–15 August 2009, Dublin, Ireland.

Campestrini, Martin, and Peter Mock. *European Vehicle Market Statistics*. Washington, D.C.: The International Council on Clean Transportation, 2011. http://www.theicct.org/sites/default/files/publications/Pocketbook_LowRes_withNotes-1.pdf.

Center for Climate and Energy Solutions. "Renewable and Alternative Energy Portfolio Standards." *www.c2es.org*. Last updated 2013. http://www.c2es.org/us-states-regions/policy-maps/renewable-energy-standards.

Davis, S. C., S. W. Diegel, and R. G. Boundy, *Transportation Energy Data Book, ORNL-6989* (Edition 32 of ORNL-5198). Oak Ridge, TN: Oak Ridge National Laboratory, 2013.

Eberhardt, James J. "Fuels of the Future for Cars and Trucks." Presentation made at the 2002 Diesel Engine Emissions Reductions (DEER) Workshop, San Diego, August 25–29, 2002. http://www1.eere.energy.gov/vehiclesandfuels/pdfs/deer_2002/session1/2002_deer_eberhardt.pdf.

Ehsani, Mehrdad, Yimin Gao, and Ali Emadi. *Modern Electric, Hybrid Electric, and Fuel Cell Vehicles: Fundamentals, Theory, and Design* (2nd Ed.). Boca Raton, FL: CRC Press, 2010.

Energy Information Administration. *Annual Energy Outlook 2013*. Washington, D.C.: US Energy Information Administration, 2013. http://www.eia.gov/forecasts/aeo/.

Energy Information Administration. *Annual Energy Review*. September 2012. http://www.eia.gov/totalenergy/data/annual/.

Energy Information Administration. "Petroleum & Other Liquids: Product Supplied." *EIA.gov*. Last updated September 27, 2013. http://www.eia.gov/dnav/pet/pet_cons_psup_dc_nus_mbbl_a.htm.

Environmental Protection Agency. *Unit Conversions, Emissions Factors, and Other Reference Data*. Washington, D.C.: Environmental Protection Agency, 2004. www.epa.gov/cpd/pdf/brochure.pdf.

European Commission. *European Commissions' and Member States' R&D Programmes for the Electric Vehicle*. Report on a European Commission Workshop (November 15, 2009). http://www.green-cars-initiative.eu/workshops/Report_WS_EC-MS_Electric_Vehicle_R-D.pdf.

European Commisson. *Flightpath 2050: Europe's Vision for Aviation*. Luxembourg: Publications Office of the European Union, 2011.

Federal Aviation Administration. "Continuous Lower Emissions, Energy, and Noise (CLEEN) Program." *FAA.gov*. Last modified January 9, 2013. http://www.faa.gov/about/office_org/headquarters_offices/apl/research/aircraft_technology/cleen.

Federal Aviation Administration. *Destination 2025*. Washington, D.C.: Federal Aviation Administration, 2011. http://www.faa.gov/about/plans_reports/media/Destination2025.pdf.

Global Change. "What's the Carbon Footprint of Building Your Car, and How does that Compare to Tailpipe Emissions?" *Global Change Blog* (September 24, 2010). http://www.globalchangeblog.com/2010/09/whats-the-carbon-footprint-of-building-your-car-and-how-does-that-compare-to-tailpipe-emissions/.

Gomes, Carlos. "Global Auto Report." *Scotia Bank* (December 5, 2013). http://www.gbm.scotiabank.com/English/bns_econ/bns_auto.pdf.

International Air Transport Association. *IATA 2012 Report on Alternative Fuels*. Montreal: IATA, 2012. http://www.iata.org/publications/Documents/2012-report-alternative-fuels.pdf.

International Civil Aviation Organization. *Annual Report of the Council: 2011*. Montreal: ICAO, 2012. http://www.icao.int/publications/Documents/9975_en.pdf.

International Civil Aviation Organization. "Carbon Emissions Calculator." *ICAO.int* (2013). http://www.icao.int/environmental-protection/CarbonOffset/Pages/default.aspx.

International Civil Aviation Administration. *ICAO Annex 16: Environmental Protection, Volume II; Aircraft Engine Emissions*. Montreal: ICAO, 2008.

International Civil Aviation Organization. *ICAO Environmental Report, 2013: Aviation and Climate Change*." Montreal: ICAO, 2013. http://cfapp.icao.int/Environmental-Report-2013/#210/z.

International Energy Agency. *Energy Technology Perspectives 2010*. Paris: IEA, 2010.

International Energy Agency. *EV City Casebook*. Paris: IEA, 2012. https://www.iea.org/publications/freepublications/publication/EVCityCasebook.pdf.

International Energy Agency. *Global EV Outlook: Understanding the Electric Vehicle Landscape to 2020*. Paris: IEA, 2013. http://www.iea.org/publications/freepublications/publication/GlobalEVOutlook_2013.pdf.

International Energy Agency. *Technology Roadmap: Electric and Plug-in Hybrid Electric Vehicles.* Paris: IEA, 2009.

International Energy Agency, *Transport, Energy and CO2: Moving Toward Sustainability.* Paris: OECD/IEA, 2009.

International Transport Forum. *Reducing Transport Greenhouse Gas Emissions: Trends & Data 2010.* Paris: OECD Publishing, 2010.

Laurent, Fabien and Elisabeth Windisch, "Triggering the Development of Electric Mobility: A Review of Public Policies," *European Transport Research Review* 3, no. 4 (2012): 221–35. http://dx.doi.org/10.1007/s12544-011-0064-3.

Leggett, Jane A., Bart Elias, and Daniel T. Shield. *Aviation and the European Union's Emission Trading Scheme.* Washington, D.C.: Congressional Research Service, 2012.

McKinsey & Company. *Roads Toward a Low-carbon Future: Reducing CO_2 Emissions from Passenger Vehicles in the Global Road Transportation System.* New York: McKinsey & Co., 2009.

Ministry of Agriculture and Food. *Energy Factsheet: Propane as a Fuel for Vehicles and Tractors*, Order No. 210.510–2. Victoria, BC: Ministry of Agriculture and Food, British Columbia, 1982. http://www.agf.gov.bc.ca/resmgmt/publist/200Series/210510-2.pdf.

Moran, Michael J., and Howard M. Shapiro. *Fundamentals of Engineering Thermodynamics*, 5th ed. New York: John Wiley and Sons, 2004.

Mustoe, Howard. "Airbus Raises A380 Price by 8.4% as Model Beats Airline Cost-Savings Goal." *Bloomberg* (January 18, 2011). http://www.bloomberg.com/news/2011-01-18/airbus-raises-a380-price-by-8-4-as-model-beats-airline-cost-savings-goal.html.

Office of the Press Secretary. "Fact Sheet: Energy Independence and Security Act of 2007." *Whitehouse.gov* (December 19, 2007). http://georgewbush-whitehouse.archives.gov/news/releases/2007/12/20071219-1.html.

Organisation International de Construceurs d'Automobiles (OICA). "Production Statistics." http://www.oica.net/category/production-statistics/.

Pasaoglu, Guzay, Michel Honselaar, and Christian Thiel, "Potential Vehicle Fleet CO_2 Reductions and Cost Implications for Various Vehicle Techology Deployment Scenarios in Europe," *Energy Policy* 40 (2012): 404–21. http://dx.doi.org/10.1016/j.enpol.2011.10.025.

Pike, Ed. *Calculating Electric Drive Vehicle GHG Emissions.* Washington, D.C.: The International Council on Clean Transportation, 2012. http://www.theicct.org/calculating-electric-drive-vehicle-ghg-emissions.

Pollet, Bruno G., Iain Staffell, and Jin Lei Shang. "Current Status of Hybrid, Battery, and Fuel Cell Electric Vehicles: From Electrochemistry to Market Prospects." *Electrochemica Acta* 84 (2012): 235–49. http://dx.doi.org/10.1016/j.electacta.2012.03.172.

Research and Innovative Technology Administration (RITA). "Airline Fuel Cost and Consumption (U.S. Carriers)." *US Department of Transportation*, 2012. http://www.transtats.bts.gov/fuel.asp?pn=1.

Serra, João Vitor Fernandes. *Electric Vehicles: Technology, Policy and Commercial Development.* New York: Earthscan, 2012.

Sperling, Daniel, and Deborah Gordon. *Two Billion Cars: Driving Toward Sustainability.* New York: Oxford University Press, 2009.

Westbrook, Michael H. *The Electric Car: Development and Future of Battery, Hybrid, and Fuel-cell Cars.* London: The Institution of Electrical Engineers, 2005.

Webster, Ben. "Electric cars may not be so green after all, says British study." *The Australian* (June 10, 2011). http://www.theaustralian.com.au/news/health-science/electric-cars-may-not-be-so-green-after-all-says-british-study/story-e6frg8y6-1226073103576.

Chapter 10
Policy Challenges for the Built Environment: The Dilemma of the Existing Building Stock

MARK SHAURETTE

Abstract

The built environment accounts for approximately forty percent of the total energy consumption in developed countries. Because buildings have a long life, the greatest opportunity for energy reduction in the built environment will come from energy conservation in the existing building stock. An overview of the policy challenges presented by the built environment, with an emphasis on existing facilities, is accompanied by a discussion of specific technologies that may have the potential to reduce energy use. To illustrate the degree of complexity associated with shifts to new technologies, lighting, a major consumer of electric energy in the built environment, is described in an expanded narrative by Kevin Kelly. The chapter concludes with a discussion of recent policy schemes that have been employed in the United States and the European Union to promote energy conservation in the built environment. Both voluntary and public policy programs are included, along with an examination of resultant successes and failures. Based on this discussion, a series of recommendations and opportunities for future solutions is provided for each of the challenge areas presented earlier in the chapter.

10.1. Introduction

Energy use in the built environment comprises approximately forty percent of the energy consumed in developed countries. In 2004, the emissions resulting from direct energy use in the built environment were about 8.6 Gt of CO_2 per year. Through the use of mature technologies, building energy use can be reduced substantially. Due to the long life of buildings, energy policy aimed at promoting building energy conservation in existing buildings is confronted by significant technical challenges. New construction produces substantially more efficient structures, but there are obvious limits to the rapid replacement of inefficient buildings. In addition, the complexity of design and construction required for the multitude of unique structures makes a one-size-fits-all solution impossible.[1] These and other barriers make public policy and the integration of energy conservation measures in the built environment an interesting and challenging area of study. The intersection of technology and public policy in reducing energy use and environmental impact of buildings necessitates integrated thinking by technologists, designers, managers, the business community as well as those who control and influence public policy.

The Urban Land Institute (ULI), a nonprofit research and education organization in the United States, examined some of the complex issues surrounding climate change, energy use, and the way leaders working in real estate and the built environment are challenged in this complex area. The authors emphasize the fact that "Even at the peak of recent building cycles, only 2 percent of the total existing floor space annually is added by new commercial building construction. In the years ahead, this portion is likely to remain below 1 percent. . . . It is the balance of buildings—the overwhelming majority of the existing building stock—that remains the dominant untapped market opportunity to invest in energy efficiency."[2]

The sheer size of this challenge can be appreciated by examining an identifiable segment of the built environment. In 2003, a total of nearly 4.9 million commercial buildings in the United States, comprising more than 71.6 billion square feet of floor space, consumed more than 6,500 trillion Btu of energy.[3] The critical portion of existing commercial buildings is the considerable inventory of structures constructed during the years between World War II and the late 1970s (about forty percent of the total commercial space). These structures were built prior to the energy use reduction efforts that became more common through changes in design, materials, and construction practice brought about by increasing energy costs in the mid to late 1970s. Most of the buildings from this era are reaching the end of their designed economic life (forty to sixty years in the US) and are prime candidates for retrofit or reconstruction.

Residential buildings also contribute significantly to the energy crisis. In addition to the challenge of replacement over an extended time period, homes also present complications due to the variety of owners and occupants. The type and degree of challenges can also vary by climate, age of the structures, and ownership patterns. This is a global challenge. For example, in 2006 more than fifty percent of Ireland's existing domestic buildings were built before the first thermal energy insulation standards were put in place. As a result Irish homes consume thirty-one percent more energy per dwelling than the EU-15 average and thirty-six percent above the EU-27 average. Since residential structures represent twenty-three percent of total final energy consumption in Ireland, improvements in residential energy efficiency holds significant potential for reducing overall energy use.[4] This is a major concern because Ireland has relied on imported energy for approximately ninety percent of its total energy needs between 2001 and 2011.[5]

While the pursuit of technological solutions to the energy crisis through renewable energy production has long-term merit, a more immediate step is to minimize building operational energy consumption. The US National Academy of Sciences concluded that well-designed policies by nations and states can result in substantial energy savings.[6] What remains is for nations and states to reach agreement on the best combination of standards, design, research, financing, incentives, and education as well as the optimal combination of technologies needed to maximize energy conservation efforts in the built environment.

10.2. Energy Conserving Building Retrofit Technologies

A thorough discussion of the technological options available for energy conserving retrofit of existing buildings is beyond the scope of this book. Nevertheless, some basic understanding of the options available is necessary as an introduction to policy discussion. This section is devoted to basic details about several categories of technology for building energy conservation. The complicating factors of human behavioral impact on technology performance are excluded to simplify the discussion. Readers should review this material with the understanding that consumer decisions to adopt or not adopt energy conserving technologies are not always rational and may be influenced by factors other than cost and benefit evaluations.[7]

10.2.1. Temperature, Ventilation, and Humidity

10.2.1.1. Insulation

One of the most universal categories of energy conserving technologies is the use of insulating products to reduce the thermal losses through the exterior structure of a building. Insulation can range from very simple to install to very complex. Material and labor costs also vary widely. This variability is intensified by the diversity of building types and the inaccessibility of the space within the walls and ceilings of existing buildings where insulation is typically installed. In addition, a significant lack of understanding of the building science behind the performance of insulating products by building trades often leads to inefficiency or even damage due to condensation resulting from improper installation.

Insulation of existing structures may require damage repair or replacement of interior or exterior finish surfaces. In addition, there is a very wide range of choice and performance of insulating products that can be confusing for contractors and consumers. Fortunately, insulation products have a long life. Any existing building insulation project should strive to maximize the insulation level achieved. This will avoid the need for future disruptions and a repeat of surface finish damage caused by attempts to increase the insulating value at a later date. Under the appropriate conditions, a properly implemented addition of insulation is an important component of any energy conserving building retrofit. In addition to the reduced energy loss and elimination of thermal bridges that can result, occupants can expect increased comfort when insulation is properly installed.

10.2.1.2. Air Sealing

An often overlooked but critical component of any attempt to reduce building energy consumption is air sealing the structure. Although there are many reasons that buildings need to be ventilated, the ventilation process is more effective and efficient if outside air is introduced in a controlled manner. The penalty for uncontrolled building ventilation in cold climates can be as high as, or higher than, the energy lost from heat conducted through the building structure. As buildings become more highly insulated, the energy cost penalty for air leakage becomes an even higher percentage of the building operating energy load. Ventilation as part of the heating and cooling strategy for the structure is preferable to letting the building breathe through uncontrolled air infiltration and exfiltration.

Techniques for air sealing are typically low-technology caulks, tapes, and air barriers. A critical component of air sealing is assuring that the tradesmen completing the air sealing understand how to apply the appropriate technology, as well as quality control of the final installation. An understanding of building science principles can help mitigate the potential for moisture problems that may be created by extensive air sealing. Unsealed forced-air heating and cooling ductwork located outside of the heated and cooled portion of the structure is a common source of air leakage. Inaccessible ductwork can be a challenge to seal. Recent advances in the use of adhesive coated particles sprayed into leaky ductwork to seal leaking joints shows promise for commercial application.

10.2.1.3. Heating and Cooling Systems

Systems for heating and cooling a structure use different technologies based on climate, location, available energy sources, building type, cultural norms, regulatory standards, contractor experience, or even regional availability of system components. Systems can be serviced by centralized heating and cooling plants that supply a district of buildings or all occupancies of a single large building. Alternatively systems are available to serve a single occupancy independent of others in a building or district. District systems offer efficiencies of scale but may suffer from distribution losses. District systems typically utilize steam or water to distribute heating or cooling to buildings being supplied. Increasing boiler or chiller efficiency, distribution efficiency or advanced control systems are key retrofit opportunities in these distributed systems. Advanced technical solutions for district systems include efforts to maximize the output from fossil fuel consumption, for example by simultaneously extracting both heat and energy. The consistent temperature deep in the ground can also be used as a heat energy source or heat sink for cooling to increase system efficiency.

Heating and cooling systems for single occupancies typically use forced air circulation or water circulation to heat or cool the occupied space. Many of these systems burn fossil fuels in a central furnace or boiler. Burning biomass is an alternative. The thermal efficiency of new equipment can exceed ninety-five percent. The fact that older equipment operates at sixty percent efficiency or less can be justification for the replacement of currently operational equipment. Increasing the efficiency of forced air or hydronic (water) distribution systems is often limited by restricted access to the distribution system. Nevertheless, fan motor upgrades, variable fan speed control, pump replacements, insulation of distribution ducts or pipes, and sealing duct leakage all offer energy efficient retrofit opportunities.

Heating by electricity is also used in single occupancy systems. The simplest systems are inexpensive, easy to control, and capable of supplying heat directly to any space through electrically heated elements. Unfortunately these basic systems suffer from distribution losses inherent with electricity. As a result they have high operating costs and impose significant loads on electric grids. Heat pumps, which use compressed gas as a means to transfer energy from the outside air to the inside air (or the reverse for air conditioning), are an alternative employed to reduce energy use in electric systems. Heat pump systems are capable of utilizing energy from outside air until the outside air temperature drops well below the freezing point of water. The efficiency of energy capture for a heat pump will fall as the outside temperature decreases, making them impractical in extremely cold climates. Heat pumps can also be installed as *ground source heat pumps* that use the warmer and more consistent deep ground temperature rather than the outside air as the energy source. Heat pump systems are desirable where fossil fuels are hard to supply to the building. As a retrofit option, heat pumps have limitations because they require the installation of a distribution system, usually using forced air, for occupancies that contain multiple rooms.

10.2.1.4. Control Systems

Control of temperature, ventilation, and humidity are typical features of modern heating and cooling systems. Opportunities for control retrofits are presented as a separate category to emphasize their unique importance. Temperature control in older buildings is usually limited to maintaining a set temperature. Because many spaces are only occupied for a portion of the day, retrofit opportunities exist in control systems with automatic temperature set-point adjustments that maintain the most comfortable conditions only when needed. Ventilation, a requirement for human habitation and health, is best controlled through automatic mechanical systems that capture the heat available in the exhaust air and use it to warm the incoming ventilation air. Heat recovery ventilation systems can capture as much as eighty percent of the exhaust heat. Adequate humidity control is important for occupant comfort. Humans prefer relative humidity in the forty to sixty percent range. When humidity is maintained in this comfort range, energy savings can be obtained because occupants are usually satisfied with cooler air temperatures in the winter and warmer air temperatures in the summer. In addition, adequate humidity control can help prevent building damage from excess moisture that can accumulate in well insulated and sealed structures.

10.2.2. Windows

The thermal performance of windows has improved substantially through the use of multiple layers of glass, inert gases sealed between the glass layers, low emissivity glass coatings, thermal break spacers between the glass layers, thermal break window frames, and improved gasketing on operable window sashes. Window replacement is often part of building retrofit strategies, but high cost can limit the viability of window replacement.

Window replacement may also be limited by architectural requirements in historic buildings. When cost or the desire to maintain the original appearance of a window is a concern, window repair using double glazing or the use of a separate storm sash can be considered. Because some older windows do not have adequate thickness to accept double glazing, the addition of a storm sash may be needed. An internally mounted storm sash can provide substantial energy loss reduction and improved occupant comfort by raising the interior surface temperature of the glass. The benefits of a storm sash can be maximized when the storm sash glazing has a hard coat low emissivity (Low E) coating.

Windows are also impacted by radiant energy that passes through the glass. Radiant energy passes through the window glass, striking interior surfaces where it is converted to heat energy. This may be beneficial in winter months but detrimental during summer time. Shading strategies can be employed to maximize the passive solar heat gain in the winter but shade the windows in the summer when the heat gain would be undesirable. Existing buildings may be limited in the ability to effectively capture passive solar heating because window locations are fixed and horizontal shading may be difficult to add architecturally. Window coverings and tinted or reflective glass coatings are an option in these cases. Advanced technologies which darken the window glazing based on an applied electric current, change in temperature or light intensity are available but have limited commercial availability.

10.2.3. Domestic Hot Water

Hot water production for domestic use is a significant component of energy use in residential buildings as well as in commercial buildings that require hot water for washing or industrial processes. The most beneficial hot water retrofits reduce both energy and water quantity demand through improved washing machines, dishwashers, and process equipment. Additional savings can come from more efficient production and storage of hot water. Hot water use tends to be intermittent. To meet the fluctuating demand, water heaters typically use a storage tank to maintain a ready supply of hot water. Heavily insulating the storage tank is often a simple and inexpensive retrofit option. Efficient natural gas water heaters are available for replacement of older electric water heaters. Other alternatives are tank-less water heaters which avoid the standby tank losses but may have limitations in hot water capacity under heavy demand. When an electric water heater is the only option, a heat pump water heater can be used. Heat pump water heaters can utilize the heat obtained from the central heating heat pump or from a stand-alone heat pump which extracts energy from its surroundings to heat the water. Stand-alone heat pump water heaters which operate within the building enclosure provide some limited air conditioning for the occupants.

10.2.4. Lighting

Lighting is often the largest consumer of electricity in a commercial building and can also be a significant energy draw for a residential dwelling. Long reliance on incandescent lighting in residences allows easy retrofit to more efficient compact fluorescent lighting (CFL) systems. In commercial buildings ,more efficient fluorescent fixtures have been used for years. Unfortunately the T-12 fluorescent fixtures with magnetic ballasts that were installed in the past are no longer a reasonable choice. Easily obtained T-8 or T-5 high efficiency fixtures with electronic ballasts can be used to reduce energy consumption by as much as fifty percent. Use of natural lighting for illumination (day-lighting) in combination with task lighting, proper levels of background illumination, and sensor controlled supplemental lighting can be useful in reducing energy consumption when windows are available. Exterior lighting and areas not used continuously can benefit from timer controls or occupancy sensors to turn off lighting when not needed. Retrofit using high efficiency light emitting diode (LED) lighting has recently become available (see sidebar on the next page for expanded lighting discussion).

10.2.5. Additional Options

Improvemetns in energy efficiency can be obtained via additional means such as replacement or retrofit of commercial refrigeration systems, home appliances, passive design principles, passive cooling techniques, the introduction of thermal mass, impacts of building orientation or vegetation, as well as building operations and maintenance improvements.

It is appropriate to emphasize the importance of retro-commissioning. Commissioning is the terminology used to describe the process of quality assurance for new systems to ensure they are operating at design parameters and in turn at their highest efficiency level. Older systems frequently become misadjusted over time and therefore require retro-commissioning. Retro-commissioning can be applied to most if not all of the non-passive energy related technologies in a building. Occupant training should be a part of this process. Occupant behavior has a major impact on energy consumption. Even if the original occupant understood the most efficient operation of the building systems, it is very likely that over time retraining will become necessary as occupants change. Retro-comission of heating and cooling systems, for example, should always be completed as part of any building retrofit

10.3. Complexity of Energy Efficiency Retrofit Strategies

A host of issues influence and complicate decisions regarding how, when and where to implement energy efficiency retrofits, including a wide array of non-technical factors. These factors relate to building characteristics as well as to the impact on owners, occupants, and stakeholders involved in the retrofit process. Even a well-planned community-wide retrofit scheme may entail extensive examination or nearly every building as an individualized custom project. This requires understanding, environmental focus, and monetary incentives linking yearly energy savigns to investment in green and efficient technologies.

The choice of which technologies to utilize is highly dependent on climate and the physical characteristics of the structure. Even residential structures built using standardized materials and designs are prone to modification through their lifetime. The

Energy Saving Developments in Lighting

Kevin Kelly

Recent developments in artificial lighting design, lamp technology and control options provide potential for significant energy savings going forward. Historically, equal illuminance across the whole *working plane* was the goal of lighting designers, however this is now considered wasteful of energy. For example, in an office setting the working plane was interpreted as the whole plan area of the room at desk height; 300 to 500 lux was specified, depending on whether work was mainly PC based or paper based. This resulted in arrays of lights that provided high levels of lighting throughout the space, whether needed or not, and often for periods extending beyond the working day, as evidenced in large cities where empty office blocks had lights switched on well into night hours. This criterion of near equal illuminance across a working plane also tended to lead to rather boring and monotonous interiors. Today such energy inefficiency is unacceptable. LED lamp development also provides potential for energy savings as these lamps replace less efficient lamps.

New recommendations, such as those specified in the SLL Code for Lighting 2012,[8] offer pragmatic design advice to ensure adequate and efficient lighting while maintaining balance in financial outlay (purchase, energy cost, and end-of-life disposal) and environmental impact (electricity load, chemical pollution, and light pollution at night). The code is based on quantitative recommendations that meet minimum lighting requirements but also acknowledges that there is a need to target lighting more carefully and address quality issues. For example, modelling of people in offices to ensure good visual interaction becomes important and good quality lighting and energy efficiency are now as important as quantitative specifications about light levels.

Good quality and efficient lighting in buildings starts with the need to maximize daylight penetration. Maximizing daylight offers opportunities to lift the spirit with natural light and so daylight must be carefully designed into a building in tandem with the artificial electric lighting and controls to create good quality efficient lighting in the space. Human beings have a preference for natural light over artificial light and side lit interiors often automatically offer good modelling by providing a strong cross vector of light. This means that people can see other people more easily as light falls on their faces from the side windows. More recently, the need to maximize daylight is also driven by the necessity to reduce energy used by electric lighting. Maximizing daylight and minimizing energy used by electric lighting must take place in a way which minimizes overall energy consumption in the building. It is counterproductive to maximize daylight in order to reduce light energy consumption if thermal energy requirements increase due to the need for extra heating or cooling. It should be noted that extra glazing will increase heating load in winter and cooling load in summer, whilst electric lighting can also contribute significantly to building cooling load requirements. A balance needs to be sought with building type, method of construction, orientation, and occupation, usage and location.

Daylight availability charts can be used to conclude that there is an external illuminance of in excess of 10,000 lux for seventy percent of the office working day in London.[9] This suggests that a room with a five percent daylight factor would have an average illuminance of 500 lux minimum for seventy percent of the working day. The artificial lighting in a space with this level of daylight might be turned off or at least dimmed without any significant disadvantage to work efficiency in such an area. A room with this level of daylight factor (above five percent) would merit consideration of daylight detection. This should be incorporated into an automatic control system. Experience to date indicates that without such an automatic control system, the potential energy saving benefit of daylight is unlikely to be fully realized. Ensuring user satisfaction throughout the working day would require integration of the lighting control system in an acceptable way to ensure lights are on when needed and off or dimmed at appropriate times. It is important that clients and facilities managers are adequately briefed about the operation of the automatic control system, in order to ensure optimal operation while realizing effective energy savings.

While standards, demands, and design methodologies change, major change is also underway in lamp technology. It is notable that the development of solid state lamp technology is revolutionizing the lighting industry. As with many revolutionary step changes in development and use of new technologies, there has been collateral damage to early adaptors of poor-quality light emitting diode (LED) lamps. However, the pace of growth of this technology is exponential and it is still at an early stage in development. In a study by Philips Lighting it is estimated that while only six percent of lighting was solid state in 2010, seventy-five percent of lighting is expected to be LED lighting by 2020.[10] At present the biggest applications of LED lighting are for stage, external lighting, architectural lighting, retail, cold rooms, transport, and hospitality. Going forward, LED lamp technology is expected to impact office and general interior lighting, but what is the current status? Exaggerated performance of LEDs by some newly emerging companies has resulted in disappointment among clients who have expressed growing skepticism. Lighting designers complain that there are not sufficient and reliable specifications underpinning LEDs, which places risk on the designers who specify them and the contractors who install them. Lighting manufacturers respond that the technology is evolving at such a fast rate that it is pointless to create specifications that are out of date as soon as they are printed. They also point out that it is impossible to reliably guarantee and measure lamp life-cycle; LED lamps should typically last in excess of eleven years (up to 100,000 hours) of constant use. At present, measurements are recorded over a time period of 9000 hours and life expectancy results are based not on lamp failure but on an accepted minimum level of lux depreciation, with data extrapolated for longer periods of time.

Present development of LED technology suggests that the efficacy of these lamps is soon to surpass even the most efficient fluorescent lamps; in the near future it is also likely to surpass the monochromatic Low Pressure Sodium (SOX) lamp used on motorways and in similar applications. McKinsey estimates that global revenue for LED lighting will be €65 billion by 2020 and LED usage will be over sixty percent of the entire market.[11] This is consistent with similar forecasts by Philips above. It is proposed there will be a focus shift from lamp replacement to fixture replacement. With fluorescent lamps, the luminaire is likely to last for a couple of decades and lamps will be replaced very cheaply every couple of years. LEDS on the other hand come hand in hand with the luminaire and if one needs replacing, usually the other does also. This raises questions about life cycle and replacement cost considerations. When replacing the whole luminaire, it is unlikely the same unit will still be manufactured due to the rapid developments in this area. This will mean that all luminaires in a space must be replaced once lamps begin to fail or their output drops markedly. The question must also be raised as to why one would replace a highly efficient fluorescent luminaire, whose lamps are providing in excess of 100 lumens per watt, with a much more expensive LED luminaire with lamps of a similar efficacy, especially when they are so expensive to buy at present. Interior lighting relies on inter reflected lighting to create an acceptable visual ambiance. Considerable light falls on walls and ceilings through reflection. However, some direct application of light onto an object or surface can create a more visually appealing and stimulating environment. At present it is this directional light characteristic of LEDs, providing color variation and visual stimuli, which provides great potential for indoor use. However, as previously mentioned, poor quality, relatively cheap LED lamps have fallen short of expectations to date. Poor heat dissipation has also been a limitation. Low-cost, modern T5 fluorescent lamps provide 100 lumens per watt, with very good color rendering and a variety of color temperatures. The long history and successful application of these fluorescent luminaire lamps enables them to retain the pole position for the general interior lighting market at present.

The cooler color temperature of many LED lamps is deemed unacceptable by many home owners and other users. The generally more appealing warmer color LEDs are available but are usually much more expensive. The present high cost of good quality LED lamps and luminaires along with the above may delay their widespread use for interior lighting. LEDs may be the future for interior lighting but they are not yet the optimal choice. However, owing to their directional accuracy, LEDs may be more suitable for many applications including outdoor use. There is a lack of reliable research in this area at present, and this needs to be addressed going forward. LEDs may also form a useful alternative to traditional lighting in future indoor applications particularly as the tendency to flood light onto a general working plane is replaced by more individual targeting of light on a specific set of task areas.

> This is an exciting and challenging time for the lighting industry with good potential for LED lighting and improved lighting controls generally. The challenge is to provide robust solutions that will maximize the benefits of new technologies whilst protecting clients from poor quality products and installations. A further goal will be to maximize light quality and minimize energy use by integrating daylight with appropriate artificial light in a way that lifts the spirit of those using the space with easy facilitation to operate and override automatic lighting controls when required. Product reliability and integrated standards will be required in order to leverage the benefits of new technologies and in so doing help reduce energy use, improve upon energy efficiency, and contribute to reduction of greenhouse gas emissions.

resulting variation from structure to structure limits one-size-fits-all solutions. In addition, the selection of one technology can impact the performance or specification of a completely separate part of a whole-building retrofit strategy. An example of this phenomenon is the influence of the level of insulation on the design and performance of mechanical heating and cooling systems. Increasing the insulation level reduces the required capacity of a building's heating, ventilation and air conditioning (HVAC) system. Installing an oversized HVAC system can negatively impact the system's efficiency, equipment life and occupant comfort. The timing or staging of installation can introduce additional complexities and potentially lead to inefficient choices. Failing to increase insulation levels at the same time the HVAC equipment is replaced results in equipment that does not match the actual heating and cooling needs of the building. The best retrofit technology upgrade strategies seek to implement all possible technologies on a whole-building basis.

Goals and expectations of owners, occupants, policy makers and taxpayers, who may subsidize or incentivize retrofit activity, can vary. What level of occupant comfort is expected? To what degree are energy use, CO_2 emissions, and property value important? Are project costs justified by cash flow, payback period, investment return or carbon reduction goals? How are the interests of those who own a rented building, and presumably pay for the retrofit, cost balanced with the interests of occupants who will benefit from the reduced cost of energy used? The answer to each of these questions is probably different depending on which stakeholder group is questioned. Some useful guidance for policy makers in examining the cost vs. benefit impacts on various stakeholders in building energy efficiency programs is provided by the US Environmental Protection Agency, which offers five principal approaches to guide public utility ommissions, city councils, and utilities. They are careful to state that "there is no single best test for evaluating the cost-effectiveness of energy efficiency." If a single cost-effectiveness measure is used it may not balance the costs and benefits of all stakeholders.[12]

In addition to payback or cost vs. benefit considerations, the source of funds can influence retrofit decisions. Grants to promote energy conservation typically require some form of decision oversight by the funding agency in order to maintain adequate control over the appropriate use of the funds. Funding agencies often lack sufficient understanding of the complex technologies involved in energy related retrofits to adequately provide oversight. This can lead to rigid guidance which may not meet the specific conditions of every structure. The use of partial funding through tax incentives, rebates or bank loans can permit greater flexibility for more building-specific solutions.

Regardless of the source of funds, the fragmented nature of construction contracting for residential and small commercial buildings provides challenges to successful energy conserving retrofits. Building contractors typically do not have a thorough understanding of all available technology solutions or the underlying science. Even when trained designers, such as architects or engineers are involved, retrofits intended to reduce air leakage often result in problems with elevated moisture levels in the structure. The challenge is complicated by the fact that building contractors typically hire an assortment of subcontractors to work with the specialized crafts involved. Each of these crafts lack an understanding of how their use of technology interacts with the work performed by other subcontractors. An outside party to supervise technical specification and quality control may be needed.

The strategies described to this point have concentrated on the reduction of energy use during the occupancy of the building, or the *use phase.* Although a significantly smaller component of the lifetime environmental impact of the building, many experts advocate that the energy associated with the building's *construction phase* be included in energy efficiency decision making. This *embodied energy* represents the sum of all the energy required to construct the building, including energy consumed to mine raw materials, create the building materials, construct the building, and to demolish and dispose of the building itself at the end of its life. By including embodied energy, the cost vs. benefit considerations can encompass the full life cycle cost of the building. Currently the availability of data required to accurately calculate embodied energy for complex buildings is limited, but the body of knowledge for life cycle cost analysis is developing rapidly.

10.4. Policy Challenges to Energy Efficiency Retrofit Success

The multi-dimensional and complex influences on energy retrofit decisions create many barriers to the creation and implementation of policies designed to promote and facilitate building energy efficiency. In 2010 the US Department of Energy (DOE) published a report from the National Renewable Energy Laboratory (NREL) titled *Summary of Gaps and Barriers for Implementing Residential Building Energy Efficiency Strategies.* The report's authors outline barriers and opportunities for market transformation for future reduction in building energy use.[13] Some of the significant challenges for residential energy conservation include:
- Limited training and certification for contractors,
- Difficulty in evaluating the quality or quantity of benefits other than simply by cost,

- Complexity created by regulations and incentives intended to promote conservation,
- Lack of reliable energy savings information to build stakeholder confidence,
- The disconnect between cost and building value created by energy conserving interventions,
- The need to verify performance based on sound building science principles,
- Incomplete understanding of homeowner motivation for energy conservation investments,
- Limited energy conservation knowledge of those who influence residential investment, and
- Lack of standardization in technologies and building standards.

A follow-up report published by DOE in 2011 provides additional detail, especially on the relationship between technical and non-technical barriers. The report points out that the building science behind whole-building energy performance is complex, requiring extensive education for homeowners, contractors, regulators, appraisers, real estate agents, material suppliers, architects, engineers, program managers, lawmakers and those involved with financing.[14]

Commercial buildings have many of the same barriers to energy retrofit as residential buildings with the addition of greater scale and complexity. Not only are the buildings more variable in design and material use, but additional non-technical factors must be considered. The US Zero Energy Commercial Buildings Consortium (CBC), a voluntary group of professionals from the commercial real estate industry formed at the request of the DOE, compiled a report in 2011 that outlines many of the challenges they face in working toward significant energy use reduction in commercial buildings.[15] In addition to the impediments identified by NREL for residential buildings, the CBC report highlights the following:

- Conflicting retrofit costs and benefits for building owners and occupants,
- Lack of individual tenant utility metering,
- Significant impact of occupant plug loads,
- Disincentive created by energy upgrades that trigger added upgrades to meet minimum codes,
- Frequent failure to maintain equipment at originally installed efficiency,
- Traditional design and construction practice does not promote whole-building efficiency,
- No facility for accurate low-cost performance comparison between similar buildings,
- Lack of accountability for energy performance and consistency of performance,
- Utility profits and pricing linked to energy sales, limiting utility incentives for energy reduction,
- Competition for owner's capital to be made available for energy conservation investment, and
- Energy cost savings often have limited impact on the financial position of the building asset.

These reports make it clear that the process involved with planning, implementing and assuring performance of energy related building improvement requires significant management. To begin the process, a thorough understanding of the science involved with building energy performance must become commonplace to architects, engineers, contractors, subcontractors, equipment and material suppliers as well as building facilities managers, purchasing agents and maintenance personnel.

By examining the simple example of adding insulation to an existing structure, the potential for unintended consequences becomes clear. In Figures 10.1 and 10.2, simplified brick masonry wall diagrams before and after insulation is added, are shown. The previously uninsulated brick wall normally absorbs moisture from the exterior. In Figure 10.1 moisture in vapor form is free to pass in and out of the wall either from the interior or the exterior surface because all of the building materials used allow the passage of moisture by diffusion. This unrestricted moisture flow allows the wall to dry to the interior in the winter. In Figure 10.2 a thin layer of insulation was added to the interior. A thin layer was chosen to avoid losing valuable useable floor space while reducing the heat flow through the wall. The insulation layer reduces heating energy consumption and increases occupant comfort by raising the surface temperature of the interior wall during cold winter days. A common insulation choice for this purpose is a product with a high resistance to heat flow per unit thickness that maximizes the thermal performance. These insulating products are frequently made from foam materials that do not allow the passage of moisture to the interior through diffusion.

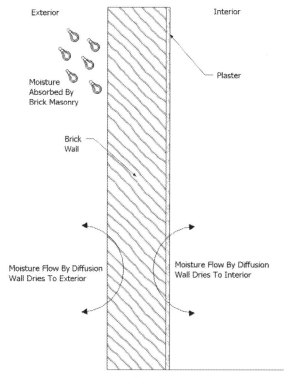

Figure 10.1. Uninsulated Masonry Wall Dries to Both Interior and Exterior.

With the addition of internal wall insulation, the internal wall temperature is cold because it is no longer heated from within the building. Figure 10.2 shows the resulting condensation of water on the inside of the masonry when exterior temperatures are low enough to reach the dew point or when warm moist air (typically above thirty percent relative humidity) leaks into the wall. The dew point is the temperature at which moisture will condense for a given level of moisture in the air. The two conditions previously described have a dew point of 3°C (37°F) or below, a common winter condition in much of the world. The moisture absorbed in the masonry by the upgraded wall now does not dry to the interior as necessary to avoid deterioration as it once did.

In addition to possible liquid moisture flowing into the structure, Figure 10.3 shows the result of the brick masonry wall reaching sub-freezing temperatures. The now consistently moist masonry suffers from wintertime formation of ice crystals within the brick, causing the expanding frost to deteriorate the exterior surface of the brick.

Over and above the universal education needed to promote the desired performance in design, there is a significant need for education and enforcement in implementation of energy retrofit. The fragmentation of the construction contracting and procurement process compounds the problems that result from poor understanding by those responsible for retrofit design.

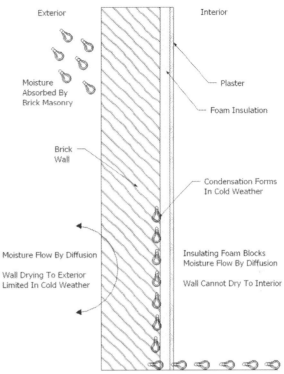

Figure 10.2. Potential for Condensation Resulting from Addition of Interior Wall Insulation.

Figure 10.3. Surface Deterioration from Sub-freezing Consistently Moist Brick.

Unfortunately, program management of policy schemes put in place to promote and incentivize energy improvement can also add to the disconnect between science based decisions and achievement of successful retrofit in practice. The best executed design will fail if the procurement and construction process is not properly managed. Choice of retrofit measures or materials which are inappropriate for the building's characteristics, poor availability of a trained workforce and/or construction management by unqualified energy services providers will all add to the uncertainty of energy saving outcomes. It is easy to understand the limited willingness of building owners to invest in energy improvements.

Even when technical and implementation challenges of energy services contracting are solved, there are still barriers to funding the retrofit work. Curtin and Maguire in a report for the Institute of International and European Affairs note that in Ireland there is a lack of attractive loan offerings for homeowners to finance residential retrofits. Homeowners are also hesitant to invest in energy improvements because of the risk of inadequate performance to justify the investment.[16] This observation is not unique to Ireland. Low participation rates and performance uncertainty have stymied financial innovation. For commercial building owners there is a divergence in strategic planning between the capital costs needed to implement retrofits and the resulting operating cost reduction. Owners frequently take a short-term view of return on investment while energy upgrades are a long-term investment. Long-term investments typically require financing. Lenders have not been able to agree on the method needed to value the effect that improvements in energy performance have on a building's marketability.[17] Without an increase in value, lenders are hesitant to provide vehicles for energy services lending. Utility company participation is often suggested, but utilities need encouragement to make changes in the revenue/profit structure of their business to justify investments that effectively result in reduced sales of their service.

There also appears to be an incomplete understanding of the most effective methods to market energy conserving retrofits. There is little agreement on the approach or message that is likely to cause differing groups to take action. Foremost in this discussion is the manner in which costs and benefits are considered. *Payback* (the time threshold for energy savings to repay retrofit costs), *Return on Investment* (the investment analysis common to business investment), *Net Present Value* (analysis using time value of money to compare projected energy savings to retrofit costs in current dollars) and *Internal Rate of Return* (discount rate that makes the net present value of all retrofit cash flows equal to zero) can be logically applied to investments in energy conserving retrofit projects, but may lead to different conclusions for different project characteristics. Payback is commonly used because of its simplicity, but is it the most appropriate? Benefits to society are also appropriate in marketing messages to stakeholders. What should the priorities be? Carbon reduction, economic development outcomes, job growth, new business development, advancement of redevelopment/regeneration can all result from energy related retrofit activities. In the final analysis, difficulty in determining the best approach to take stems from a lack of agreement on how to value the benefits.

The discussion which has been presented to this point revolves around use of existing technologies to improve the energy performance of buildings. There is some advocacy for advancement of new technologies to provide simple solutions that either dramatically reduce energy consumption or provide low-cost renewable sources of energy for building operations. While there are promising developments in many new products, barriers to their widespread use exist. To an even greater extent than the automotive sector, the built environment has a great deal of inertia. Thus the status quo represents one of the industry's greatest barriers to rapidly adopting new technology. Substantial investment is required on many levels to change the way things are done. The complexity of interactions and the many players involved with the manufacture, distribution, adoption and implementation of new technologies leads to challenges for any new building technology. Often what is lacking is cooperation between researchers, energy laboratories, manufacturers and end-users to help promote technology transfer through planning, assessment and trial implementations that provide the needed outreach to building owners and contractors.

Despite the potential for both direct cost savings and long-term societal benefits, the end-user is frequently a barrier to wide acceptance of energy related upgrades. Personal values can play a strong role in homeowner reluctance to upgrades. Conflicting values can range from resistance to potential changes that impact the appearance of the residence to concerns about the quality of workmanship and the perceived value of improvements. In some cases the inconvenience of workers being present in an occupied home or the need to clear out an attic space used as storage to make room for increased insulation can become a deterrent. As a result, there is a clear need to understand the priorities, values and long-term goals of homeowners when policy toward energy retrofit is considered. In the commercial building sector, additional end-user challenges exist because of the split incentives between building owners and tenants. Some experts have suggested that because of the many barriers, market forces alone are not adequate to motivate a significant increase in energy efficiency even with escalating energy prices. Based on the results of a multi-year simulation study focusing on the six markets of Brazil, China, Europe, India, Japan and the US sponsored by the World Business Council for Sustainable Development (WBCSD), the WBCSD advocates for a broad set of interventions by industry, governments, and building owners to make meaningful market transformation a reality. The WBCSD conclusions are noteworthy because their study area represents nearly two-thirds of the world's energy consumption.[18]

To achieve the market transformation needed to move beyond these barriers, programs with government support may be needed. The development of government programs must avoid creating barriers that further prevent wide acceptance of energy related retrofit as a result of program design or management. Program and regulatory barriers can come from both unexpected and anticipated program consequences. Incentive program design problems can arise from incentives that are too small to change behavior or too complicated to manage either by program managers or potential recipients. Regulatory restrictions based on conflicting social concerns such as building conservation or historic preservation planning constraints must be carefully considered to avoid significantly diminishing the program outcomes.

10.5. Building Energy Reduction Programs Recently Employed in the US and EU

Attempts have been made to devise programs which reduce the operating energy consumption of the built environment. As was noted in the previous section, participation in building energy reduction programs has often been lower than expected, even when they involved limited financial investment by the building owner. This fact reflects recognition of the many non-monetary costs of building energy retrofits that complicate technical and non-technical barriers. Policy commentary has exposed the substantial conflict between programs that base energy reduction investments solely on the monetary energy cost saving for the existing building owner and the longer-term societal imperative imposed by the global energy crisis.

For the European Union, the societal obligations created by the global energy crisis were recognized in the Energy Performance of Buildings Directive (EPBD), first implemented in 2002. The EPBD was updated in 2010 with more ambitious provisions creating obligations for EU countries to implement programs that achieve the requirements for building certification, inspections, and training with the ultimate goal to improved energy performance. While these directives are creating future obligations for EU countries, solutions will come only from transformation of markets which in many ways are localized. A large body of public policy commentary and research has been devoted to program design. Brenda Boardman, in a policy framework for the UK titled *Achieving Zero: Delivering future-friendly buildings,* makes many suggestions for the future based on successes and failures of programs implemented in the UK.[19] Many of the policy schemes presented by Boardman are useful for stimulating creative public policy discussion, but cannot be applied without careful consideration for the conditions specific to the market where the policy will be put to work. The following sections provide a small sample of past attempts to reduce energy consumption in buildings.

10.5.1. Voluntary Schemes to Promote Reduced Energy Consumption

Numerous voluntary schemes to promote energy reduction in the built environment have been offered on a local level. These voluntary programs have for the most part been targeted to owners of domestic buildings. While having the advantage of easy contact with local conditions and end-users, these programs have limitations in their ability to influence wide market transformation. Program incentives of many types are used with the 'reward' for participation as variable as the program sponsors.

Some programs simply provide ratings or labels in an attempt to enhance the recognition of value for the building owner created by the energy conserving investment. Enhanced ratings can be based on established standards such as construction specifications that are beyond code or when buildings achieve certified levels for recognized sustainability programs. Schemes based on sustainable building ratings such as Leadership in Energy and Environmental Design (LEED), the BRE Environmental Assessment Method (BREEAM) and Green Globes provide a framework for best practice in energy conservation as well as additional areas of environmental sustainability. Rating programs specific to building energy use such as Energy Star developed and promoted by the US government, Passivhaus which is a rigorous standard for low energy building first developed in Germany and the Home Energy Rating System (HERS) developed by Residential Energy Services Network (RESNET), a group of US energy professionals, offer a more energy conservation targeted evaluation.

Utility companies have developed programs to promote energy use reduction through rebates that reward owners who upgrade to appliances, water heaters or heating and cooling systems that operate at a high efficiency. Tax reduction is also a common form of reward for voluntary programs. The tax reward may be a credit against income tax due or an abatement of real estate property tax increases that might otherwise result from energy related building improvements. A type of reward for voluntary participation that often targets low income owners takes the form of a grant. Subsidized loans to complete the work have been suggested for higher income owners, but as described in a later section, lack of certainty that the energy savings will be adequate to repay the loan limits their use.

Overall, participation in voluntary programs has not been overwhelming. Demonstration programs that bring the potential for energy use reduction to the attention of building owners have been one approach to increase participation. The US Department of Energy has launched a program called the Better Buildings Challenge which works in coordination with communities, community based organizations, education providers and corporate partners to demonstrate residential, commercial and industrial projects that are successful in dramatically reducing energy use. Prior programs sought to achieve a similar outcome and future programs will be needed to provide adequate market penetration. Education is also an area where governments both local and national play a role. In the US the Environmental Protection Agency (EPA) and Department of Energy (DOE) provide building energy related educational materials through the Energy Star program and the Energy Efficiency and Renewable Energy (EERE) program. These educational programs are supported by the research efforts of the Oak Ridge National Laboratory (ORNL), the National Renewable Energy Laboratory (NREL), Lawrence Berkeley National Laboratory (LBNL) and the Pacific Northwest National Laboratory (PNNL). The laboratories are located in diverse regions of the US which allows them to concentrate on energy research specific to differing climates.

10.5.2. Public Policy Schemes Recently Employed and Proposed

A review of some national or regional schemes to promote building energy use reduction through public policy intervention is informative. An example of a regulatory mandate is the EU requirement for member nations to institute a national building energy performance assessment scheme. The impetus for universal building energy rating is to develop a publicly available uniform measure of the energy performance for every building. These ratings, typically referred to as *asset ratings*, are based on the theoretical performance of the building under a specified set of environmental conditions. Although asset ratings do not reflect the actual energy consumed by a building, they are preferable to collecting past utility use data because asset ratings eliminate the influence of occupant behavior.

Through the uniform collection of building ratings, consumers can compare buildings based on energy performance. Ratings can also provide a basis for building values assessments that reflect investments in energy performance. Some compare building energy performance ratings to miles-per-gallon ratings for automobiles.

In 1997 the first national building energy rating program was put into effect in Denmark. In 2002 the EU enacted the Energy Performance of Buildings Directive (EPBD). As part of the EPBD, member states were required to implement a national building energy rating and disclosure program beginning in 2009. In recent years Brazil, China and Australia have instituted national building rating schemes, although China's program is only mandatory for government owned or funded buildings. The US has not yet developed mandatory state or national level building rating programs, but several major cities have mandatory rating and disclosure programs underway.

Building rating and disclosure is not without its challenges. Some argue that the simulated energy use developed for the asset ratings is not an adequate reflection of the building energy consumption and is subject to lack of uniformity based on the accuracy of the building data which is used to generate the rating. Enactment of universal building ratings in the US has been hampered by disagreement on a standardized methodology for collecting building data. In an article published by the Bureau of National Affairs in 2011, the authors describe a consensus standard being developed by the American Society for Testing and Materials International (ASTM) which may provide the uniformity needed in the US for collecting and analyzing building performance information for regulatory compliance.[20] In addition to quality control in the data collection and analysis, many issues remain to be resolved. Rating programs must make performance data easy to understand yet meaningful, translate the performance data into building valuation information for lending and investment decision makers, preserve the privacy of sensitive building data while permitting adequate public disclosure of building energy performance as well as maintaining and updating the large database required to assure that ratings remain consistent over time.

A recently introduced program called the Green Deal is a more market or investment driven concept. The Green Deal, authorized by the UK Energy Act 2011, is an attempt to seamlessly combine energy company obligations to support building energy use improvements along with a financing mechanism. Upgrades for buildings under the Green Deal must meet the *Golden Rule*. The basic concept of the Golden Rule is that all energy improvement investments must produce savings that equal or exceed the cost of the improvements. In addition, the time period required for repayment of the improvement cost must not exceed the expected life of the improvement. Repayment is to be administered through the energy utility billing process. The regulatory complexity of the Green Deal is beyond the scope of this chapter, but several of the program management issues are noteworthy. For instance if this market driven program is to be successful, Green Deal Providers must be able to negotiate the complexity of program guidelines, obtain sufficient financial backing for up-front costs, work with energy companies to facility loan payback and implement energy conserving retrofits that apply appropriate building science and energy cost saving analysis to assure a quality retrofit that meets the Golden Rule payback requirements. A tall order based on the barriers previously introduced. Who will bear the cost if Green Deal Providers fail to adequately plan or implement their retrofits? What risks do Green Deal Providers accept by making the upfront investments needed to become a viable Provider if several years into the program the Green Deal is determined to be too costly for the UK government to maintain?

10.5.3. Program Successes and Failures

Descriptions of successful building energy programs are common in the mass media, but how success is defined can radically change the assessment of a program. Many programs describe their success by providing details of work accomplished or percentage reduction in energy consumption for a group of buildings. While these accomplishments are noteworthy in demonstrating the potential for reduction in energy use and carbon production, the magnitude of the problem truly requires many millions of structures to be retrofit every year. The UK has committed that zero carbon emissions will be associated with all buildings in the entire country by the year 2050. While zero carbon emission does not mean zero energy use, it does mean that most if not all of the twenty-six million homes and two million business structures must undergo some form of energy retrofit by 2050. Simple math tell us that for success in the UK alone several hundred thousand retrofits per year will be required immediately, growing to a million or more per year over time.

Failures are easier to identify because they are characterized by lack of participation. A notable example is in the area of financing building energy improvements. Energy upgrades require funds to complete the work, so financing is critical. Energy efficiency financing programs have failed to attract participation from both lenders and borrowers. Neither is willing to take the risk that the energy performance will justify the debt. Some have suggested that energy improvement loans should be tied to the real estate through on-bill financing that is paid for by current owners as well as future owners if the expected life of the improvement exceeds average years of ownership. While this may solve some owner concerns, lenders and investors continue to worry about the loan risk because the increase in building value resulting from the energy improvement may not adequately reflect the improvement cost. Energy Services Companies (ESCO) have been successful in reducing risk for entities with large portfolios of buildings by providing up-front capital in return for a share in the energy savings. This process can reduce building owner risk. A US government program issuing Energy Savings Performance Contracts (ESPCs) to ESCOs aims to achieve extensive recuctions in energy use, emissions, and cost in large installations where economies of scale can be leveraged. However, in its current form the ESCO process is not well suited to small projects. Nearly all of the ESCO projects to date have been for public projects or very large scale private projects.

Another notable failure is that the incremental learning that takes place with every new program is seldom passed on to future stakeholders. This lack of continuity of knowledge as programs expire is just one example of challenges to policy making. Limits on the life of a program also limit the possibility of sustained success. Energy retrofits are complex and require training for many levels of stakeholder. Many of the stakeholders who participate in the process are small business entities that depend on a financial return for continued existence. Will the large number of small and medium size businesses needed to carry out the energy improvements invest the time and resources needed for a program with an uncertain future? Will the lessons learned from previous short-lived programs be available to them as preparation? How do energy retrofit services become a trusted process implemented by proven entities without continuity? Will an efficient low-cost material and equipment supply chain capable of meeting high volume demands develop without continuity?

10.6. Recommendations and Opportunities for Future Solutions

While no single solution or ideal combination of recommendations will provide an answer to the challenge of excess energy consumption by the built environment, a number of opportunities for improvement exist. The following is an amalgamation of suggestions from current literature on the subject grouped by related topic.

10.6.1. Education, Motivation, and Marketing

- Use demonstration projects open to the public to make building owners aware of energy conserving retrofit potential. Include government owned buildings in the demonstration.
- To increase confidence in the potential energy savings, make simplified kitchen-table discussion information about energy conserving retrofit available to building owners.
- Work toward branding of energy retrofit solutions based on sound building science.
- Create education programs for builders, developers, architects, engineers, lenders and appraisers.
- Education programs should include building science in addition to standard practice so that designers and builders fully understand the technical ramifications of their retrofit decisions.
- Home improvement contractors drive adoption of technologies. Use educated service providers to promote energy retrofit to improve participation rates.
- Experiment with new measures and messengers to inform and engage people toward an energy-aware culture.
- Develop mechanisms that make energy performance improvement visible to guests, friends, and neighbors.
- Help building owners distinguish between dubious and real energy savings claims.
- Develop marketing to promote energy retrofit whenever any renovation activity is considered to take advantage of the economies available at the time.
- Market all benefits of energy conserving retrofit including improved occupant comfort, safety, indoor air quality, and overall environmental quality.
- Target populations need to be exposed to energy conserving retrofit communication and marketing three or more times before showing recognition of the message.
- Promote the value of energy audits.
- Educate the building construction industry about the pending growth opportunities in energy services contracting.

- Utilize workforce re-education funding to support education of needed energy services workers and managers.
- Develop building science elementary and secondary school curriculum to aid in understanding of building energy performance for future building owners and occupants.

10.6.2. Technology, Building Codes, and Implementation

- Support government funded research to develop and commercialize energy conserving technologies.
- Expand and strengthen support for energy improvement contractors that promote integrated design through whole-building retrofit options.
- Include technologies with longer payback periods such as energy recovery ventilation and triple glazed windows to achieve ambitious building energy use reduction goals.
- National building codes and enforcement need to be strengthened to improve building energy performance.
- Eliminate disincentives to energy conserving retrofits created by code triggers that require other code compliance upgrades when energy retrofits are completed.
- Combine building energy performance codes and renewable energy regulations to help promote whole-building performance analysis.
- Develop standardized practices when possible to streamline decision making and completion of energy services work.
- Encourage integrated design approaches and innovations.
- Develop standardized green lease agreements that help sort the split incentives between building owners who pay for energy conserving retrofits and renters.
- Develop simple to use energy retrofit quality criteria and checklists.
- Develop standardized energy audit procedures, tools and reports.
- Increase indoor air quality research to develop mechanical ventilation requirements for buildings retrofits that dramatically reduce building air leakage.
- Encourage the development of very low-capacity high-efficiency heating and cooling equipment that will be needed for high performance homes.
- Simplify the complexity of programmable thermostats.
- Promote community or district wide volume purchases of materials for energy related retrofits.
- Promote one-stop-shop business models that combine design, financing and contracting for energy conserving retrofit.

10.6.3. Building Performance Rating and Labeling

- Based on a national asset rating assessment standard, implement building ratings and labels that are uniform, simple to understand, normalized to eliminate inconsistencies and adequately reflect value related to building marketability.
- Rating and labeling programs should adopt mechanisms to assure that label information remains current and accessible to the public.

- Research is needed to ascertain the market value impact of building energy performance labels.
- Develop simplified energy performance simulation tools that don't require extensive knowledge or training for use by building owners and managers interested in building energy upgrades.
- Life Cycle Energy Consumption is beyond the scope of near-term comparisons, but work should continue to perfect the analysis for future building energy performance comparisons.
- Create simple benchmarking and goal setting tools to assist commercial building owners in making energy performance decisions thru comparison to similar high performance buildings.

10.6.4. Financing

- Concentrate efforts on those most in need of financing.
- Align monthly repayment cost to energy savings to avoid net increase in monthly cost.
- Allow the financing of non-energy upgrades to promote energy conserving retrofit at the same time as other improvements.
- Loan terms in excess of five to seven years will be needed to promote comprehensive whole-building retrofits.
- Enable lending tied to the asset such as on-bill financing repaid through utility billing systems or municipal financing funded through bond sales that are repaid through real estate taxes.
- Develop and mandate standardized building appraisal procedures that reflect building energy performance for use in risk analysis for energy improvement mortgages.
- Create protocols for reporting of building energy use that can be used during commercial real estate due diligence and underwriting for building loans.
- Investigate turn-key practices which combine financing and performance guarantees or shared savings through contracts with ESCOs.

10.6.5. Program Design

- Programs and policies should strive for integrated market transformation targeting barriers and contextual challenges specific to a target population in a community or region.
- Programs should employ rigorous assessment including measurement and verification of the installed energy savings measures to validate program goals.
- Program assessment and validation data collection should begin immediately so that mid-program corrections can be made.
- Consider action research that collects and builds on programs so that lessons learned from past programs are not lost as new programs are developed.
- Subsidies may be required for demonstration and roll-out support to implement residential programs.

- Design incentive programs where the cost of access does not exceed the value of the benefit.
- Incentive programs should provide increasing benefit levels for more aggressive whole-building retrofits.
- Consider tax incentives such as reduced real estate taxes for owners completing energy retrofit.
- To increase the scale and scope of utility sponsored programs, decouple the connection between reduced sales resulting from energy conservation and utility revenue.
- Develop certification programs to help assure the efficacy of installed energy savings measures.
- Design residential retrofit programs that align with and take into account personal values and decisions of homeowners. Design programs around people not buildings.
- Design programs around specific communities, contexts and populations.
- Partner with and showcase renovation and regeneration contractors with adequate training and proven performance that will include energy retrofit at the same time as other building improvements.
- Provide low-risk entry points for building contractors interested in business development for energy services contracting.
- Utilize trusted third parties such as non-profits, charities, religious institutions, schools, or community organizations as referral agents for retrofit programs.
- Programs should be kept in place for a decade or more to avoid the reluctance of business participants fearful of developing new practices for short-lived programs.

Obviously, not all of these suggestions can be adopted in every situation. Some are even in conflict with others. Policy must adapt to the community and context in which it is implemented. For new construction and developing countries policy should concentrate on new construction codes and code enforcement. In the developed world more complex solutions will be necessary.

Discussion Problem

In Happy Hills, a planned community of approximately 250,000 that serves a bedroom community to a nearby major city, the residents live in predominantly multi-family housing which is comprised of buildings with twenty to forty living units in buildings three to four stories in height. Residents enjoy a good standard of living and can commute by rail to jobs in the nearby major city.

The multifamily residential buildings are privately owned and most residents rent their apartments paying monthly rent to a wide diversity of landlords. The buildings were constructed in the 1960s and most of the residents have been living in their homes for many years. Because of a lack of new construction in the area and the excellent physical condition of the existing buildings, the majority of the residents would like to remain in their current apartments.

> Happy Hills operates a local utility that supplies both electricity and natural gas to the community. Community leaders are well aware of the challenges of the global energy crisis and the local impacts of the crisis because of their position as a public utility. Utility operators have noted that on a per capita basis Happy Hill residents use substantially more energy than the average based on national statistics. Because the local buildings were metered for energy consumption by the building or grouping of buildings when they were constructed, the utility managers have no easy method to track individual apartment energy use. The local government would like to establish policies and programs that promote improved residential energy conservation in Happy Hills. What public policy or program recommendations would you recommend?
>
> As you make your recommendations be sure to consider energy conserving technology selection and integration, program standards and implementation, financing for incentives or grants, building owner acceptance, building occupant reaction, and how to monitor program success and metrics. After you have your initial recommendations, consider the following questions: What would be different if the apartments were owned individually by the occupants? What would be different if the residences were detached from each other rather than in a multiple occupancy building?
>
> This problem is likely to have a multitude of solutions, none of which would be ideal due to the complex interaction involved. The case example should stimulate thought and discussion about the range of issues encountered at the nexus of technology and a common set of circumstances with both private and public implications. The best solution will integrate the concerns of the stakeholders resulting in measurable reductions in building operating energy as an outcome.

Notes

The principal author wishes to express his thanks to Mr Lloyd Scott for his facilitation in researching the Irish and European landscape in contribution to this chapter.

1. Mark Levine et al., "Residential and Commercial Buildings," in *Climate Change 2007: Mitigation. Contribution of Working Group III to the Fourth Assessment Report of the Intergovernmental Panel on Climate Change*, ed. Bert Metz, Ogunlade Davidson, Peter Bosch, Rutu Dave, and Leo Meyer (Cambridge: Cambridge University Press, 2007).
2. Urban Land Institute (ULI), *New Tools. New Rules: Climate Change, Land Use, and Energy 2010* (Washington, D.C.: Urban Land Institute, 2010), 7.
3. Energy Information Administration, "Commercial Buildings Energy Consumption Survey," *EIA.gov*, last modified 2013, http://www.eia.gov/consumption/commercial/index.cfm.
4. Fergal O'Leary, Martin Howley, and Brian Ó Gallachóir, *Energy in the Residential Sector* (Dublin: Sustainable Energy Ireland, 2008), http://www.seai.ie/News_Events/Press_Releases/Energy_in_the_Residential_Sector_FNL.pdf.
5. Sustainable Energy Authority of Ireland (SEAI), *Energy Security in Ireland: A Statistical Overview* (Dublin: SEAI, 2011), http://www.seai.ie/Publications/Statistics_Publications/EPSSU_Publications/Energy_Security_in_Ireland/Energy_Security_in_Ireland_A_Statistical_Overview.pdf.
6. National Academy of Sciences (NAS), *Real Prospects for Energy Efficiency in the United States* (Washington, D.C.: National Academies Press, 2010), http://www.nap.edu/catalog.php?record_id=12621.

7. Marius C. Claudy, Claus Michelsen, and Aidan O'Driscoll, "The Diffusion of Microgeneration Technologies: Assessing the Influence of Perceived Product Characteristics on Home Owners' Willingness to Pay," *Energy Policy* 39, no. 3 (2011): 1459–1469, http://dx.doi.org/10.1016/j.enpol.2010.12.018.
8. The Society of Light and Lighting, *SLL Code for Lighting* (London: CIBSE, 2012).
9. The Society of Light and Lighting, *SLL Lighting Handbook* (London: CIBSE, 2009).
10. Philips, "The LED Lighting Revolution: A Summary of the Global Savings Potential," *asimpleswitch.com*, last modified May 2012, http://www.lighting.philips.com/pwc_li/main/lightcommunity/assets/the-led-lighting-revolution-facts-figures.pdf.
11. McKinsey & Company, *Lighting the Way: Perspectives on the Global Lighting Market* (McKinsey & Co., 2012).
12. Environmental Protection Agency (EPA), *Understanding Cost-Effectiveness of Energy Efficiency Programs: Best Practices, Technical Methods, and Emerging Issues for Policy-Makers* (Washington, D.C.: EPA, 2008), http://www.epa.gov/cleanenergy/documents/suca/cost-effectiveness.pdf.
13. Department of Energy (DOE), *Summary of Gaps and Barriers for Implementing Residential Building Energy Efficiency Strategies* (2010), http://apps1.eere.energy.gov/buildings/publications/pdfs/building_america/49162.pdf.
14. DOE, *Technical Barriers, Gaps, and Opportunities Related to Home Energy Upgrade Market Delivery* (2011), http://apps1.eere.energy.gov/buildings/publications/pdfs/building_america/tech_barriers.pdf.
15. Zero Energy Commercial Buildings Consortium (CBC), "Analysis of Cost & Non-Cost Barriers and Policy Solutions for Commercial Buildings," *Zeroenergycbc.org*, 2011, http://zeroenergycbc.org/pdf/CBC_Market-Policy_Report_2011.pdf.
16. Joseph Curtin and Josephine Maguire, *Thinking Deeper: Financing Options for Home Retrofit* (Dublin, Ireland: Institute of International and European Affairs, 2011).
17. Mark Chao, *Recognition of Energy Costs and Energy Performance in Real Property Valuation*, 2nd ed. (Washington, D.C.: Institute for Market Transformation, 2012). http://www.imt.org/uploads/resources/files/Energy_Reporting_in_Appraisal.pdf.
18. World Business Council for Sustainable Development (WBCSD), *Transforming the Market: Energy Efficiency in Buildings* (2009), http://www.wbcsd.org/pages/edocument/edocumentdetails.aspx?id=11006.
19. Brenda Boardman, *Achieving Zero: Delivering Future-Friendly Buildings* (Oxford, UK: Environmental Change Institute, 2012), http://www.eci.ox.ac.uk/research/energy/achievingzero/achieving-zero.pdf.
20. Mark J. Bennett, Anthony J. Buonicore, and David J. Freeman, "New ASTM Standard May Serve as Safe Harbor for Emerging Building Energy Labeling, Transactional Disclosure, and Benchmarking Regulations," *Bureau of National Affairs* 231:2271 (2011), http://www.bepinfo.com/images/pdf/ASTMBEPABennett.pdf.

Bibliography

Bennett, Mark J., Anthony J. Buonicore, and David J. Freeman. "New ASTM Standard May Serve as Safe Harbor for Emerging Building Energy Labeling, Transactional Disclosure, and Benchmarking Regulations." *Bureau of National Affairs* 231:2271 (2011). http://www.bepinfo.com/images/pdf/ASTMBEPABennett.pdf.

Boardman, Brenda. *Achieving Zero: Delivering Future-Friendly Buildings*. Oxford, UK: Environmental Change Institute, 2012. http://www.eci.ox.ac.uk/research/energy/achievingzero/achieving-zero.pdf.

Chao, Mark. *Recognition of Energy Costs and Energy Performance in Real Property Valuation, 2nd ed.* Washington, D.C.: Institute for Market Transformation, 2012. http://www.imt.org/uploads/resources/files/Energy_Reporting_in_Appraisal.pdf.

Claudy, Marius C., Claus Michelsen, and Aidan O'Driscoll. "The Diffusion of Microgeneration Technologies: Assessing the Influence of Perceived Product Characteristics on Home Owners' Willingness to Pay." *Energy Policy* 39, no. 3 (2011): 1459–1469. http://dx.doi.org/10.1016/j.enpol.2010.12.018.

Curtin, Joseph, and Josephine Maguire. *Thinking Deeper: Financing Options for Home Retrofit.* Dublin, Ireland: Institute of International and European Affairs, 2011.

Department of Energy (DOE). *Summary of Gaps and Barriers for Implementing Residential Building Energy Efficiency Strategies.* (2010). http://apps1.eere.energy.gov/buildings/publications/pdfs/building_america/49162.pdf.

Department of Energy (DOE). *Technical Barriers, Gaps, and Opportunities Related to Home Energy Upgrade Market Delivery.* (2011). http://apps1.eere.energy.gov/buildings/publications/pdfs/building_america/tech_barriers.pdf.

Energy Information Administration. "Commercial Buildings Energy Consumption Survey." *EIA.gov.* Last modified 2013. http://www.eia.gov/consumption/commercial/index.cfm.

Environmental Protection Agency (EPA). *Understanding Cost-Effectiveness of Energy Efficiency Programs: Best Practices, Technical Methods, and Emerging Issues for Policy-Makers.* Washington, D.C.: EPA, 2008. http://www.epa.gov/cleanenergy/documents/suca/cost-effectiveness.pdf.

Levine, Mark, Diana Ürge-Vorsatz, Kornelius Blok, and Luis Geng, Danny Harvey, Siwei Lang, Geoffrey Levermore et al. "Residential and Commercial Buildings." In *Climate Change 2007: Mitigation. Contribution of Working Group III to the Fourth Assessment Report of the Intergovernmental Panel on Climate Change.* Edited by Bert Metz, Ogunlade Davidson, Peter Bosch, Rutu Dave, and Leo Meyer. Cambridge: Cambridge University Press, 2007.

McKinsey & Company. *Lighting the Way: Perspectives on the Global Lighting Market.* McKinsey & Co., 2012.

National Academy of Sciences (NAS). *Real Prospects for Energy Efficiency in the United States.* Washington, D.C.: National Academies Press, 2010. http://www.nap.edu/catalog.php?record_id=12621.

O'Leary, Fergal, Martin Howley, and Brian Ó Gallachóir. *Energy in the Residential Sector.* Dublin: Sustainable Energy Ireland, 2008. http://www.seai.ie/News_Events/Press_Releases/Energy_in_the_Residential_Sector_FNL.pdf.

Philips. "The LED Lighting Revolution: A Summary of the Global Savings Potential." *asimpleswitch.com.* Last modified May 2012. http://www.lighting.philips.com/pwc_li/main/lightcommunity/assets/the-led-lighting-revolution-facts-figures.pdf.

Society of Light and Lighting, The. *SLL Code for Lighting.* London: CIBSE, 2012.

Society of Light and Lighting, The. *SLL Lighting Handbook.* London: CIBSE, 2009.

Sustainable Energy Authority of Ireland (SEAI). *Energy Security in Ireland: A Statistical Overview.* Dublin: SEAI, 2011. http://www.seai.ie/Publications/Statistics_Publications/EPSSU_Publications/Energy_Security_in_Ireland/Energy_Security_in_Ireland_A_Statistical_Overview.pdf.

Urban Land Institute (ULI). *New Tools. New Rules: Climate Change, Land Use, and Energy 2010.* Washington, D.C.: Urban Land Institute, 2010.

World Business Council for Sustainable Development (WBCSD). *Transforming the Market: Energy Efficiency in Buildings.* 2009. http://www.wbcsd.org/pages/edocument/edocumentdetails.aspx?id=11006.

Zero Energy Commercial Buildings Consortium (CBC). "Analysis of Cost & Non-Cost Barriers and Policy Solutions for Commercial Buildings." *Zeroenergycbc.org,* 2011. http://zeroenergycbc.org/pdf/CBC_Market-Policy_Report_2011.pdf.

"Energy and persistence conquer all things"

Benjamin Franklin

Epilogue: Reflections on Our Path Forward

What do the trajectories of energy supply and demand, population growth, and climate change suggest about the future? Can sustainable, practical, affordable solutions be brought to bear upon these challenges? As we deepen our understanding of the questions, we come to realize that the solutions will not be easy ones at all. These are wicked problems that will require trade-offs and tough decisions. They will lead to new questions, some more complicated yet. As we noted in Part I, it is unlikely a single technology or a single country will swing the needle entirely by itself. Nor will any single policy, however robust and timely. Lasting solutions will rely on creative interdisciplinary collaboration, and not just within academia but from the furthest reaches of the public and private sectors as well. They will involve more active participation from consumers and civil society, be accompanied by a growing social consciousness, and demand further technological innovation and coordinated policy dialogue.

When considering tomorrow's energy supply and demand trends, it is important to recognize that compelling market factors will continue to heavily influence these. In the near term, if one expects that consensus opinion will overwhelmingly embrace large price premiums in return for low carbon energy, then one has not fully considered the adaptability and market efficiency of today's current energy matrix. That said, with increasing concern over resource scarcity and environmental impacts, it is now more widely appreciated that existing energy resources, natural gas and nuclear power as examples, can represent a viable bridge to a more diverse, sustainable, and lower carbon energy future. This will not happen by accident, nor will the bridge have any value if it does not have an opposite shore on which to land. Thus, technology, policy and consumer response must become more synchronized and flexible to spawn robust energy and climate strategies in the face of changing variables, evolving economies and electorates, and new scientific data and discoveries.

With world population having reached seven billion in late 2011, "medium band" projections indicate a world population of approximately nine billion by 2050, increasing relatively modestly to ten billion by 2100. Thus, by 2050 the planet will be home to thirty percent more people than it is today. Asia is the most populated continent today with 4.3 billion inhabitants (sixty percent of world population), largely comprised of the combined populations of China and India. Africa hosts one billion inhabitants. These statistics confirm the growing life force of the planet and global energy requirements will continue to increase accordingly. And while expansion of developing regions often welcomes a rising middle class, nations must thoughtfully grapple with the resource implications of broader economic prosperity. Challenges in meeting the social, economic, and technological requirements for a larger emerging world force are therefore considerable, not least owing to the fact that the world is struggling today with oil prices in excess of $100 per barrel, coupled with an awareness of the historic dependence of global economic growth on oil production.

On climate, it is widely considered that a global temperature increase in excess of 2°C will result in considerable variation to habitats, which in turn will affect food production and livelihood for populations in many regions of the world. Impact predictions are well beyond the scope of this book, however experts note that a number of geographical locations including the United States, Australia, and Africa may be increasingly susceptible to drought. Other regions are expected to experience increasing levels of rainfall, including Asia, parts of Europe and the US. Some regions will suffer on account of crop failure while others may benefit through warmer weather pattern shift from southern to northern regions.

Part II of our book has laid a strong foundation from which we can draw some salient observations about prospects for energy production and conversion technologies. Carbonaceous fossil fuels—principally coal, oil, and gas—remain the major constituents fueling our growing energy demands as the search for new and unconventional resources, once considered beyond reach, gains pace via new and unprecedented mining and extraction technologies. The debates relating to developments in carbon capture and storage (CCS) and extraction of natural gas from shale continue to gather pace, while coal is expected to continue providing significant baseload power in many countries. A serious conundrum remains that national climate objectives cannot be achieved without the effective demonstration and implementation of CCS, which is not expected to be achievable at sufficient scale prior to 2030.

Natural gas can be viewed as a microcosm of the types of issues that may increasingly accompany the world's energy future. Consider that the tremendous quantities of low cost natural gas from shale formations via new hydrofracking techniques have been made possible by innovative technologies pioneered over the past several decades, with both government *and* private support. From 2005 to 2013, the share of US natural gas from shale formations has increased from about four percent to more than thirty percent and will continue to climb. Estimates of gas quantities that may be recoverable through hydrofracking vary; some analysts suggest that up to forty trillion cubic meters is attainable in the US alone. This, if achievable, could satisfy a large portion of energy demand for upwards to one hundred years at about half the emissions of coal. The increase in emissions to Earth's atmosphere through greater deployment of gas will nevertheless be substantial; there is no room for complacency in assuming natural gas will be a solution to a warming atmosphere and climate change.

And yet while some see unconventional natural gas as the solution to national if not global energy problems, public opinion is divided on the subject. Critics fear environmental damage may result through widespread exploitation of gas by hydraulic fracturing and are concerned about the effectiveness and standardization of regulatory frameworks. There is credible concern of a significant increase in carbon footprint owing to venting and leakage of both carbon dioxide and methane during mining, and through the combustion of natural gas to produce energy following extraction and processing. A further point of controversy relating to fracking comes from concerns about increased pressure that underground fluid injection can exert on seismic faults, thus making them more likely to slip. Reports by seismologists in the US note a number of small quakes which have been triggered by far off larger earthquakes; it is apparent that wastewater injection critically loads local faults, placing them on the verge of rupture. Amid these challenges stand great opportunities for technologists, policy makers, and informed consumers to collectively balance the trade-offs and adopt lasting strategies with regard to natural gas.

Additional challenges and opportunities surround the deployment of renewable energy resources as well. The European Wind Energy Association is intent on achieving a target of twenty percent total EU electricity consumption powered by wind, while the United States claimed the number one spot in wind connectivity from China for 2012. Combining with a very strong year in Europe, the annual world market grew by ten percent, equivalent to forty-five gigawatts, and with total combined global wind installation of 282.5 GW. Other regions of the world are seriously engaged in wind energy development and installation as well, including Latin America, Africa and Asia. Policy at the national level is a critical factor in driving both domestic and regional markets. Having greatly incentivized record US installations in 2012, the US Federal Renewable Electricity Production Tax Credit (PTC) was extended for one year on 1 January 2013. The message is clear that wind energy is growing and that renewable energy technologies can indeed be deployed into existing markets at scale. However intermittency, grid integration and paybacks without subsidy support remain key hurdles to overcome going forward.

Throughout the first decade of the twenty-first century solar photovoltaic energy (PV) has been the fastest growing renewable energy technology globally. The cumulative total reached 100 GW by the end of 2012, of which seventy-five percent was installed since 2009. Concentrated solar power (CSP) which generates heat and electricity, is also gaining in popularity, particularly in the United States and in Spain. IEA estimates that solar energy could provide up to one-third of the world's energy demand after 2060. Though optimistic, it signals the potential for growth in what is perhaps the most natural energy available to planet Earth. Costs have fallen and in sunny countries, solar thermal energy (STE) and solar photovoltaic electricity (PV) are competitive against oil-fueled electricity generation, and are very effective in helping meet demand peaks. Rooftop solar water heating has soared in China, and IEA believes that rooftop PV in sunny countries could be a way forward in meeting local generation requirements. The tough reality is that further support by way of feed-in tariffs, including market-based reverse auction mechanisms, will be required for some time before the industry can stand on its own and eventually sustain itself. It is worth noting that solar power has been accompanied by certain unintended consequences over the past decade, for example by guaranteeing artificially high electricity rates. The hope is that these growing pains and lessons learned help make utilities, investors and policy makers more effective at commercializing renewable energy.

Hydro remains the world's largest source of renewable electricity. It was second to wind power in new installations between 2005 and 2010. Although more suited to some countries than others based on geography and natural resources, there is room for significant growth in the decades ahead. In pumped storage hydroelectricity, Japan has successfully pioneered a seawater energy storage facility at Kunigama, Okinawa, a non-trivial challenge due to corrosion considerations. A consortium in Ireland has proposed development of a large-scale pumped storage scheme which, coupled with wind turbines whose energy pumps seawater to upper storage lakes, could provide the majority of the country's electricity needs. Energy storage remains a strategic need that can enable renewable energy generated from intermittent sources, and one where innovation, investment and multi-stakeholder dialogues are in great demand.

While Ocean Energy developments, including tidal power, tidal marine currents, wave power, gradients of temperature and salinity, are at an early stage, there is promise that breakthroughs will enable deployment, particularly in island and remote locations. An array of proposed, planned and operational wave and tidal energy projects and marine test sites in Europe, span the continent from Norway to the northern coasts of Africa. Technological advances will be critical in demonstrating the viability of this resource. Similarly, there is also potential for a scaled increase in the global production of geothermal heat and electricity outwards to 2050. IEA estimates that geothermal energy can comprise 3.5% of global electricity and 4% of heat energy by 2050, perhaps optimistic, but again indicative of what technology appropriately deployed may be capable of achieving.

Biofuels are similar to shale gas, in that they represent another intriguing microcosm of the interplay between technologies and policies with regard to economic and environmentally sustainable energy solutions. As we graduate from first generation to second, agricultural land area, second-generation conversion technologies, environmental promises, and practical policies are all central to the evolving debate. Sustainable deployment of high-impact, second-generation biofuels will undoubtedly require balanced policy measures and an improvement in the economics of advanced biofuel production. With experience as our guide, active participation and compromise from engineers, scientists, economists, law makers, and consumers can also be expected to figure in.

With regard to the future of nuclear energy, we again see that a delicate balance is imperative to address the social and political complexities associated with sensitive technology. Societal concerns and the consequential realities of nuclear accidents continue to reappear at Fukushima, following the March 2011 nuclear accident. Among these were concerns relating to recorded radiation levels in sea water near the power plant. Strategies to address these issues remain ongoing, and represent another area in which the technological and policy communities must cooperate and counsel each other to facilitate best possible outcomes. Another takeaway from this ongoing dilemma is the reality that although accidents at nuclear power plant are rare, the resulting consequences can be far reaching. This adds to public disquiet and mistrust of nuclear energy and factors into societal acceptance of regional nuclear plant construction. Yet for many countries, strategic national targets call for meaningful contributions from nuclear power to help meet key CO_2 reduction goals by mid-century.

As with energy production where the easy oil has been consumed, and decisions have become much more complex and interconnected, future energy use will present some equally difficult challenges. The world has grown accustomed to a cheap and readily available supply, creating a behavior on the consumption side that is as predictable as it is unsustainable. Perspectives drawn from Part III speak to the future of energy consumption.

We have seen that the introduction of renewable energy sources into existing markets is a complex socio-techno-economic challenge that depends upon alignment of stakeholder objectives in order to truly succeed. Thus far, complete alignment has generally proven elusive. While the economics of energy distribution and use currently favors traditional sources, growing government and private sector investment in renewable energy systems is reflecting broader awareness and acceptance of the need to address fossil fuel dependence and climate change. A key takeaway is that civil society, and more specifically the individuals that comprise it, have important roles to play. The best renewable technology under the most appropriate policy must still be met by a market where individuals advocate and engage in its adoption and use. Timing and communication among the stakeholders are tantamount to the technology and policy themselves.

The broad categories of transportation and buildings collectively account for about seventy percent of global energy consumption. While each is unique, some common trends apply including near-term efforts aimed at improving energy efficiency. Admittedly, this may not sound as exciting or cutting edge as other emerging technologies, but energy efficiency may in fact be among the most critical near-term means of reducing energy demand and emissions in relatively immediate and affordable ways. The technologies largely exist, there is substantial policy and regulatory precedent, and consumers can see the direct economic benefits from their investments. More efficient vehicles, including optimization of internal combustion engines and aggressive national fuel economy standards have combined to accelerate efficiency efforts in transportation. Implemented effectively, this will have the effect of dramatically reduced oil demand and emissions from the sector. New technologies including hybrid and electric vehicles, and advanced fuels will increase in market share, but their commercial viability will depend on development, validation and production costs. Fleet turnover will be gradual, but new technologies in automotive and aviation show great promise and have done well to ensure alignment within industry and government entities, to streamline the eventual policy rollout.

In the built environment, insulation, lighting, controls, heating/cooling, and other conservation and efficiency improvements from off-the shelf technologies are contributing to dramatic gains in energy savings and reduced environmental impact. This is a success story in real time whose value should not be understated. For new buildings, the outlook is promising and will increase compliance to stringent standards. That said, older buildings comprise a dominant share of the stock, exceeding ninety percent in many regions. This dilemma means assessing the benefits and costs associated with upgrades becomes more difficult. Longer term solutions are case dependent, and will likely involve innovation related to more intelligent design and use of HVAC, lighting, water, and advanced building materials, among others. An exciting aspect of built environment energy analysis is the more clearly defined interaction of the consumer with energy decisions. Though more data-informed and rational responses are desirable in consumer decisions, lessons learned from this sector may have far reaching application to others, and provide avenues for more proactive energy involvement by lay society.

We hear you saying, "I want to make a difference." Consider three closing themes as we conclude our reflections on the path forward.

For the future of energy supply, diversity in the energy matrix is imperative. By all means, there is room to scale-up renewable technologies, but it will happen alongside efforts to optimize conventional and fossil resources. This means that wind, biofuels, solar, and other clean alternatives will become truly competitive; and it also means the momentum in natural gas and nuclear must be leveraged in order to reduce the environmental impact. It means managing the tough realities of coal and oil. For now, all cost effective kilowatt-hours (and BTUs) will find a market; and in the very near future, there will be an ever-increasing demand for the clean ones.

For the future of energy consumption, efficiency and conservation must not be overlooked. In the next decade, these areas will potentially do more to reduce energy costs, use, and emissions than even the cleanest new energy sources. While the innovation may look different from R&D on new conversion technology, it will be no less impactful. This is about a relentless focus to extract the most value from every watt. Intelligent controls, advanced materials, cutting edge optimization technologies for mature industries, and more informed and proactive consumers—these will become the agents of change.

For the next generation of energy leaders, social awareness and technological innovation must go hand in hand. Grand global challenges like energy and climate demand that problem solvers interact across boundaries, and outside of comfort zones. Tomorrow's policy makers, for instance, must more deeply appreciate the complexity of engineering and economic challenges. Tomorrow's scientists and engineers must not only be aware of social implications, but conversant enough to speak candidly and affect them favorably. The world is in need of better data to be sure. But equally critical are educated experts willing to interpret data and communicate complex issues throughout the world.

We have not found a perfect source of energy. Nor have we resolved the difficult issues of consumption. For the foreseeable future, we thus have some compromises to make. Since the first man-made fire, draft animals, steam engines, jet airplanes, personal computers, and smart phones we have opted to leverage our natural resources to reduce our physical burden, improve our quality of life, and accelerate our journey in this world. A world that is an organism, of which we are a critical and wonderful part. We have pioneered great achievements and witnessed many natural and man-made disasters. Yet we adapt, learn, innovate, and try to improve. Taking on board all this new data affects everyone differently—some retreat, many carry on as usual, others take up action. In crisis, action is often needed; the engine does not run itself, so to speak. And yet, even action alone may not guarantee positive change. The action must be informed, coordinated, flexible; it must be backed by a critical mass; and it must leverage the best of technology and policy, of science and the human heart. The physical laws and data that describe our world to us are impersonal and without conscience. Not so for us. And therein lies perhaps our greatest challenge and our greatest opportunity.

Index

Achieving Zero: Delivering future-friendly buildings, 272
adoption, of technology, 200
advanced biofuel, sugarcane ethanol as, 56–57
advanced power networks, 117
advanced technologies, in achieving clean coal, 22
advanced vehicle targets, by region, 239t
agents, in systems, 77
Agricola, George, on biogenic of fossil fuel creation, 15
agricultural commodities, price drivers for, biofuels as, 154–155
Air Force (US), alternative fuel blend goals of, 247
air sealing, in retrofitting building, 258
aircraft
    hybrid, 245
    solar-powered, 245
alkaline fuel cells, 240
Alternative Aviation Fuel Experiment (AAFEX II), 246
Alternative Fuel Effects on Contrails and Cruise Emissions (ACCESS), 246
alternative fuels. *See also* biofuel(s)
    aviation
        development and use of, GFAAF and, 246
        emissions and, 244–246
    fuel cells as, 240–241
    research efforts on, 220
American Electric Power v. Connecticut, 39
American Reinvestment and Recovery Act (ARRA, 2009), 38–39
America's Climate Choices (2011), responsibility for CO_2 emissions and, 3
Anderson, Robert, battery electric vehicle of, 228
Arab oil embargo (1973)
    UK energy response to, 206
    US energy policy and, 34
Ardnacrusha hydroelectric power plant, 100
Arrhenius, Svante, on heat-absorbing atmospheric gases and ground temperature, 14
Asia, rooftop solar water heaters in, 128
asset ratings, of building energy performance, 273
associated gas, 18, 19

atmosphere, earth's, greenhouse gases and, 13–23
atomised gas fuels (AGF), 219–220
attenuator devices, for wave energy, 108
attitudes, individual behavior and, 81
automobile. *See also* vehicles
automobile, petroleum industry and, 18
avgas, for aircraft, 242
aviation
    fuels for
        alternative
            development and use of, GFAAF and, 246
            emissions and, 244–246
        overview of, 242–244
        regulation of, 239–248
    greening of, policy frameworks and initiatives for, 246–248

bargaining power
    consumer, as barrier to taking new technology to market, 195
    supplier, as barrier to taking new technology to market, 195–196
batteries, nanotechnology and, 140–141
battery electric vehicles (BEVs), 228–229
    internal combustion engine vehicle domination over, 229
behavior
    consumer, in reducing oil use, 223–225
    factors affecting, 81
    planned, theory of, 81, 81f
belief systems, individual behavior and, 81
benefits, and costs of energy conserving retrofits in marketing, 270
Benz, Karl, internal combustion engine vehicle developed by, 229
Bernoulli, Daniel, on hydrodynamic theory, 100
Better Buildings Challenge, 273
Betz, Albert, on energy from wind, 93
Betz's Law, 93–94
biochemical conversion, of biomass, 159
bioenergy, 6

biofuel(s)
- advanced, sugarcane ethanol as, 56–57
- agricultural commodity prices and, 154–155
- in Brazil, flexible dispensing infrastructure for, 153–154
- cellulosic
  - economics of, 157
  - feedstocks for, availability and cost of, 157–158
  - in US Renewable Fuel Standard, 152, 152f
- environmental issues on, 160–161
- ethanol as (*See* ethanol)
- in EU, biodiesel as, 154
- feedstocks and conversion technologies for, 157–159, 159–160
- first-generation, 153–155
  - environmental issues on, 160–161
- government policy on, 161–162
- greenhouse gas emissions of, 156t, 157
- history of, 152–153
- major challenges and opportunities for, 162–163
- mandate *vs.* subsidy for, 153
- nanotechnology and, 140
- oil price uncertainty and, 157, 158f
- production of, global share of, 153, 153f
- rural incomes and, 154–155
- second-generation, 155–157
  - environmental issues on, 161
  - reverse auctions for, 162
- in uncertain world, 151–166
- in US, ethanol as, blend wall and, 154

biogenic gas, creation of, 19

biomass
- conversion to biofuels, second-generation technologies for, 159–160
- global capacity of, 92
- in India's energy supply, 59

blend wall, for ethanol, 154

Blueprint for a Secure Energy Future (2011), 40

brand recognition, as barrier to taking new technology to market, 194

Brazil
- in biofuels history, 152
- energy and climate change policy of, 54–58

BRE Environmental Assessment Method (BREEAM), 272

building(s)
- commercial (*See* commercial buildings)
- embodied energy of, 266
- energy performance of, rating systems for, 273–274
- existing
  - commercial, contribution of, to energy crisis, 256
  - residential, contribution of, to energy crisis, 256
- retrofit of, energy conserving, 257–261 (*See also* retrofit technologies, energy conserving)
- residential (*See* residential buildings)

built environment, 255–282
- energy reduction in, voluntary schemes to promote, 272–273
- energy savings and reduced environmental impact in, 287–288
- energy use in, 7

bulk-heterojunction solar cells, 132–133

Bush, George H. W. (President), refusal of, to sign Convention on Biodiversity, 37

Bush, George W. (President), on Kyoto Protocol, 38

CAFE program. *See* Corporate Average Fuel Economy (CAFE) program

Canada, as confederation, energy policy in, 196–197

cancer, radiation exposure and, 185–186

cap-and-trade principle, in EU Emissions Trading System, 49
- in aviation emissions reduction, 248

capital investment, as barrier to taking new technology to market, 194

carbon
- in coal, 16
- industrial revolution and, 17–18

Carbon Calculator, 246

carbon capture and storage (CCS) technologies, 22–23

carbon dioxide, atmospheric lifetime of, 14–15

carbon dioxide (CO_2) emissions
- aviation growth rate projections and, 241–242
- as challenge for transportation, 227
  - technology in reducing, 227–228
- from coal-fired power plants, 16
- of electric *vs.* gasoline vehicles, 230
- of internal combustion engine and electric vehicles compared, 228
- international standard for reducing, 24
- nuclear power and, 173
- from production of vehicle, 222, 222f
- regulation of, EPA in, 39
- trend in, 1, 2
- from vehicle fuels, 221, 221t

carbonaceous fuels, 14–15

casinghead gas, 19

CCS. *See* carbon capture and storage (CCS)

cellulosic biofuels
- economics of, 157
- feedstocks for, availability and cost of, 157–158
- technology for, challenges to development of, 162–163
- in US Renewable Fuel Standard, 152, 152f

Central Intelligence Agency (CIA), Center on Climate Change and National Security of, 40

INDEX

charcoal, evolving uses of, 12
chemical conversion pathways, for biomass, 159–160
Chernobyl accident (1986), 178–179
    radiation-induced illness/death from, 185–196
China
    energy and climate change policy of, 50–52
    energy relationship of, with Russia, 54
    hydroelectricity production in, 101
    nuclear power generation in, 171
    oil consumption growth in, 19
    vehicle fleet growth in, 216
choice, and preference in social action, 78
Clean Coal Technologies (CCTs), 21–23
climate
    long-term effects of greenhouse gases on, 14
    variability of, pioneering thought leaders on, 13–14
climate change
    Brazil's energy policy and, 57–58
    China's energy policy and, 51–52
    combating, Obama administration in, 39–41
    EU energy policy and, 45, 47–48
    future role of nuclear power and, 183
    India's energy policy and, 60
    Russia's energy policy and, 53–54
    US energy policy and, 37–41
        American Reinvestment and Recovery Act and, 38–39
        Earth Summit and, 37
        Kyoto Protocol Conference and, 37–38
closed fuel cycle, future of nuclear energy and, 184–185
coal
    clean, achieving
        advanced technologies in, 22
        carbon capture and storage in, 22–23
        control of pollutants and emissions in, 22
    Clean Coal Technologies and, 21–23
    composition of, 16
    as energy source
        in China, 50–51
        global, 16
            challenges to, 16
            declining supply of, 182
        in India, 58
    formation of, 16
    in India's energy policy, 58
    industrial revolution and, 17–18
    mining of, problems accompanying, 17–18
    through history, 15–18
coal rank, 16
coke, uses of, 12
combustion, overview of, 220–222
Commercial Alternative Aviation Fuels Initiative (CAAFI), biofuels and, 248

commercial buildings
    contribution of, to energy crisis, 256
    retrofitting
        challenges to, 267
        lighting in, 260–261, 262–263
commissioning, definition of, 261
common authentic values and principles, 78, 80
common good, 80
    in decision on ethanol production, case study on, 85–86
    definition of
        individual fulfillment in, 82–83
        religious, 82
        utilitarian, 82
    globalization and, 84
    individual rights and, 81–83
commons, tragedy of the, 83–84
compact fluorescent lighting systems, in retrofitting residential buildings, 260
Comparative Aircraft Flight Efficiency's CAFE Foundation, emission elimination and, 245
complex hybrid drivelines, 232f, 236
compressed natural gas vehicles, market share of, 226
computer aided design (CAD) tools, for microelectronic industry, 131
concentrated solar power (CSP), 126
concentrating photovoltaics, commercial development of, 133–134
confederate political systems, 196–197
consumer bargaining power, as barrier to taking new technology to market, 195
consumer behavior, in reducing oil use, 223–225
consumption, products of, as challenge for transportation, 227
Continuous Lower Energy, Emissions and Noise (CLEEN) program, 247
control rods, in nuclear reactor reactivity control, 176
control systems, in retrofitting building, 259
cooling system, in retrofitting building, 258–259
corn, as biofuel, 154, 155
    environmental issues on, 160–161
Corporate Average Fuel Economy (CAFE) program, in 1975 and 2011, 35–36
    in regulation of vehicle fuel emissions, 222–223
    technology improvements for efficiency and, 217–218
cost(s)
    and benefits of energy conserving retrofits in marketing, 270
    of feedstocks for cellulosic biofuels, 157–158
    of geothermal energy, 187
    levelized
        in comparing economics of energy sources, 193, 204, 205t, 206t
        of wind energy in UK, 208

of nuclear energy, components of, 173, 173f
of nuclear reactors, 187
switching, as barrier to taking new technology to market, 194–195
of thermoelectric power materials, 139–140
crude oil. *See* petroleum
Cugnot, Nicolas, steam-powered vehicles of, 228
customer loyalty, as barrier to taking new technology to market, 194

Darracq, M. A., regenerative braking technology demonstrated by, 229
Davenport, Thomas, DC electric motors for vehicles of, 228
daylight, in building lighting, 262–263
de Bélidor, Bernard Forest, *Architecture Hydraulique* of, 100
Denmark, in wind turbine development, 93, 207
Department of Defense (DOD)
renewable energy goals of, 247
renewable jet fuel and, 162
Department of Energy (DOE), creation of, 34
developing countries
energy and climate change balance for, 31–33
globalization and, challenges for, 28–29
as greenhouse gas polluters, 3–4
rural incomes in, biofuel industry and, 155
diesel fuel
for aircraft, 242
calorific value of, 220–221
combustion of, emissions from, 221, 221t
diffusion, of technology, 200, 201
direct methanol fuel cells, 240
diurnal variation, climate stability and, 14
Drake, Edwin Laurentine, Colonel, oil first drilled for by, 18
driveline configurations, in modern vehicle configurations, 231, 232f
drivelines
for electric vehicles, 232f, 234
for hybrid electric vehicles, 232f, 236
for internal combustion engine (ICE) vehicle, 232f, 233
dry-cask storage, of nuclear waste, 181

Earth Summit, 37
economic challenges, to solar power growth, 127–128
economic factors, in taking renewable energy technologies to market, 194–195
economic growth
energy and, 28
expected, energy demand and, 28–34
in India, energy policy and, 58, 60–61
economic impact, of hydroelectricity, 101–102

efficiency improvements, under American Reinvestment and Recovery Act, 38
Electric Vehicle Act of 1976, 229
Electric Vehicle Initiative (EVE), 238
electric vehicles (EVs)
battery, 228–229
charging of, greenhouse gas emissions and, 218
CO_2 emissions of, compared with internal combustion engine, 228
drivelines for, 232f, 234
early, 228–229
market share of, 226
modern, developments in, 230
regenerative braking technology and, 229
targets for, by region, 239t
electricity
base-load, nuclear power as source of, 173–174
heating systems using, 259
nuclear power in production of, 171–172
electricity grid developments, impact of renewable technologies on, 117–118
Electrostatic Precipitators, 22
embargo
Arab oil, US energy policy and, 34
oil, after Yom Kippur War, EU economic consequences of, 44
emerging economies, globalization and, challenges for, 28–29
emission factor (EF_c), 22
emissions
carbon dioxide (*See* carbon dioxide (CO_2) emissions)
greenhouse gas (*See* greenhouse gas (GHG) emissions)
Emissions Trading System (ETS), of EU, 49
in aviation emissions reduction, 248
energy
conservation of, retrofit technologies for, 257–261 (*See also* retrofit technologies, energy conserving)
demand for
global, 5–6
growth expectations and, 28–34
distribution and use of, 7, 191–282
taking renewable energy technologies to market in, 193–213 (*See also* market, taking renewable energy technologies to)
transportation in, 215–254 (*See also* transportation)
economic growth and, historical background of, 28
economics of, for alternative energy sources, 202–204
embodied, of building, 266

food as source of, 12
man's quest for, 12–13
transitions in, inertia of, 28
for transportation, 7
use of, in built environment, 7
reduction of, voluntary schemes to promote, 272–273
world statistics on, for 2011, 30t
energy conversion devices, microelectronics in, 131
energy conversion technology, 89–190
geothermal energy as, 114–117 (*See also* geothermal energy)
hydroelectric energy as, 99–102 (*See also* hydroelectric energy)
impact of, on electricity grid developments, 117–118
solar power as, 125–150 (*See also* solar power)
tidal energy as, 102, 111–114 (*See also* tidal energy)
wave energy as, 102–111 (*See also* wave energy)
wind energy as, 92–99 (*See also* wind energy)
energy crisis
commercial buildings' contribution to, 256
residential buildings' contribution to, 256
energy density, of fuels, compared, 172–173, 173f
Energy Independence and Security Act (EISA 2007), 36, 117
Energy Information Administration (EIA), 204, 205t, 206t
Energy Performance of Buildings Directive (EPBD), 274
energy policy
Brazil, 54–58
authority over, 55
climate change and, 57–58
corporate influence over, 55–56
Canada, 196–197
China, 50–52
EU, 42–50
climate change and, 45, 47–48
Emissions Trading System in, 49
Kyoto Protocol and, 47
solar photovoltaic growth and, 126–127
twenty-first-century, 44–45
global, solar industry predictions and, 127
government structure and, 197–198
India
administration of, 59
economic growth and, 58, 60–61
poverty and, 58–59
inducements and, 198–199
mandates and, 199
Russia, 53–54
agencies involved in, 53
UK, wave energy in, 207–208

US
American Reinvestment and Recovery Act (2009) and, 38–39
Arab oil embargo and, 34
climate change and, 37–41
Earth Summit and, 37
Energy Policy Act (2005) and, 38
federalist political structure and, 197
Iranian revolution and, 35
Kyoto agreement and, 37–38
natural gas in, 36
under Obama administration, 38–41
in 1970s and 2007- compared, 35
shale gas in, 36
Three Mile Island nuclear reactor meltdown and, 35
vehicular fuel efficiency in, 35–36
Energy Policy Act (2005), provisions of, 38
energy security, future role for nuclear power and, 182–183
Energy Star, 273
energy storage, in modern vehicle configurations, 231
energy storage devices, nanotechnology and, 140–141
engineer(s)
in reducing greenhouse gas emissions, 4–5
social engagement by, 6, 73–88
engineering
definition of, 73
nuclear, 6–7
history of, 171
enrichment process, in nuclear power, 168–169
environment, built, 255–282. *See also* built environment
environmental impact
of geothermal energy, 117
of hydroelectricity, 101–102
environmental issues
on biofuels, 160–161
on oil recovery from shale, 20–21
on wind energy, 98–99
environmental pollution, in Chinese energy policy, 51
Environmental Protection Agency (EPA)
in carbon dioxide emissions regulation, 39
carbon pollution standards for power plants proposed by, 39
stratospheric ozone protection plan of (1986), 37
ethanol. *See also* biofuel(s)
corn-based, 154–155
production of, environmental and economic issues for, case study on, 85–86
sugarcane, in Brazil, 56–57, 153–154
ethanol industry, Brazilian, 56–57
ethyl vinyl acetate (EVA), encapsulating photovoltaics, problems with, 135
EU. *See* European Union (EU)

Euler, Leonard, on theory of hydraulic machines, 100
European Atomic Energy Community (EURATOM), establishment of, 43
European Aviation Safety Agency (EASA), 247
European Coal and Steel Community (ECSC), formation of, 42
European Economic Community (EEC), establishment of, 43
European Green Cars Initiatives, research and, 220
European Strategic Technology Plan, 45, 46b
European Strategy for Sustainable, Competitive and Secure Energy, 45
European Technology Platform for Electricity Networks of the Future, 117
European Union (EU)
    building energy reduction programs recently employed in, 272
    early biofuel subsidies in, 152
    energy generation policies in, structuration and, 78
    Energy Performance of Buildings Directive of, 274
    energy policy in, 42–50 (*See also* energy policy, EU)
    European Commission of, in regulation of vehicle fuel emissions, 223
    oil embargo of 1973 and, 44
    Russian energy exports to, 52
EVs. *See* electric vehicles (EVs)
exergetic efficiency, 135

fast reactors, future of nuclear energy and, 184–185
Federal Aviation Administration (FAA), 247
federalist political structure, 197
feebate system, to support market share growth for electric vehicles, 237
Feynman, Richard, on nanotechnology, 128
fiber optic applications, of microelectronics, 131
Fischer-Tropsch (FT) process
    for biomass conversion, 159, 160
    in jet fuel production, 19
        liquid drop-in alternative, 244
fissility, of shale, 20
fission, nuclear, 169
fission products, creation of, 169, 170
Flightpath 2050, 247
floating offshore wind turbine platforms (FOWT), 97–98, 98f
flue gas desulfurization, 22
fluidized bed combustion technology, 22
fluorescent fixtures
    compact, in retrofitting residential buildings, 260
    in retrofitting commercial buildings, 260–261
Ford, Henry, Model T Ford introduced by, 229

fossil fuel emissions. *See also* greenhouse gas (GHG) emissions
fossil fuels. *See also* coal; gas(es); Petroleum
    clean, 21–23
    through history, 15–21
450 Scenario, 2–3
    components of, 48
Fourier, Jean Baptiste Joseph, greenhouse effect and, 13
fracking. *See* hydrofracking
Frisch, Otto, in history of nuclear engineering, 171
fuel(s)
    alternative (*See* alternative fuels; biofuel(s))
    aviation
        alternative, emissions and, 244–246
        overview of, 242–244
    calorific value of, 220–221
    energy density of, compared, 172–173, 173f
    higher-heating value of, 220–221
    nuclear
        disposal of, 170
        uranium as, 168–169
fuel assembly, in nuclear reactor, 169
fuel cells, 240–241
fuel efficiency, vehicular, in US energy policy, 35–36
fuel power, for internal combustion engine vehicle, 232
fuel rods, in nuclear reactor, 169
Fukushima accident (2012), 179–180
fusion, nuclear
    in electricity generation, efficiency of, 172
    future for, 188–189

gas
    associated, 18, 19
    atomised, as transport fuel, 219–220
    biogenic, creation of, 19
    casinghead, 19
    to liquid (GTL), 19
    natural, 19–20 (*See also* natural gas)
        increased demand for, environmental concerns on, 4
    propane, liquified, as transport fuel, 220
    shale, 20–21 (*See also* shale gas)
    solution, 18
    thermogenic, creation of, 19
gas centrifuge, in uranium enrichment, 169
gaseous diffusion, in uranium enrichment, 169
gasification, in biomass conversion, 160
gasoline
    calorific value of, 220
    combustion of, emissions from, 221, 221t
geothermal energy, 114–117
    cost of, 116–117
    demand and supply management for, 116

global installed capacity of, 92
    overview of, 114–116
    social and environmental impacts of, 117
Germany, nuclear power in, 172
    phase out of
        Case Study on, 79–80
        risk in, 188
global energy crisis, 5
global energy outlook, future role for nuclear power and, 182–183
global energy policy, 27–72
Global Framework for Alternative Aviation Fuels (GFAAF), 246
global warming, first use of, by EPA, 37
globalization
    common good and, 84
    developing economies and, challenges for, 28–29
"Golden Rules for a Golden Age of Gas" (IEA), 4
government policies and initiatives, for change in automobile market, 237–239, 238t, 239t
government regulations, as barrier to taking new technology to market, 195
government structure, energy policy and, 197–198
Grand Coulee Dam, 100
Great Britain, genetically modified food adoption and diffusion in, 200
Green Deal, 274–275
Green Globes, 272
greenhouse effect, 14
    early contribution on, 13
greenhouse gas (GHG) emissions
    in aviation, reduction of
        approaches to, 247–248
        policies and initiatives for, 246–248
    in Brazil, 57
    as challenge for transportation, 227
    in China, 51
    electric vehicle charging and, 218
    global
        by 2035, estimated, 30–31
        from 1971-2012, 30
    global average temperatures and, 2
    global responsibility for, 3–4
    in India, reduction goals for, 60
    negative impacts of, growing public awareness of, 31
    reducing, Kyoto Protocol on, 47
    for renewable fuel categories, 156t, 157
    in Russia, 53
    for selected countries
        total and per capita, 32f, 33
        trends in, 2005-2011, 33, 33f
    from vehicle fuels, 221–222, 221t
        combustion and, 220–222, 221t, 222f
        regulation of, 222–223

greenhouse gases
    earth's atmosphere and, 13–23
    listing of, 14
    long-term effects of, on climate, 14
grid, in wave energy utilization in UK, 209
Grid Modernization Commission, 117
grid parity, of solar photovoltaic, challenges to, 127–128
ground source heat pumps, in retrofitting building, 259
growth, economic. See economic growth
Guri Dam, 101

Hahn, Otto, in history of nuclear engineering, 171
hard dry rock (HDR), for electricity generation, 115–116
heat gain, solar, passive, through windows, 260
heat pump water heaters, in retrofitting building, 260
heat pumps, in retrofitting building, 259
heat recovery ventilation systems, in retrofitting building, 259
heating system, in retrofitting building, 258–259
HEVs. See hybrid electric vehicles (HEVs)
high voltage direct current (HVDC) transmission, in future grids, 118
Hoover Dam, 100
horizontal axis tidal current turbines, 113
horizontal axis wind turbine (HAWT), 95
humidity control, in retrofitting building, 259
hybrid aircraft, 245
hybrid electric vehicles (HEVs), 235–237
    drivelines for, 232f, 236
    early, 229
    market share of, 226
    modern, developments in, 230
hydroelectric energy, 99–102
    in Chinese energy policy, 51, 101
    generation of, 101
    global installed capacity of, 91, 100
    historical perspective on, 99–101
    in India's energy policy, 59
    production of, early, 100
    social, environmental, and economic impact of, 101–102
hydrofracking, of shale gas
    development of, 20
    in energy future, 182–183
    environmental issues on, 20–21, 182–183
    in US energy policy, 36
hydrogen
    production of, advanced reactors in, 184
    as transport fuel, research needs for, 220
hydropower, history of, 99–100
Hydrotreated Esters and Fatty Acids (HEFA) method, in liquid drop-in alternative jet fuel production, 244

ICE vehicle. *See* internal combustion engine (ICE) vehicle
IEA. *See* International Energy Agency (IEA)
India
    energy and climate change policy of, 58–61
    nuclear power generation in, 171
individual
    behavior of, factors affecting, 81
    rights of, common good and, 81–83
inducements, energy policy and, 198–199
industrial revolution, carbon and, 17–18b
information processing devices, microelectronics in, 130–131
information transfer, microelectronics and, 131
insulation, in retrofitting building, 257
    unintended consequences of, 268–269, 268f, 269f
integrated gasifier combined cycle, 22
Intergovernmental Panel on Climate Change, China's participation in, 51
intermediate-band solar cells, for energy conversion efficiency, 133
internal combustion engine (ICE) vehicle, 233–234
    CO_2 emissions of, compared with electric vehicles, 228
    domination of, over battery electric vehicles, 229
    driveline for, 232f, 233
    improvements in, 230
    inefficiency of, 234
    market share of, 226
International Civil Aviation Organization (ICAO)
    fuel efficiency goals of, 247
    in reducing GHG emissions in aviation, 246
International Energy Agency (IEA)
    formation of, oil crisis of 1973 and, 35
    in historical energy data compilations and analysis, 29–30
    projections of, for future energy trends, 36–37
    on reducing atmospheric emissions trend, 1
    World Energy Outlook of, 3
International Energy Agency Task 29 working group on wind energy, 98–99
International Thermonuclear Experimental Reactor (ITER) project, 189
Ioffe, Abraham, thermoelectric figure-of-merit introduced by, 136
Iran, gas field in waters off, 20
Ireland, existing homes in, energy consumption by, 256
irrigation, hydropower for, 99
Itaipu Dam, 100, 101

Japan, Fukushima accident in, 179–180
jet fuel, 242
    alternatives to
        conversion pathways for, 244, 245t
        emissions and, 244–246, 246t
        converting gas to, 19
        liquid drop-in alternative, 244

Kettering, Charles, in evolution of internal combustion engine vehicle, 229
Kyoto Protocol (1997)
    amendment to (2013), 48–49
    Canada's participation in, confederate political structure and, 197
    EU and, 47
    US and, 37–38
Kyoto units, 47

lamp, LED, in building design, 263–265
Leadership in Energy and Environmental Design (LEED), 272
LED lighting. *See* light emitting diode (LED) lighting
levelized costs
    in comparing economics of energy sources, 193, 204, 205t, 206t
    of wind energy in UK, 208
life cycle analysis (LCA), of fuels for environmental and cost-impact, 245–246
light, natural, in building lighting, 262–263
light emitting diode (LED) lighting
    in building design, 263–265
    in retrofitting buildings, 261
lighting
    energy saving developments in, 262–265
    in retrofitting building, 260–261, 262–265
linear wave theory, 105
liquid drop-in alternative jet fuel, 244
liquified propane gas (LPG)
    carbon dioxide emissions of, per volume consumed, 221t
    as transport fuel, 220
Low Nitrogen Oxide burners, 22

maintainability, in taking renewable energy technologies to market, 202
mandates, energy policy and, 199
market
    automobile, change in, government policies and initiatives for, 237–239, 238t, 239t
    competitive intensity of, 195
    competitive structure of, bringing new technology to market and, 194
    taking renewable energy technologies to, 193–213, 287
        barriers to, 194–195
        economic factors in, 194–196
        economics of energy and, 202–204
        government structure and, 197–198
        maintainability factors in, 202
        political factors in, 196–199

social factors in, 199–201
for wave energy in UK, 204–209
Marrakesh Accords, 47
Massachusetts v. EPA, 39
Masuda, Yoshio, as father of modern wave energy, 103
Messina Conference, in evolution of EU, 43
methane, atmospheric lifetime of, 15
methanogenic organisms, in biogenic gas creation, 19
microelectronic revolution, 130
mills
　　tide, 102
　　waterwheel, 99–100
molten carbonate fuel cells (MCFCs), 242
Moore's law, 131
Musgrove, Peter, in wind turbine design, 93

nanomaterials, classification of, 129, 129f
nanostructure, in solar photovoltaics, 132–135
nanostructured thermoelectric material, 137–139
nanotechnology, 128–132
　　biofuels and, 140
　　definition of, 128
　　energy storage devices and, 140–141
　　introduction to, 125–126
National Action Plan on Climate Change (NAPCCC), India's, 60
National Intelligence Council (NIC) report, on climate change impacts in Russia, 53–54
national security, US, energy and climate change in, 39–40
natural gas, 19–20
　　as bridge fuel, 36
　　in Chinese energy policy, 51
　　compressed, vehicles using, market share of, 226
　　in energy future, 182–183
　　as microcosm of world's future energy issues, 284
　　processing of, byproducts of, 19
　　reserves of
　　　　Brazilian, pre-salt, 56
　　　　locations of, 20
　　　　offshore, development of, by India, 59
　　　　Russian, 52
　　as transport fuel, 219–220
　　in US energy policy, 36
natural light, in building lighting, 262–263
network effects, as barrier to taking new technology to market, 195
new energy vehicles, in Chinese energy policy, 51
nitrous oxide, atmospheric lifetime of, 15
nuclear energy
　　as base-load electricity source, 173–174
　　cost components of, 173, 173f
　　cost of reactors and, 187
　　culture of, changes in, public acceptance and, 186–188

current status of, 171–174
essentials of, 168–170
future of, 167–190, 286–287
future role for, 182–185
　　advanced reactors and, 183–184
　　closed fuel cycle and, 184–185
　　global energy outlook and, 182–183
generation of
　　residual heat removal and, 176–177
　　since 1980, 172f
in Germany, phase out of, Case Study on, 79–80
in global energy mix, 12–13
in India's energy policy, 59
peaceful utilization of, 171
public acceptance of, 185–188
　　change in nuclear energy culture and, 186–188
　　radiation risk and, 185–186
safety of, 175–177
US energy policy and, 40
　　Three Mile Island nuclear reactor meltdown and, 35
nuclear engineering, 6–7
　　history of, 171
nuclear fission, 169
　　chain reaction in, as safety feature, 175–176
　　in electricity generation, efficiency of, 172
nuclear fuel
　　disposal of, 170
　　　　challenges in, 180–181
　　spent, inventory of, in 2007, 180f, 181
nuclear fusion
　　in electricity generation, efficiency of, 172
　　future for, 188–189
nuclear power plants
　　accidents in, impacts of, 177–180
　　　　at Chernobyl in 1986, 178–179
　　　　at Fukushima in 2012, 179–180
　　　　at Three Mile Island in 1979, 177–178
　　defense-in-depth concept of, 175, 175t
　　reactor reactivity control in, 176
nuclear reactors
　　advanced, 183–184
　　cost of, 187
Nuclear Regulatory Commission (NRC) response, to Three Mile Island nuclear reactor meltdown, 35
nuclear waste, as fuel for advanced reactors, 84
nuclear waste management, 170
　　challenges in, 180–181

Obama administration
　　in combating climate change, 39–41
　　international energy cooperation initiatives of, 41
　　National Security Strategy of, new energy sources and combating climate change in, 39–40

offshore natural gas reserves, development of, by India, 59
offshore wind energy
    future potential for, 96–98, 97f
    turbine maintainability and, 202
oil
    crude (*See also* Petroleum)
        supply *vs.* demand for, 4
    prices of
        biofuel demand and, 155
        uncertain, cellulosic biofuel investment and, 157
        wave energy patents and, 106–107, 106f
    production of
        inelastic, energy security and, 182
        in Russia, 53
    reliance of transportation sector on, 216
        for aviation, 239, 241–244
        for vehicles, 226–227
    reserves of
        Brazilian, pre-salt, 56
        finite, as challenge for transportation, 226–227
        Indian, 59
        of UK in North Sea, development of, 206–207
OPEC. *See* Organization of the Petroleum Exporting Countries (OPEC)
operational effectiveness, of renewable energy technologies, 202
Organization of Arab Petroleum Exporting Countries (OAPEC), oil embargo proclaime4
Organization of the Petroleum Exporting Countries (OPEC)
    creation of, 43–44, 205–206
    members of, 43–44
    oil reserves of, 19
oscillating water column (OWC), as wave energy converter, 103, 107
oscillating wave surge converter, 107–108
overtopping devices, for wave energy, 108–109
Oxy-Coal combustion process, 23

parallel hybrid driveline, 232f, 236
Peltier effect, 136
perceived behavioral control, 81
petroleum. *See also* oil
    consumption of
        Chinese, 50
        global, 19
    foreign, India's reliance on, 59
    products derived from, 18
    reliance of transportation sector on, 216
        for aviation, 239, 241–244
        for vehicles, 226–227
    through history, 18–19

phosphoric acid fuel cells, 240
photosynthesis, artificial, nanotechnology in, 140
photovoltaics, solar, 132–135. *See also* solar photovoltaics (PV)
Planté, Camille, rechargeable battery invented by, 228
point absorber devices, in wave farms, 108
policy(ies)
    energy (*See* energy policy)
    government, for change in automobile market, 237–239, 238t, 239t
    on greening of aviation industry, frameworks and initiatives for, 246–248
    on retrofitting for energy conservation, challenges to, 266–267
    transport, 225
policy instruments, energy policy and, 198–199
policy perspectives, on emerging renewable energies, 7
policy specialists, in addressing climate change, 5
political factors, in taking renewable energy technologies to market, 196–199
political systems, 76–77
    confederate, 196–197
    interactions of, in social environment, 76
    non-federalist *vs.* federalist, 196–197
    unitary, 196
polymer electrolyte membrane (PEM) fuel cells, 240
Poncelet, Jean Victor, in turbine and waterwheel design, 100
population, world, trends in, environmental implications of, 283–284
post-combustion capture, 23
poverty, in India, energy policy and, 58–59
power, fuel, for internal combustion engine vehicle, 232
power plants, carbon pollution standards proposed by EPA for, 39
pre-combustion capture, 23
preference, and choice in social action, 78
President's Climate Action Plan (2013), 40
principles, common authentic, 78, 80
propane, as transport fuel, 220
proton exchange membrane fuel cells, 240
pyrolysis, in biomass conversion, 160

Qatar, gas field in, 20
quantum dots, for concentration of diffuse solar energy, 133–134
quantum wells, 130

radiation risk, public acceptance of nuclear energy and, 185–186
radioisotope thermoelectric generators (RTGs), 136
Rayleigh distribution, in analyzing wave amplitude and heights, 105

recycling, of nuclear spent fuels, 170, 181
 future of nuclear energy and, 184–185
regenerative braking technology, introduction of, to electric vehicles, 229
renewable energy, under American Reinvestment and Recovery Act, 38–39
renewable energy resources, challenges and opportunities of, 285–286
renewable energy technologies, 6
 in Chinese energy policy, 51
 impact of, on electricity grid developments, 117–118
 taking to market, 193–213 (*See also* market, taking renewable energy technologies to)
Renewable Fuel Standard (RFS 2007 and updates), 36
 cellulosic biofuels in, 152, 152f, 161–162
residential buildings
 contribution of, to energy crisis, 256
 retrofitting
  challenges to, 266–267
  lighting in, 260–261, 262–265
retro-commissioning, in retrofitting buildings, 261
retrofit technologies, energy conserving
 air sealing as, 258
 complexity of, 261, 265–266
 cost *vs.* benefit considerations on, 265
 discussion problem on, 279–280
 domestic hot water production as, 260
 embodied energy of building and, 266
 end-user as barrier to acceptance of, 271
 fragmented nature of contracting for residential/small commercial buildings and, 266
 funding of
  barriers to, 270
  source and, 265–266
 government programs supporting
  potential pitfalls of, 271
  in US and EU, recently employed, 272
 heating and cooling systems as, 258–259
 insulation as, 257
 lighting as, 260–261
 marketing of, 270
 occupant training as, 261
 programs for, successes and failures of, 275–27
 recommendations and opportunities for future solutions in, 276–279
  in building performance rating and labeling, 277–278
  in education, motivation, and marketing, 276–277
  in financing, 278
  in program design, 278–279
  in technology, building codes, and implementation, 277

 retro-commissioning as, 261
 success of, policy challenges to, 266–267
 window replacement as, 259–260
reverse auctions, for second-generation biofuels, 162
Ridley, Matt, on energy sources for Industrial Revolution, 17
Rio+20 Summit (2012), 41
Rockefeller, John D., in developing petroleum industry, 18
Royal Dutch Company, in history of petroleum, 18
Russia. *See also* Chernobyl accident (1986)
 Chernobyl accident in, 178–179
 energy and climate change policy of, 52–54
 energy relationship of, with China, 54
 natural gas reserves in, 20

safety
 of advanced reactors, 183–184
 nuclear energy, 175–177
Sakharov, Andrei, in tokamak development, 189
Schuman, Robert, in evolution of EU, 42
scrubber technology, 22
Selective Catalytic Reduction, 22
semiconductors, microelectronics in making of, 130
series hybrid driveline, 232f, 236
series-parallel hybrid driveline, 232f, 236
Severn Barrage Tidal Power system, 102
shale gas, 20–21
 deposits of, locations of, 20
 extraction of (*See also* Hydrofracking, of shale gas)
  environmental issues on, 20–21
 history of, 20
 in US energy policy, 36
Shell, in history of petroleum, 18
SLL Code for Lighting 2012, 262
smart energy, 118
smart grids, 117
SmartGrids ETP, 117
social engagement, by engineer, 73–88
social environment(s)
 components of, 74–75, 74f
 description of, 74
 individual experience with, 75
 social systems in, 75–78
social factors, in taking renewable energy technologies to market, 199–201
social impact
 of geothermal energy, 117
 of hydroelectricity, 101–102
social institutions, adoption and diffusion of emerging technologies and, 200
social issues, with nuclear energy, 185–188
social scientists, in addressing climate change, 5

social systems, 75–78
    political systems as, 76–77
    structuration theory and, 77–78, 77f
solar cells
    bulk-heterojunction, 132–133
    hybrid, organic-inorganic, 132
    intermediate-band, for energy conversion efficiency, 133
    monolithic tandem stacks of, efficiency of, 133
solar heat gain, passive, through windows, 260
solar photovoltaics (PV), 132–135
    concentrating, commercial development of, 133–134
    failure of, 134–135, 134f
    global installations of, growth of 2003-2012, 126, 126f
solar power, 6
    for aircraft, 245
    in Chinese energy policy, 51
    concentrated, 126
    conversion applications of, nanoscience in, 129
    enabling role of nanotechnology and, 125–150
    global installed capacity of, 92
    growth of, 126–127, 126f
    introduction to, 125
    in Middle East, future plans for, 118
    overview of, 126–128, 126f
    technological barriers to adoption of, 127
    in wave production, 103
solar water heaters, rooftop, in Asia, 128
solid oxide fuel cells (SOFCs), 241
solution gas, 18
Standard Oil Company, in history of petroleum, 18
steam, heating systems using, 258
Strassmann, Fritz, in history of nuclear engineering, 171
Strategic Energy Technology Plan (SET Plan), European, 45, 46
Strategic Petroleum Reserve (SPR), establishment of, 34
structuration, 77–78, 77f
structures, in systems, 77
subjective norm, individual behavior and, 81
Suez Crisis, impact on Western Europe, 43
sugarcane ethanol, in Brazil, 56–57, 153–154
Summary of Gaps and Barriers for Implementing Residential Building Energy Efficiency Strategies, 266
supplier bargaining power, as barrier to taking new technology to market, 195–196
Sustainable Energy for All campaign, 41
Sustainable Way for Alternative Fuels and Energy for Aviation (SWAFEA), 247
switching costs, as barrier to taking new technology to market, 194–195

Tamm, Igor, in tokamak development, 189
tariffs, as barrier to taking new technology to market, 195
technology(ies)
    acceptance or resistance to, 201
    adoption of, 200
    diffusion of, 200, 201
    in reducing CO_2 emissions from vehicles, 227–228
    renewable energy, 6, 193–213 (*See also* renewable energy technologies)
    retrofit, energy conserving, 257–261 (*See also* retrofit technologies, energy conserving)
temperature, global
    average, climate change and, 2
    increase in, consequences of, 284
temperature control, in retrofitting building
    air sealing in, 258
    control systems for, 259
    heating and cooling systems in, 258–259
    insulation in, 257
thermochemical conversion, of biomass, 160
thermodynamic system(s)
    of internal combustion engine vehicle, 232f, 233
    for vehicle propulsion, 231, 231f, 233
thermoelectric coolers, 136
thermoelectric material, nanostructured, 137–139
thermoelectric (TE) power, 135–140
    material cost and efficiency in, 139–140
    nanostructured thermoelectric material and, 137–139
    overview of, 135
thermogenic gas, creation of, 19
Three Gorges Dam, 100, 101
Three Mile Island nuclear reactor meltdown (1979), 177–178
    US energy policy and, 35
tidal barrages, 111
    tidal range power from, 111, 112
tidal current turbines, 111, 112–114, 113f, 114f
    future potential of, 114
tidal currents, 111
tidal energy, 111–114
    from currents, locations for, 113
    economics of, 203
    global installed capacity of, 92
    overview of, 111
    tidal currents and, 111
    tidal range power as, 111, 112
        locations for generation of, 112
    tidal variations and, 111
tidal range power, 111, 112
tidal stream power, 111
Tokamak nuclear fusion reactor, 188–189
tragedy of the commons, 83–84
transistors, invention of, 130–131

transport policy, 225
transportation
    challenges facing, 226–227
    energy for, 7, 215–254
        alternative fuels as, 219–220
        consumer behavior and intelligent transit and, 223–225
        efficiency and vehicle technology and, 217–218
        electric and hybrid vehicles and, 225–239
        emissions from
            combustion and, 220–222, 221t, 222f
            regulation of, 222–223
        overview of, 216–225
        policy overview on, 225
Trevithick, Richard, steam-powered vehicles of, 228
turbines
    tidal current, 111, 112–114, 113f, 114f
        future potential of, 114
    water, 101
    wind
        development of, Denmark in, 93, 207
        history of, 92–93
        modern, power grid and, 95–96, 96f
        noise from, acceptance/resistance and, 201
        offshore, maintainability of, 202
turbulence, in wind flow, 94–95
Tyndall, John, on climate variability, 13–14

unitary political systems, 196
United Kingdom (UK), taking wave energy technologies to market in, challenges for, 204–209
United Nations Climate Change Conference
    in Cancún, Mexico (2010), 48
    in Copenhagen (2009), 47–48
    in Doha, Qatar (2012), 48
United Nations Conference on Environment and Development (UNCED), 37
United Nations Framework Convention on Climate Change (1997), 37–38
United States (US)
    building energy reduction programs recently employed in, 272
    climate and energy policy of, 34–42
    early biofuel subsidies in, 152
    energy policy in (*See* Energy policy, US)
    as federalist political structure, 197
    genetically modified food adoption and diffusion in, 200
    nuclear waste management in, 181
    Three Mile Island nuclear reactor meltdown in, 35, 177–178
uranium
    energy density of, 172, 173, 173f
    in nuclear power, 168–169
    underutilization of, nuclear waste recycling and, 184–185
US. *See* United States (US)
utility companies, programs to promote energy reduction of, 273

values
    common, across societies, globalization and, 84
    common authentic, 78, 80
    individual behavior and, 81
Vehicle Technologies Program of Department of Energy, research and, 220
vehicles
    advanced, targets for, by region, 239t
    alternative fuels for, 219–220
        fuel cells as, 240–241
    compressed natural gas, market share of, 226
    early power sources for, 228–229
    electric (*See* electric vehicle(s))
    fuel efficiency of, in US energy policy, 35–36
    global production figures for, 226f
    hybrid electric (*See* Hybrid electric vehicles (HEVs))
    internal combustion engine (*See* internal combustion engine (ICE) vehicle(s))
    market for, change in, government policies and initiatives for, 237–239, 238t, 239t
    modern, configurations of, 231, 231f, 233
    new energy, in Chinese energy policy, 51
    production of, carbon dioxide emissions from, 222, 222f
    propulsion systems for, effectiveness of, 231, 231f, 233
    steam driven, early, 228
    technology of, efficiency and, 217–218
ventilation, in retrofitting building, 258
vertical axis tidal current turbines, 113
vertical axis wind turbine (VAWT), 95

waste, in India's energy supply, 59
water heaters, domestic, in retrofitting building, 260
waterwheel mills, 99–100
wave(s)
    freak, 105–106
        wave energy converter design and, 110
    nature of, 103–106
wave energy
    benefits and disadvantages of, 109–111
    developments in, 106–107, 106f
    devices and technologies for, 107–109, 108f
        arrays of, 109
        control of, 109
        model testing of, 109
    economics of, 203
    emerging technologies for, challenges for, 204–209

nature of waves and, 103–106
overview of, 103
patents in, oil prices and, 106–107, 106f
resources of, 106
in UK energy policy, 207–208
well-to-wake method, of environmental and cost-impact analysis of fuels, 245
wheel-to-wake approach, to environmental and cost-impact analysis of fuels, 245
wind
characteristics of, 93–95
speed of
power curve for, 95, 96f
variability of, 94
in wave generation, 103
wind energy, 92–99
bringing to market in US, federalist political structure and, 197
in Chinese energy policy, 51
environmental concerns on, 98–99
global installed capacity of, 91–92
historical overview of, 92–93
offshore
future potential for, 96–98, 97f
maintainability of turbines and, 202
social acceptance of, 98–99
wind characteristics in, 93–95
wind farms, 95
resistance to, 201
wind turbines
development of, Denmark in, 93, 207
history of, 92–93
modern, power grid and, 95–96, 96f
noise from, acceptance/resistance and, 201
offshore, floating platform, 97–98, 98f
window replacement, in retrofitting building, 259–260
World Energy Outlook (2012, IEA), 3
Wouk, Victor, as "godfather of hybrid," 229

Yom Kippur War, oil crisis resulting from, 44
Young, James, in shale gas processing, 20
Yucca Mountain project, for nuclear waste storage, 181

ZT (thermoelectric figure of merit), 136–137